# SPACE
# FOR THE
# 21ST CENTURY

·

# DISCOVERY
# INNOVATION
# SUSTAINABILITY

# SPACE
# FOR THE
# 21ST CENTURY

·

# DISCOVERY
# INNOVATION
# SUSTAINABILITY

MICHAEL SIMPSON
RAY WILLIAMSON

AND

LANGDON MORRIS

EDITORS

FOREWORD BY
BRUCE MCCANDLESS II
NASA ASTRONAUT

AN AEROSPACE TECHNOLOGY WORKING GROUP BOOK
VOLUME 5

IN PARTNERSHIP WITH

SECURE WORLD FOUNDATION

AND

INTERNATIONAL SPACE UNIVERSITY

AND

THE INTERNATIONAL INSTITUTE OF SPACE COMMERCE

Cover design by Langdon Morris

Photo Information

Front Cover:
Starfield: Hubble Image (NASA)

Front Cover Insets, Top to Bottom:

NASA astronaut Scott Kelly and ESA astronaut Tim Peake. aurora photograph taken from the International Space Station, January 20, 2016. (NASA)

SpaceX Dragon berthing at ISS, March 3, 2013. (NASA)

High Resolution Imaging Science Experiment (HiRISE) camera aboard NASA's Mars Reconnaissance Orbiter shows the western side of an elongated pit depression in the eastern Noctis Labyrinthus region of Mars. Along the pit's upper wall is a light-toned layered deposit. Nov. 24, 2015. (NASA)

Possible variations in chemical composition from one part of Saturn's ring system to another are visible in this Voyager 2 picture as subtle color variations that can be recorded with special computer-processing techniques. This highly enhanced color view was assembled from clear, orange and ultraviolet frames obtained Aug. 17, 1981 from a distance of 8.9 million kilometers (5.5 million miles). August 28, 1981. (NASA)

NASA's Solar Dynamics Observatory image in extreme ultraviolet light shows an active region of the sun's coronal loops taken over about a two-day period, from February 8 - 10, 2014. Coronal loops are found around sunspots and in active regions. These structures are associated with the closed magnetic field lines that connect magnetic regions on the solar surface. (NASA)

Deep Space Harvester Concept. (Brian Versteeg/Deep Space Industries)

ISBN-13:
978-1532784422

# Table of Contents

The Aerospace Technology Working Group                    V

Secure World Foundation                                   VI

The International Space University                        VII

The International Institute of Space Commerce            VIII

Foreword                                                  IX
BY Bruce McCandless II


1.  Introduction:                                         3
    Benefits for the 21ST Century
    BY Michael K. Simpson


Part I:                                                   9
The Multilateral Effort to Assure
Space Sustainability

2.  Reflections on Space Sustainability                   11
    BY Dr. Karl Doetsch

3.  The Long Road to a Sustainable Use of                 31
    Outer Space
    BY Gérard Brachet

4.  Sustainability, COPUOS and the LTSSA                  47
    Working Group
    BY Dr. Peter Martinez

5.  Strengthening Stability In Outer Space From           61
    The Bottom Up: The Role Of Guidelines For
    The Long-Term Sustainability Of Outer Space
    Activities
    BY Kenneth D. Hodgkins, Richard H. Buenneke, and
    Amber Charlesworth

6.  Forwarding Multilateral Governance of                 75
    Outer Space Activities:
    Next Steps for the International Community
    BY Theresa Hitchens

PART II:                                                      119
ECONOMIC AND LEGAL FOUNDATIONS OF SPACE
SUSTAINABILITY

7.   THE ECONOMICS OF SPACE SUSTAINABILITY        121
     BY  BRIAN WEEDEN AND TIFFANY CHOW

8.   ASSURING A SAFE, SECURE AND SUSTAINABLE      137
     SPACE ENVIRONMENT FOR SPACE ACTIVITIES:
     CONSIDERATIONS ON A CONSENSUAL ORBITAL
     DEBRIS REMEDIATION SCHEME
     BY PROF. DR. LESLEY JANE SMITH

9.   LEGAL ISSUES RELATING TO UNAUTHORIZED        157
     SPACE DEBRIS REMEDIATION
     BY JOYEETA CHATTERJEE

10.  DIRECTED GENETIC MANIPULATION AND            183
     ENGINEERING: REGULATORY ISSUES FACING
     AN INCIPIENT INDUSTRY SHAPING TRANSHUMANS
     AND POST HUMANS IN SPACE
     BY DR. GEORGE S. ROBINSON


PART III:                                                     193
NATIONAL PERSPECTIVES ON SPACE SUSTAINABILITY

11.  NATIONAL SPACE POLICIES AND THEIR IMPORTANCE  195
     IN ENSURING THE LONG TERM SUSTAINABLE USE OF
     SPACE
     BY VICTORIA SAMSON

12.  ESA CLEAN SPACE INITIATIVE                    213
     BY JESSICA DELAVAL

13.  CREATION OF A SUSTAINABLE SPACE               229
     DEVELOPMENT BOARD AND OTHER INITIATIVES FOR
     ENVIRONMENTALLY SUSTAINABLE LAUNCH ACTIVITIES
     BY OLGA ZHDANOVICH

PART IV:                                                    269
SUSTAINING AND EXPANDING SPACE OPERATIONS AND
COMMERCE

14. INNOVATIVE MODELS FOR PRIVATE FINANCE              271
    INITIATIVES FOR SPACE SCIENCE MISSIONS
    BY JEFFREY NOSANOV, NORAH PATTEN, MICHAEL POTTER, AND
    CHRISTOPHER STOTT

15. AN ALTERNATIVE MODEL FOR SPACE COMMERCE           295
    SUSTAINABILITY
    BY THOMAS E. DIEGELMAN

16. HELIUM 3: THE FIRST TRUE COMMERCIAL SPACE         313
    ACTIVITY AND RECOVERY OF LUNAR RESOURCES BY
    MINING AND SPACE TRANSPORTATION INDUSTRIES
    BY THOMAS TAYLOR

17. AN EMERGING BUSINESS MODEL FOR COMMERCIAL         329
    SPACE
    BY DR. MICHAEL WISKERCHEN

18. FIREFLY                                           337
    BY MICHAEL BLUM

19. ARE SOLAR POWER SATELLITES SITTING DUCKS          343
    FOR ORBITAL DEBRIS?
    BY ALFRED ANZALDUA, BRAD BLAIR AND DAVID DUNLOP

20. UPDATE ON COMMERCIAL AND ACADEMIC SPACE           361
    ENDEAVORS
    BY THE EDITORS

PART V:                                               369
SPACE AND SUSTAINABILITY ON EARTH

21. THE MEANING OF SPACE FOR DEVELOPING               371
    COUNTRIES
    BY CIRO ARÉVALO-YEPES

22. SPACE SYSTEMS FOR SUSTAINABLE DEVELOPMENT         387
    ON EARTH
    BY FELIPE DUARTE SANTOS

23. CONCLUSION: MULTILATERAL INITIATIVES FOR          403
    SPACE SUSTAINABILITY
    BY RAY WILLIAMSON

INDEX                                                 421

# The Aerospace Technology Working Group

The Aerospace Technology Working Group, also known as ATWG, is an independent space policy research and innovation group led by seasoned professionals in aerospace and other fields who seek to further humanity's exploration of space while simultaneously benefiting people on Earth.

The ATWG was instituted by NASA Administrator Richard Truly in 1990 as an independent body to perform future planning for the nation's space efforts. Initially, the ATWG began identifying and seeking improvements in both existing and developing space systems through the planned application of emerging technologies and the development of new ways of doing business, including the application of distributed missions and innovative strategic concepts in operations under the direction of Dr. Ken Cox.

Today, the ATWG is an independent entity, using semi-annual and regional Forums, technical and strategic dialogs, personal interactions, books, articles, and speeches to explore topics pertinent to developing a space faring people and prepare policy recommendations for national and global leaders.

Using the organization's substantial base of management, engineering and scientific expertise, the ATWG also provides strategic and technical advice, public speakers, and consulting teams to address specific aerospace tasks and broad conceptual and philosophical questions. The ATWG collaborates actively with other space-related national and international organizations.

In addition, the ATWG places special emphasis on promoting and stimulating education in the sciences, mathematics, the engineering disciplines, and other technical areas.

Participants in ATWG include experts from throughout NASA, ISU, IISC, Secure World Foundation, aerospace contractors, systems suppliers, entrepreneurial businesses, professional societies, universities, and government agencies including the DOD, FAA, and DOE.

You can learn more about the ATWG at www.atwg.org.

# SECURE WORLD FOUNDATION

Secure World Foundation envisions the secure, sustainable and peaceful uses of outer space contributing to global stability on Earth. The Foundation works with governments, industry, international organizations and civil society to develop and promote ideas and actions for international collaboration that achieve the secure, sustainable, and peaceful uses of outer space.

With the end of the Cold War and rapid spread of access to information, more of humanity is seeking to obtain the security and socioeconomic benefits that space systems can provide. This trend, in turn, is promoting rapid growth in the number of space actors. The growth in stakeholders benefitting from space systems has exposed the limitations of existing global legal, policy, technical, and operational regimes to preserve the space environment.

At this point, just one half century into the Space Age, the Foundation believes it has a unique opportunity to play a role in establishing the secure and sustainable use of the space domain. Central to this opportunity are: increasing the knowledge about the space environment and the need to maintain it; promoting international cooperation and dialogue; and helping all space actors realize the benefits that space can provide.

For more information  on Secure World Foundation, please visit http://swfound.org/.

PROMOTING COOPERATIVE SOLUTIONS FOR SPACE SUSTAINABILITY

# THE INTERNATIONAL SPACE UNIVERSITY

The International Space University provides graduate-level training to the future leaders of the emerging global space community at its Central Campus in Strasbourg, France, and at locations around the world.

In its two-month Space Studies Program and one-year Masters program, ISU offers its students a unique core curriculum covering all disciplines related to space programs and enterprises – space science, space engineering, systems engineering, space policy and law, business and management, and space and society. Both programs also involve an intense student research team project providing international graduate students and young space professionals the opportunity to solve complex problems by working together in an intercultural environment.

In addition, the school offers an Executive MBA focusing on the space sector, a Technical Masters in Space Systems Engineering in cooperation with The Stevens Institute of Technology, several short courses including the Executive Space Course, and an innovative 5 week program focusing on space issues of particular concern to the Southern Hemisphere in Adelaide, Australia in cooperation with the University of South Australia.

Since its founding in 1987, ISU has graduated more than 3000 students from over 100 countries. Together with hundreds of ISU faculty and lecturers from around the world, ISU alumni comprise an extremely effective network of space professionals and leaders that actively facilitates individual career growth, professional activities and international space cooperation.

You can learn more about ISU at www.isunet.edu/.

# THE INTERNATIONAL INSTITUTE OF SPACE COMMERCE

The International Institute of Space Commerce, or simply 'the Institute,' has been established on the Isle of Man through a partnership between the International Space University (ISU) and the Manx Government.

The Institute's mission is to become the leading think-tank in the study of the economics of space. It is intended to be the intellectual home for the industry and space academia around the world for which it shall perform studies, evaluations and provide services to all interested parties with the ultimate aim to promote and enhance world's space commerce to the general public. The vision for the Institute is to act as a resource for all, being an international and nonpartisan think-tank, drawing upon new ideas and solutions to existing and future problems the space industry faces by drawing together experts from academia, government, the media, business, international and non-governmental organizations, most notably those from the ISU and its extended network of people and resources. The aim of the Institute is to broaden the professional perspective and personal understanding of all those involved in the study, formulation, execution, and criticism of space commerce.

The Institute is a Not for Profit Foundation and has been located at the site of the International Business School (IBS) on the Isle of Man to capitalize on the Isle of Man's growing importance and position in the world's space industry.

You can learn more about the IISC at www.iisc.im/about.asp.

INTERNATIONAL INSTITUTE
OF SPACE COMMERCE

# FOREWORD

## Bruce McCandless II
### Former NASA Astronaut

This volume is the fifth in the series on contemporary space topics by the Aerospace Technology Working Group with support from Secure World Foundation, the International Space University, and the International Institute of Space Commerce. It deals principally with the topic of sustainability of space operations. In all fields of challenging endeavor actually accomplishing an objective (e.g., putting a satellite into orbit) comes first, followed by exploitation or commercialization, and lastly by a realization that the resource is finite. Such "finite-ness" may come from considerations of pollution (e.g., space debris, propulsion effluent) or of actual limitations on the availability of the resource (e.g., crowding of Geostationary Earth Orbit – GEO). Both of these topics are among those discussed in detail in this volume. Developing countries, in particular, may find such considerations too burdensome, and this begs the need for regulation to avoid the classic "Tragedy of the Commons" situation.

In the case of orbital debris we have collectively arrived at a point where tens of millions of tiny pieces of debris are currently in orbit, decaying at diverse rates in a situation where a single flake of paint has been demonstrated to be capable of causing damage when impacting at high relative velocities. At the other end of the spectrum, defunct satellites (e.g., ESA's Envisat) present discrete problems worthy of individual retrieval/disposal efforts but fraught with complications arising from ownership to potentially still effective ITAR constraints on access to onboard technology.

And, of course, the managers of the International Space Station are absolutely paranoid about higher altitude orbital debris eventually decaying to and ultimately impacting their very large orbiting facility.

While space may realistically be dubbed "infinite," very specific orbits, or sets of orbits, have practical capacity limits. In GEO, for example, spacing of satellites along it are subject to constraints arising from use of the same radio frequency spectra and the size of ground based antennas required to spatially discriminate between adjacent satellites. In popular high inclination sun-synchronous Earth imaging orbits, these all converge near the poles, creating a traffic management concern arising from the risk of collision.

The subject of "green propellants" is treated from several aspects. The Liquid Oxygen / Liquid Hydrogen system, while yielding only water vapor from combustion, may have a significant carbon footprint associated with the manufacture of the LH2 from methane or methanol. Aluminum oxide, an exhaust product of common solid propellant boosters is generally regarded as inert, but the inhalation of fine particles of it can cause pulmonary fibrosis or other lung damage in humans. Additionally the need for oxidizer depletion shutdown in the family of hydrazine/oxidizer booster stages results in significant quantities of UDMH (for example) being dispersed upon impact of the early stages.

No Foreword can do adequate justice to the carefully developed material within the publication itself. For a detailed and thought provoking coverage of the principal topics associated with the sustainability of space operations, this book is highly recommended, authoritative, and "a good read."

•••

## BRUCE McCANDLESS II

Bruce McCandless II is a 1958 graduate of the U. S. Naval Academy and holds advanced degrees from Stanford University and the University of Houston. He first married the former Alfreda Bernice Doyle of Roselle, New Jersey, now deceased. They had two children and two grand-children. Subsequently, he married the former Ellen Conkling Shields, and adopted two additional children with six grand-children. McCandless retired from the Navy in 1990 with the rank of Captain, and is a Designated Naval Aviator Astronaut. He flew in a Navy fighter squadron (VF-102) for four years (including the Blockade of Cuba) prior to being assigned to graduate school, and, ultimately, to duty as an astronaut.

In the latter capacity he served as CAPCOM for Apollo 10, 11, and 14. He was a backup crew member for the first SKYLAB mission and flew on two Space Shuttle missions: STS 41-B, in the course of which he made the world's first untethered "space walks" in 1984 using the Manned Maneuvering Unit, which he helped design, develop and qualify (See photo above); and STS-31, that deployed the Hubble Space Telescope, which he helped to adapt for on-orbit serviceability. McCandless was awarded the NAA Collier Trophy and the National Air and Space Museum Trophy for his accomplishments on STS 41-B, holds one patent for an in-space tool tethering system in use today, and received numerous awards from both NASA and the Navy.

Following his retirement, McCandless took a middle management position with Lockheed Martin Space Systems Company in Denver, Colorado, and managed a variety of small contracts, retiring in 2005 into an "on call" capacity. He was inducted into the Astronaut Hall of Fame in 2005, and designated a Distinguished Graduate of the Naval Academy in 2012.

# SPACE
# FOR THE
# 21ST CENTURY
•
# Discovery
# Innovation
# Sustainability

CHAPTER 1

# INTRODUCTION: BENEFITS FOR THE 21ST CENTURY

MICHAEL K. SIMPSON, PH.D.
SECURE WORLD FOUNDATION

The Space Age began in the 20th Century with monumental efforts that transformed our understanding of ourselves, our planet, and our place in the cosmos. Those efforts combined the pragmatism of hard core science and technology with the idealism and aspiration to venture beyond our home world, with the opportunism to seize upon the realities in geopolitics and commerce to go beyond GEO and begin our endeavors in exopolitics. Together, all of this enables us to see our planet and ourselves much differently now than we did a century ago.

Now, at the outset of the 21$^{st}$ Century we are continuing to build upon that composite foundation of pragmatism, idealism, and opportunism and we have progressed to the point that we now understand that accessing the benefits of space are rightfully the province of all humans, and we are steadily increasing our capacity to make such an understanding into reality.

There have been many lessons along the way.

Perhaps we have learned from the rabid excess of self-inflicted environmental damage on Earth to avoid duplicating that experience in space.

Hopefully we have learned from our increasingly successful efforts to coordinate the flows of traffic on roads, in the air, and around busy harbors on Earth, to recognize the need for similar coordination in orbit.

And ideally we would have learned from our experiences on Earth with innovation and with rising expectations, to recognize that the potential benefits of our activities in space will only be sustained and fulfilled through expanded efforts.   These efforts should engage the active participation of more nations and more enterprises, as they will bring creative new applications and technological advances along with broader engagements and expanded benefits.

Practically, we also need to acknowledge that as incomprehensibly enormous as the cosmos may be, the critical orbits around our planet are becoming crowded with both active satellites and the orbital remnants of earlier space activity, and that overcrowding will be detrimental to all present and future users.  Satellites in orbit require orbits free of destructive debris, and they also require reliable access to radio frequency spectrum to communicate their data and deliver their services to Earth.

Hence, it would be enormously beneficial to acknowledge that Earth-bound humanity has become increasingly dependent on space capabilities and space-enabled services, and that these must be managed thoughtfully and proactively.  Communications, broadcast, remote sensing, positioning, navigation, timing, and weather forecasting are all essentials of modern life and modern commerce, and of course all are now dependent on value that is being created by assets located in space.

And since they operate in a brutally hostile environment of near total vacuum, extreme cold, intense heat, microgravity, elevated radiation, and difficult space weather, they are a testament to the knowledge and skills of scientists and engineers of all nations who transformed possibilities into realities.  A great deal of knowledge has already been accumulated to facilitate the effective operation of spacecraft in these conditions, but significant advances await us yet, and still more scientists and engineers and visionaries are stepping forward to make their own contributions.

Expanding on this list and thereby bringing the benefits of space to all humanity in the 21$^{st}$ Century, and doing so through many new and different kinds of commercial and scientific endeavors, is the common theme that unites all of the chapters in this, the fifth volume in the Aerospace Technology Working Group series. We are grateful to a distinguished set of authors who have contributed their thoughts and insights here, and we hope you enjoy reading them as much as we do.

•••

Part 1 of this volume underscores the intense international interest in the issue. The authors have all played pivotal leadership roles fostering multilateral cooperation and seeking agreement on best practices both inside the United Nations system and beyond.

Part 2 examines the conceptual foundations of space sustainability with straightforward discussions of the political and legal questions that need to be resolved, and the opportunities being suggested for resolving them.

The third part draws our attention to national perspectives and the differences that become apparent based on when countries first became actively involved in space activity. Here our authors point out not only that the differences are real, but also that idealists have reason to hope that those differences can be resolved equitably.

Part 4 demonstrates the wide range of commercial thinking inspired by space sustainability. What does it take to sustain a business or generate a new one using space technology? What does it take to sustain human life as our species gets closer to settling on celestial bodies other than Earth? Can we find new methods for escaping Earth's gravity well that make the planet's own intra-atmospheric environment more sustainable?

The last section provides both a practical look at what space technology is doing and can do to stimulate and accelerate sustainable economic development on Earth, and an idealistic insight to the hopes many developing countries have for adding space-enabled capabilities to their development toolkit.

•••

This book is concerned with both the practical and the ideal aspects of space. It illuminates the breadth of discovery, innovation and space sustainability, and the challenges confronted to maintain it, and as you read through these pages, we hope that these points will become clear and compelling:

- The importance of keeping space open for innovation
- The need to make it accessible to the growing number of new participants
- The necessity that space be capable of supporting the enormous demands placed on it
- The inevitability that space sustainability can only be achieved through international cooperation

Ultimately the sustainability of our efforts in space are dynamic.

They are about doing things better, more creatively, more cooperatively, and more sensibly.

Many years ago I heard a Clan Mother of the Oneida Indian Nation say, *We do not inherit the Earth from our ancestors; we borrow it from our children.*

That is how I feel about space.

That is how we are learning to think about space sustainability.

•••

## DR. MICHAEL K. SIMPSON

Dr. Michael K. Simpson is Executive Director of the Secure World Foundation and former President of the International Space University. He has also been President of Utica College and the American University of Paris with a combined total of 22 years of experience as an academic chief executive officer. He currently holds an appointment as Professor of Space Policy and International Law at ISU. After graduating from Fordham University, Simpson accepted a commission as an officer in the U.S. Navy, retiring from the Naval Reserve in 1993 with the rank of Commander.

His naval experience included service as a Political-Military Action Officer at US European Command in Stuttgart, Germany. He completed his Ph. D. at Tufts University, The Fletcher School of Law and Diplomacy, holds a Master of Business Administration from Syracuse University; and two Master of Arts degrees from The Fletcher School. He has also completed two one-year courses in Europe: the French advanced defense institute (Institut des Hautes Études de Défense Nationale) and the General Course of the London School of Economics.

He is a member of the International Academy of Astronautics, a member of the International Institute of Space Law and a Senior Fellow of the International Institute of Space Commerce. He is the author of numerous scholarly papers, presentations, articles and book contributions.

His practical experience includes service as an observer representative to the UN Committee on the Peaceful Uses of Outer Space, participation in the IAF committees on Commercial Spaceflight Safety and Space Security, participating organization representative to the Group on Earth Observations, Vice Chair of the Hague Space Resources Governance Working Group, participation in the UN High Level Forum preparing UNISPACE +50, the Board of Directors of the World Space Week Association, and the Board of Governors of the National Space Society.

# PART I

# THE MULTILATERAL EFFORT TO ASSURE SPACE SUSTAINABILITY

## INTRODUCTION TO PART I

In Part 1 we hear from the global experts who are working directly in and with governments, multi-national agencies and governing bodies to craft the standards and agreements that are essential to sustaining humanity's access to use and use of its unique capabilities.

CHAPTER 2

# REFLECTIONS ON SPACE SUSTAINABILITY

## DR. KARL DOETSCH
### CHAIRMAN OF THE BOARD, ATHENA GLOBAL

Sustainable development of humankind on Earth involves, among other factors, assessing and managing natural and humankind's impacts on the sustainability of Earth's environment, alleviating, if possible, the threat of collisions between Near Earth Objects (NEOs) and Earth, using space both as a safe haven and as a source for extra-terrestrial resources, and finally safeguarding the capacity of the space environment to continue to support space operations in valuable Earth orbits. Space activity is central to achieving each of these aspects of sustainable development.

## NATURAL DISASTERS HAVE SHAPED EARTH AND ITS LIFE SINCE THE BEGINNING

Earth was formed as part of the Solar System some 4.5 billion years ago. Asteroids and comets are what remain of the primordial building blocks of the Solar System's planets. During the early formation of Earth over a period of tens of millions of years there were many collisions with the primordial building blocks that shaped Earth, and at one time caused what was to become Earth's moon to break away after a collision with what must have been a Mars-sized object. Collisions continued to occur even long after the shape of Earth was well defined and had stabilized, sometimes with large objects that had a profound influence on the life forms that had developed. For example, the Chixchulub impact 65 million years ago with a 6 km–sized object caused the K-T mass extinction. Although a collision with a body of this size might occur once in several million years, collisions with smaller bodies are increasingly likely and more frequent as their size is reduced, but the resulting damage to life forms is likely to be correspondingly less.[1]

Recent small-scale but notable recorded events are the Tunguska event in Siberia, Russia in 1908, where a small 60–190 m asteroid or comet exploded at an altitude of 5-10 km and destroyed a wide, remote forested area, and the Cheyabinsk event in 2013 during which a 20 m meteor exploded 30 km above Russia, and led to a large number of injuries and building damage.

### NUMBER OF NEOS DISCOVERED TO DATE

As of February 02, 2015, 12,191 NEOs have been discovered. Some 867 of these have been asteroids with a diameter of approximately 1 km or larger. Also, 1,545 have been classified as Potentially Hazardous Asteroids (PHAs).[2]

In seeking to protect against the effects of collisions, ways need to be found to reduce the number of potential collisions with asteroids and comets, which are referred to as "Near Earth Objects" (NEOs) when their trajectories take them into the neighbourhood of Earth, and thus to safeguard and preserve the diversity of existing life forms. Doing this will

---

[1]    International Academy of Astronautics Study, *Dealing with the Threat to Earth from Asteroids and Comets*, February 2009.

[2]    NASA, NEO Program, online: http://neo.jpl.nasa.gov/faq/

inevitably involve the world's space activity in a new kind of mission, which includes engineering and managing, both politically and technically, the application of NEO collision avoidance systems. It may also involve providing alternative robust shelters and safe havens away from Earth, leading eventually to the start of migrations of humans and other biota from Earth, first into the Solar System and even then beyond.

## EVOLUTION OF LIFE ON EARTH TOWARDS DOMINANCE BY HOMO SAPIENS

Given the concern described in this chapter with the dangers of collisions with comets and asteroids, it is ironic that it was indeed the collisions between comets and asteroids and the Earth which deposited many life-supporting materials that helped initiate and are essential to life on our planet, and thus started the long evolutionary process towards the emergence of humankind.

Life did indeed emerge, and the oldest confirmed fossils of single-celled microorganisms are 3.5 billion years old; multi-cellularity evolved perhaps 1 billion years ago, and it was some 400-250 million years ago that one of the most significant evolutionary steps occurred, the development of

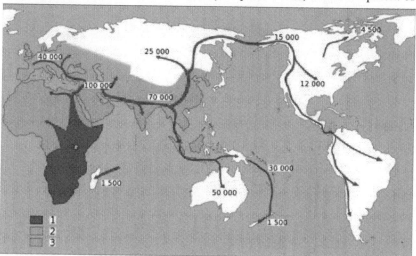

**Figure 1:**
**Homo sapiens migration out of Africa** (number of years ago)[3]

---

[3]   Wikipedia, the Free Encyclopedia, File: Spreading homo sapiens.svg, online: http://en.wikipedia.org/wiki/File: Spreading_homo_sapiens.svg

water creatures able to survive on land. Their subsequent migration from the oceans to land, and their further evolution to sentience, is one of the profound stories of life on Earth.

The Chixchulub event that occurred 65 million years ago and killed off the prevalent and dominant dinosaurs and about 60% of all other life on Earth, was the fifth known mass extinction. Mammals soon evolved in the aftermath of this event, and the first known hominid appeared on Earth some 52 million years later. The evolutionary branch we call Homo sapiens appeared some 400,000 to 250,000 years ago, residing in Africa until about 100,000 years ago when some tribes began migrating, eventually expanding the Homo sapiens sphere of influence to cover the entirety of the planet.

Neanderthals, who appear to have been better adapted to colder climates, had been settled in Eurasia from Western Europe to Central and Northern Asia some 200,000 to 250,000 years ago, but became extinct in Europe between 41,000 and 39,000 years ago, which coincided with the start of a very cold period, and was 5,000 years after Homo sapiens reached the continent.

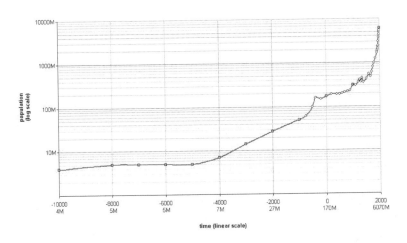

**Figure 2**
**Estimated Human World Population since 10,000 BCE[4]**

To survive, Homo sapiens became ever more capable of adapting to the environment and creating technology. He adopted and modified the tools of

---

4    http://upload.wikimedia.org/wikipedia/commons/f/f2/World_population_growth
     _%28lin-log_scale%29.png

the Neanderthals, made weapons, established complex communications patterns, worked as part of a collective and not only survived but appears to have flourished throughout the Ice Age, through which no other branch of the homo species survived. By the time that era had ended, Homo sapiens had become the dominant life form on Earth.

From the beginning around 400,000 to 250,000 years ago, human population growth accelerated quickly from a population level of about 0.8 billion at the time of the Industrial Revolution in 1760 to reach 5 billion by the late 1908s. Some current estimates of the future human population suggest that it will continue to rise from the 2014 level of 7.3 billion to reach almost 11 billion in 2100.[5]

## Homo sapiens, the Chief Operating Officer of Earth

There is no global commitment yet to dealing with the external threats faced by Earth, but what about the internal threats arising from natural forces and human activity?

Today the collective human impact on Earth is comparable to that of the forces of nature, and considerably greater than that of any other species. With such a predominant position come both demands and responsibilities, as our huge population leads to ever-increasing demands for the resources of Earth, and their processing, to meet basic needs and to satisfy human desires for increased standards of living and economic development. However, demand for an ever improving quality of life has not been satisfied uniformly; there has been a huge disparity between those living in developed nations, today comprising some 20% of the world population, and the rest.

Because of the current economic and social forces of globalization and to some extent of standardization, populations of many developing nations with rapidly emerging economies, particularly those in Asia, are striving for a much higher quality of life. At the same time, the greatest population growth and its attendant demands is occurring in developing nations, and is expected to continue to do so for several generations. Consequently, the demand for Earth's resources will continue to grow at a compounded rate, and the environmental damage caused by their processing could increase irreversibly without proper management. This will be strongly felt in

---

[5] UN, Department of Economic and Social Affairs, on line: esa.un.org/wpp/unpp/panel_population.htm

increased consumption of food, water, energy, natural resources and renewable resources. Additional adverse effects on Earth's environment will result from ever more pervasive agriculture, resource extraction and refinement processes, manufacturing processes and various other processes needed to satisfy the demands of human consumption. In addition, these activities have also precipitated mass extinctions of many other species, and are also leading to significant climate change.

## THE CONCEPT OF SUSTAINABLE DEVELOPMENT

In the 1970s, the seemingly unbounded growth in demand for resources generated by human activity, and the impact of their processing on the environment, led to the question of whether or not Earth was capable of continuing to absorb such demands.

The Club of Rome published a 1972 report entitled *The Limits to Growth*,[6] which explored scenarios relating to the impact of unlimited resource consumption in an increasingly interdependent world. It identified that choices needed to be made by society to reconcile sustainable progress within environmental constraints. As well as delivering a subsequently much-debated message on the need for collective societal action and for the imposition of constraints on societal behaviour to safeguard our future, it postulated as the underlying scientific message and warning that if the real world followed business as usual, then,

"Global society is likely to overshoot – and then be forced to decline or collapse - because of significant reaction delays in the global economy. These are the unavoidable lags in the perception and localization of global limits, the significant institutional delays involved in (democratic) decision making, and the biophysical lags between implementation of remedial action and the improvement of the ecosystem."[7]

The message was clear that action would be necessary well before adverse effects had taken an undeniable and irreversible hold. However,

---

[6]   Meadows, Donella H., Meadows, Dennis L., Randers, Jorgen, and Behrens III, William W., *The Limits to Growth: A report for the Club of Rome's Project on the Predicament of Mankind* (a Potomac Associates Book, 1972).

[7]   Jorgen Randers, *What was the message of Limits to Growth*, Version 5th, April 2010, online:
      http://connect.clubofrome.org/ecms/files/resources/What_was_the_message_of_Lim its_to_Growth.pdf

one of the unresolved arguments advanced by those not persuaded by *The Limits to Growth* was that technological advances would solve or ameliorate resource and pollution problems ahead of decline or collapse.

This difference in expectations is well illustrated by two salient examples of ecological limits being approached – the first was the degradation of Earth's ozone layer, where coordinated international regulatory action based on technological assessments appears to have reversed the decline, and the other is the cod fish stock on the Canadian continental shelf, where despite remedial action there appears to be no reversal in stock decline. Apparently technological advances coupled with regulation can resolve some issues, but not yet all, and not necessarily all of the most significant ones.

Since the publication of *The Limits to Growth* there have been significant further national and international considerations of the concept of striving for sustainable development, which responded to the warnings of this seminal report. The United Nations established the World Commission on Environment and Development in 1982, commonly known as the Brundtland Commission, to address the conflict between economic growth on a global scale and an accelerating ecological degradation. In 1987 the Commission issued its report, *Our Common Future*, to address multilateralism and interdependence of nations in search of a sustainable development path. It coined the most widely quoted definition for sustainable development as:

"Development that meets the needs of the present without compromising the ability of future generations to meet their own needs."[8]

This report was followed by a series of World Summits hosted by the United Nations that attracted many world leaders aiming to establish frameworks of cooperation and action. Each of these Summits provided carefully coordinated recommendations acceptable to the collectivity of nations represented, and identified actions to address the question of sustainable development. The Earth Summit 1992 in Rio de Janeiro, with its resulting documents, Agenda 21, the Rio Declaration on Environment and Development, the Statement of Forest Principles, the United Nations Framework Convention on Climate Change and the United Nations

---

[8]  UN, *Report of the World Commission on Environment and Development: Our Common Future* (1987), online: www.un-documents.net/wced-ocf.htm

Convention on Biological Diversity,[9] created general public awareness of the need to consider the environment and development together, with a focus on environmental protection and socio-economic development. Among many other important actions, it led directly to a study by individuals of the Indian Space Research Organisation (ISRO)[10] addressing how space activity could support a number of the key declarations of the summit.

At the Earth Summit 2002, Rio+10,[11] sustainable development was recognized as an overarching goal for society and many partnership initiatives were identified to help meet the earlier millennium development goals. More recently, at Earth Summit 2012, Rio+20,[12] emphasis was placed on building a green economy to achieve sustainable development that included the developing world, as well as on the development of an institutional framework to improve international coordination. These Summits clarified the complex interrelationship between environmental protection, social development and economic development, and sought national commitments to address various aspects of achieving sustainable development, while balancing these three aspects important to the future of all of our societies.

## THE ROLE OF SPACE IN HELPING ACHIEVE SUSTAINABLE DEVELOPMENT

Now we turn our attention to the vantage point of space, which by its very nature enables the provision of unique, continual, synoptic, temporal and spatial information about Earth and its atmosphere, and provides a vehicle for global communications and measurement in support of the various objectives defined throughout the series of Summits.

Space activity also provides an enabling capability to address potential collisions with Earth by NEOs, to increase our knowledge about the Universe, to explore parallels to our Earth system, to seek other life forms

9    UN, United Nations Conference on Environment and Development, Rio de Janeiro, 3-14 June 1992, on line: http://www.un.org/geninfo/bp/enviro.html/
10   Rao U. R., Chandrasekhar, M.G., Jayaraman, V., Space & Agenda 21, Caring for Planet Earth (Prism 1995)
11   http://www.un.org/jsummit/html/documents/summit_docs/131302
12   UN, Report of the United Nations Conference on Sustainable Development, Rio de Janeiro, Brazil 20–22 June 2012, online: http://www.uncsd2012.org/content/documents/814UNCSD%20REPORT%20final%20revs.pdf

that may exist there, and to make available extraterrestrial resources that could help future developments of the human species and Earth's biota in general.

To consider the potential roles space activity might have in addressing major issues highlighted by the concerns of global sustainability, several questions may be posed:

- Can space assets be used in a significant and continual way to detect, record and warn of the progression of unsustainable impacts of natural and human activity on Earth, and help in the effective education about and management and remediation of unsustainable activity?
- Can space assets be used to detect, warn and protect Earth from potential NEO impacts that could cause significant damage on Earth?
- Can space provide a safe haven for Earth's biota?
- Can space exploration and extraterrestrial resource exploitation augment our knowledge and develop our capacity to maintain sustainable development of life on Earth?

The answer to each of these questions is affirmative.

Activity in space is already fully integrated into the human effort to provide for the necessities of modern life in a sustainable way. It enhances our capacity to manage and provide food, potable water, resources, and energy on a global basis. It allows us to better anticipate and manage the impact of natural and man-made disasters; it enables global communications, tele-health and tele-education services to remote areas, the protection of shelters, and the management of our ever more sophisticated transportation infrastructure. It has become essential for meeting our security and sovereignty needs, our weather forecasting, and the tele-control of financial and other transactions.

Space activity is also central to our understanding of the universe and our Solar System, and we have already learned much about how humans can live in full health and remain fully functional within a protective, life-support environment in space.

Instrumented probes and robots launched from Earth have landed on planets, comets and asteroids, and NASA is currently planning an ambitious program to capture and relocate an asteroid. These are the first small but essential steps to human exploitation of extraterrestrial resources, migration from Earth, and the protection of Earth from extraterrestrial bodies.

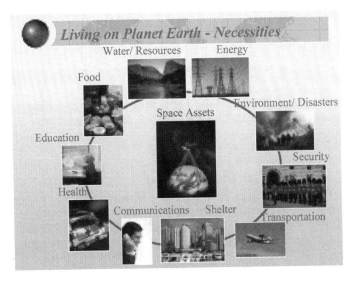

**Figure 3**
Space assets are becoming ever more central to managing the provision of the necessities of human life on Earth.

## THE EVOLUTION OF SPACE APPLICATIONS

Following the launch of the world's first orbiting satellite, Sputnik in 1957 and starting with launches of successful Earth satellites (Explorer for meteorology in 1959, Echo for communications in 1960, Transit for navigation in 1964, and Landsat 1 for Earth observations in 1972, space applications were quickly developed by several nations to provide the capability to communicate among all parts of the world, to observe and quantify different characteristics of Earth and its atmosphere, and to measure, position and navigate through GPS. Initially, each application was developed for military purposes. However, it was not long before military applications were joined by civil and commercial applications of these space-based, enabling capabilities, with each now being vigorously pursued by many nations and non-government entities. From a handful of developed nations in the early days of space, in 2015 there are some 80 nations with 1235 operating satellites applied to providing ever greater precision and more comprehensive coverage of Earth from space.

## THE EVOLUTION OF SPACE EXPLORATION

At the same time as space applications on Earth were being developed, space exploration activity was initiated to expand human knowledge of the Solar System and of the Universe, to consider whether life exists elsewhere and to consider establishing extraterrestrial human outposts and settlements, with one of its aims being to exploit extraterrestrial resources. All of our solar system planets have been visited and measured from their orbits. Landers, probes and robots have been deposited on several of them. Asteroids and comets have been visited and landed upon. Sophisticated space observatories in Low Earth Orbit (LEO) or at the Earth-Sun Lagrange point, which measure all segments of the electromagnetic spectrum as well as the forces of gravity, have been launched to better understand different characteristics of the Universe. And humans have ventured into space - after the first orbital flight of Yuri Gagarin in 1961, Neil Armstrong became the first human to set foot on an extraterrestrial body, Earth's Moon, in 1969. After this extraordinary start to human exploration of space, which culminated with the conclusion of the Apollo missions to the Moon in 1972, human presence in space has been restricted to space shuttles and space stations in LEO at altitudes of less than 500km above the Earth's surface, despite the fact that technically there exist many paths to the successful establishment of extraterrestrial outposts and settlements in the Solar System.

When the progress of space exploration to date is measured against time scales associated with past major migrations on Earth, it has been extraordinarily rapid. We would expect the migration of biota into space to take place over many human generations, and probably not be initiated on a grand scale until life on Earth came under a significant threat, perhaps because of insufficient sources of the necessities of life on Earth, or collision with a large NEO, thereby forcing an expansion into space. The first steps building for these eventualities are being taken.

In summary, activity in space through various applications can uniquely fulfill many of the demands for the information that is necessary to enhance sustainability through the management of human activity in response to natural events and human needs and disturbances. Further, space activity provides unique windows on, and entries to space that can address the threats of potential collisions of extraterrestrial objects with Earth, lay the foundation for using extraterrestrial resources to expand human capacity, and enable Earth's biota to migrate from Earth. Hence, we can further conclude that the achievement of long-term sustainability without employing space assets is unlikely.

**Figure 4**
**Will Biota from Earth migrate into our Solar System and beyond?**

## SPACE DEBRIS

The growth in space activity and its benefits has been accompanied by a deterioration of the environment of the orbits that are key to the continued efficacy of Earth applications because of the generation of space debris.

If space is to continue to play a critical role in enhancing and sustaining life on Earth, the increased use of space assets must be managed in such a way that space activity itself remains sustainable in Earth orbits that are ever more crowded with operational and defunct satellites and debris.

Efforts to control space debris are actively being pursued both at national and international levels, the latter under the auspices of the UN Committee on the Peaceful Uses of Outer Space (COPUOS). International working groups of this committee have been studying the issue and developing recommendations and guidelines since 1993.[13]

---

[13]  UN, OOSA, *Space Debris Mitigation Guidelines of the Committee on the Peaceful Uses of Outer Space*, 2010: online: http://www.iadc-online.org/References/Docu/Space_Debris_Mitigation_Guidelines_COPUOS.pdf

### SPACE DEBRIS – ONE INCIDENT

In February 2009, the first known major collision of two satellites in orbit – the non-operational Russian communications satellite, Cosmos 2251, with a mass of 900kg, and the operational US satellite, Iridium 33, with a mass of 600kg, collided at 33,000km per second about 800km above Earth – created a cloud of some 1300 pieces of space debris that will threaten orbiting spacecraft in LEO for decades. The 66-satellite fleet of Iridium currently makes about one collision avoidance manoeuvre every two weeks. The satellite collision, debris breakup and numerous near misses prompted renewed calls for international regulations or laws that would make it mandatory to dispose of defunct satellites.

## NATIONAL AND INTERNATIONAL EFFORTS TO USE SPACE ASSETS TO ENHANCE SUSTAINABILITY

The principal links between space activity and achieving sustainable development on Earth arise from the capacity of space activities to support informed decisions and actions on Earth, by providing spatial and temporal situational information about Earth and its atmosphere, real-time communications between any points on Earth, and precise positioning and movement information on Earth, as well as by providing warnings of any potential collisions with NEOs. This information is invaluable for assessing and managing environmental impacts, food production, potable water preservation, resources management, security, sovereignty, global communications, transportation, the health of Earth's flora and fauna, and man-made and natural disasters.

Many nations have addressed various aspects of sustainability through their space activities, and this is not restricted to nations with the largest space programs. For example, India has been at the forefront of specifically applying space to sustainable development. U. R. Rao and others[14] have comprehensively treated Indian work covering various aspects of space communications, tele-education, ecology, environment, disaster management, food, economic security, weather and climate. Many other nations developed similar if not such comprehensive programs, and recognized the need to share and, to some extent, integrate applicable data from their respective space systems. This was manifested through the

---

[14]   Rao U.R., *Space technology for Sustainable Development* (Tata McGraw Hill, 1996)

creation first of the Committee on Earth Observation Satellites (CEOS)[15] in 1984, and then of the Group on Earth Observations (GEO)[16] in 2005. The CEOS mission is *"to ensure international coordination of civil space-based Earth observation programs and promotes exchange of data to optimize societal benefit and inform decision making for securing a prosperous and sustainable future for humankind."* The GEO vision is *"to realize a future wherein decisions and actions, for the benefit of humankind, are informed by coordinated, comprehensive and sustained Earth observations and information."*

As the technologies for measurement and observation continue to advance, space assets have become ever more valuable as they have become further integrated with Earth-based systems to directly address many of the technical issues associated with understanding the forces of nature and the interaction of humans with our planet.

## THE UN AND GLOBAL SPACE ACTIVITY

COPUOS was established by the UN in 1959, and since the outset has addressed the issue of using space for the benefit of humankind and to support the goal of sustainability. COPUOS and its two subcommittees, the Legal Subcommittee and the Scientific and Technical (S&T) Subcommittee, provide the forum for developing international space treaties and guidelines for the international use of space, coordinate space activity within the UN system, and maintain a registry of space objects. COPUOS has developed a space applications program that emphasizes efforts in developing nations, and has sponsored and hosted many targeted conferences and special events throughout the World, generally in cooperation with other organizations. In addition, the UN has hosted 3 major general space conferences:

- Unispace 1, 1968, which succeeded in bringing an awareness of the significant potential of space benefits to all the member states;
- Unispace II, 1982, following which the UN program on space applications was considerably strengthened and expanded, resulting in increased opportunity for developing countries to participate in educational and training activities in space science and technology and to develop their indigenous capabilities in the use of space technology

---

[15]   CEOS, online: http://www.ceos.org/
[16]   GEO, online: http://www.earthobservations.org/

applications; and

- Unispace III, 1999, which provided opportunities for the international community to prepare a blueprint to maximize the benefits of space science and technology for all people into the new millennium, and led to *The Space Millennium: Vienna Declaration on Space and Human Development* (the Vienna Declaration).[17]

Unispace III and its resulting declaration addressed many of the elements of sustainable development that could be more effectively addressed by using space assets, specifically:

a) Protecting the Earth's environment and managing its resources
b) Using space applications for human security, development and welfare
c) Advancing scientific knowledge of space and protecting the space environment
d) Enhancing education and training opportunities and ensuring public awareness of the importance of space activities
e) Strengthening and repositioning of space activities in the United Nations system
f) Promoting international cooperation.

For each of these areas, specific recommendations (33 in all) were identified to advance the efficacy of using space activity for the benefit of humanity as a whole. The UN General Assembly approved these recommendations.

Starting in 2001 and continuing until 2004, the activity to advance the recommendations of Unispace III became one of the dominant considerations of COPOUS and its S&T Subcommittee, and it led to tangible results on many topics linked to sustainable development.[18]

At the same time, the Inter-agency Space Debris Coordination Committee established in 1993, and reporting regularly to COPOUS, had reached consensus on a series of guidelines for reducing the negative

---

[17] UN, *The Space Millennium: Vienna Declaration on Space and Human Development*, July 1999, online:
http://www.unoosa.org/pdf/reports/unispace/viennadeclE.pdf
[18] UN, *Review of the implementation of the recommendations of the Third United Nations Conference on the Exploration and Peaceful Uses of Outer Space*, July 2004, online http://www.oosa.unvienna.org/pdf/reports/unispace/A_59_174E.pdf

impact of the growing problem of space debris by 2002.[19]

As the Unispace III evaluation and action identification phase of the activity was coming to a close, the author, as the outgoing Chair of the UN S&T Subcommittee, was invited to address[20] COPUOS with observations on the Committee and its future directions. This address emphasized the importance of the central role of space activity in achieving sustainable development on Earth, and also noted that for this to be complete, additional emphasis needed to be placed by the Committee on the exploration of space for gaining knowledge, insights, potential resources and protection that would help achieve such sustainability on Earth.

### MAKING THINGS HAPPEN WITHIN THE UN SYSTEM

While approving the declaration emanating from Unispace III, the UN General Assembly did not assign any additional resources to its already resource-stretched Office of Outer Space Affairs (OOSA), to implement its recommendations. It fell to COPUOS to devise a process for moving forward, without additional resources from the UN system. The mechanism proposed by the author, who was then Chair of the S&T Subcommittee, was to offer for adoption and leadership by members states, various recommendations to be developed by different volunteer clusters of member states, named as Action Teams that reported to COPOUS, and worked with both government and nongovernment entities to develop more detailed recommendations and specific actions that would advance the implementation of the recommendations. This approach, in what was a novel approach for the United Nations, was developed to great effect. The resulting 12 Action Teams, involving 51 member states, 15 entities of the UN system, 10 international organizations that have observer status on the Committee and 13 other intergovernmental and non-government entities, considered 12 of the recommendations that COPUOS selected as being of highest priority. The Action Teams met regularly on the margins of the S&T Subcommittee and COPOUS annual meetings and reported progress to the committees, thereby assuring continuous and substantial progress over a three-year period.

---

[19]  UN, OOSA, *IADC Space Debris Mitigation Guidelines*, 2002, online: http://www.unoosa.org/pdf/spacelaw/sd/IADC_space_debris_mitigation_guidelines. pdf

[20]  Doetsch, K.H., *Observations on Activity of S&T Subcommittee of* COPUOS, Presented to 48th Session of COPOUS, Vienna, 8-17 Jun 2005, online: www.unoosa.org/pdf/transcripts/copuos/COPUOS_T536E.pdf

Activities of COPUOS since then have placed a consistent emphasis on such overarching goals, with a working group on the long-term sustainability of outer space being established in 2010,[21] and which is currently concluding the development of recommendations and guidelines for ensuring the sustainability of space activity.

The role of space activity towards achieving a sustainable world has been emphasized through a number of conferences and through the establishment of linkages to the main programs addressing different aspects of sustainability within the UN System.[22] Specific relationships of space to the achievement of some aspects of sustainable world have been explored,[23] for example through conferences on space and water,[24] space and the digital divide,[25] space and the global sustainability agenda,[26] space and agriculture,[27] space and climate,[28] space and disaster management,[29] space and fire,[30] space and maritime applications,[31] and space and global

---

[21]  UN, *Report of the Committee on the Peaceful Uses of Outer Space*, June 2010, online: http://www.unoosa.org/pdf/gadocs/A_65_20E.pdf

[22]  UN, Inter-Agency Meeting on Outer Space Activities: 2014, online: http://www.uncosa.unvienna.org/uncosa/en/iamos/index.htm

[23]  UN, *Report of the Inter-Agency Meeting on Outer Space Activities on its thirty-fourth session*, May 2014, online: http://www.uncosa.unvienna.org/pdf/reports/iamos/AC105_1064E.pdf

[24]  UNOOSA, Sixteenth UN/IAF Workshop on the Use of Space Technology for Water Management, Valencia, Spain, 29 to 30 September 2006, online: http://www.unoosa.org/oosa/SAP/act2006/spain/index.html

[25]  UN, *Report on the United Nations/Malaysia Workshop on Bridging the Digital Divide: Space Technology Solutions*, Kuala Lumpur, 20-24 November 2000, online: http://www.unoosa.org/pdf/reports/ac105/AC105_748E.pdf

[26]  UNOOSA, *Documents relating to Space and Socioeconomic Development in the context of Rio+20 and the Post-2015 Development Agenda*, online: http://www.unoosa.org/oosa/en/COPUOS/sustdev.html

[27]  UN, *Space for agriculture development and food security, Special report of the Inter-Agency Meeting on Outer Space Activities on the use of space technology within the United Nations system for agriculture development and food security*, April 2013, online: http://www.unoosa.org/pdf/reports/ac105/AC105_1042E.pdf

[28]  UNOOSA, United Nations/Indonesia International Conference on Integrated Space Technology Applications to Climate Change, September 2013, online: http://www.unoosa.org/oosa/en/climatechange/united-nations-indonesia-international-conference-on-integrated-space-technology-applications-to-climate-change.html

[29]  UNOOSA, *Space Technology and Disaster Management*, online http://www.oosa.unvienna.org/oosa/en/SAP/stdm/index.html

[30]  UNOOSA, International Space University Presentation to UN COPUOS STSC, *Global Wildland Fire Forecasting Using Space Technologies*, February 2006, online: http://www.unoosa.org/pdf/pres/stsc2006/tech-05.pdf

[31]  UNOOSA, 22nd United Nations/International Astronautical Federation Workshop on *Space Technologies Applied to the Needs of Humanity: Experience from Cases*

health.[32]   Particular emphasis has been placed on engaging developing nations so that the benefits of space activity could be more readily shared by all people.

In addition, educational institutes, for example the UN regional centres, the International Space University, and others have addressed many of the issues associated with space applications on Earth as well as the exploration of space, thereby actively engaging new generations of space professionals in addressing the profound influence that space can have on the development of our society.  As another example of such engagement, graduate students using public domain information to assess whether activities related to space are enhancing or reducing space security contribute strongly to the annual Space Security Index,[33] published since 2003.

## CONCLUDING COMMENTS

World recognition of the existence of limits to unconstrained population and economic growth, and of the high value of the development of space activity have advanced in parallel.  Indeed, a weak coupling between space activities and meeting basic human needs has existed since the earliest days of human utilization of space, but it is only in recent years that there is a widespread recognition that space activity, with its unique capacity to measure critical parameters and increase knowledge about the complex interaction between biota, land, water and the atmosphere, plays a unique and indispensable role in achieving the ambitious goals of humankind for a sustainable world.

It is also through space that humankind may be able to protect Earth from the collision with comets and asteroids, and it is through space that humans may find safe havens for Earth's biota and establish settlements away from Earth, and perhaps eventually replenish some of the non-renewable resources it uses on Earth.

All of the nations of the Earth can and do benefit significantly from our presence in space, and will benefit as well from a concerted and shared effort to protect and preserve this critical commons for the benefit of all.

in the Mediterranean Area , September 2012, online:
http://www.unoosa.org/oosa/en/SAP/act2012/un-iaf/index.html
[32]  UNOOSA, Space Applications for Global Health, online:
http://www.oosa.unvienna.org/oosa/en/SAP/health/index.html
[33]  Ploughshares, Space Security Index, online: http://ploughshares.ca/programs/space-security/space-security-index/

Through the efforts described in this chapter we have seen significant progress toward global alignment and agreement around the frameworks, agreements, and actions necessary to preserve this critical and unique resource that surrounds our unique planet, and we anticipate continuing diligence and progress in its care and development.

•••

## DR. KARL DOETSCH

Dr. Karl Doetsch was elected Chairman of the Board of Athena Global in 2005. In over 40 years of aerospace experience, Karl has occupied the most senior positions in his field. He worked in leadership roles on Canada's principal human space flight initiatives: the development of the Canadarm for the space shuttle, the establishment of Canada's Astronaut Program and the Canadian Space Station Program. He has served as Vice-President of Human Spaceflight, Vice President Programs, and Acting President of the Canadian Space Agency, as well as Chairman of Canada's National Space Plan Task Force, before becoming President of the International Space University.

He has served as Chairman of the Scientific and Technical Subcommittee of the United Nations Committee on the Peaceful Uses of Outer Space (UNCOPUOS), President of the International Astronautical Federation (IAF) and President of the Canadian Aeronautics and Space Institute (CASI).

Karl, who holds a Ph.D. in aerodynamics from Imperial College, London, is the recipient of numerous awards, including the Thomas Eadie Award of the Royal Society of Canada, the NASA Public Service Award, the Allan D. Emil Award of the IAF, the C.D. Howe Award of CASI and an Honorary Doctorate in Engineering from Kingston University, UK. The Executive Theatre at the International Space University is also named in his honour.

Karl's visionary leadership, extensive international experience and in-depth understanding of the power of space technologies to address environmental and social challenges have helped him make outstanding contributions to the use of space assets to meet society's needs.

CHAPTER 3

# THE LONG ROAD TO A SUSTAINABLE USE OF OUTER SPACE

## GÉRARD BRACHET

SPACE POLICY CONSULTANT

- FORMER DIRECTOR GENERAL OF CNES, THE FRENCH SPACE AGENCY (1997-2002)
- FORMER CHAIRMAN OF THE UN COMMITTEE ON THE PEACEFUL USES OF OUTER SPACE (2006-2008)
- FRENCH EXPERT ON THE UN GROUP OF GOVERNMENTAL EXPERTS ON OUTER SPACE TCBMS (2012-2013)

## 1. A WORRYING SITUATION

### 1.1 WHY WORRY?

The safety and sustainability of space activities in Earth orbit over the long term is increasingly a matter of concern. Today, most actors in Outer Space, including states, regional space organizations and commercial satellite operators, realize that our use of Outer Space since 1957 has been rather careless about its long-term sustainability. The situation is quite similar to what we have seen in the 20th Century in the exploitation of oceans for shipping and fishing: voluntary ignorance of the negative impact of pollution and generalized blindness of the long-term effects of over-fishing.

The increasing number of actors in Outer Space, both government and private, is a measure of the great success of space systems for exploration, for scientific research, and for the services they can deliver to daily life on our planet. However, this proliferation of actors requires a stronger international regulatory framework in order to guarantee every operator a fair, safe, and sustainable access to Outer Space.

The rapid increase in the population of orbital debris is the major concern, and although recent measures have been adopted at the international level to limit its future growth, the result of this will not be seen for decades.

There is also the risk associated with the potential use of weapons in Outer Space, the risk that Outer Space becomes another battlefield.

In addition, there is the difficult issue of managing the finite radio-electric spectrum available, and the orbital slots allocated to geostationary satellite operators, a challenging task for the International Telecommunication Union (ITU) which operates under United Nations rules, i.e. by consensus for most of its activities, and which also lacks any real enforcement power.

Finally, we should consider the potential effect of solar storms on the operations of orbital systems, but this is not a new issue and the risks are independent of the number of functional space objects in orbit.

These are the main factors that call into question our ability to continue operating safely and without interference in Earth orbit, and which we will explore in this chapter.

### 1.2   SOME NUMBERS

Some hard numbers are useful to help us understand the situation:

- 11 States have acquired their own space launch capability, but only 6 of them have a routine launch capability. They collectively conduct about 70 to 80 launches per year (80 in 2011, 75 in 2012, 78 in 2013 and 34 during the first half of 2014). More and more launches deliver multiple spacecraft to orbit, sometimes more than a dozen small satellites.
- 4994 space launches have taken place from 1957 till the end of

2013.

- About 20,000 objects are tracked and cataloged by the US Space Surveillance Network, of which approximately 22% are spacecraft (functional and non functional), 12% spent upper stage rocket bodies, 10% mission-related objects, 56% fragments (up from 41% before the China ASAT test of January 11, 2007).

- About 60 States and regional governmental organizations operate satellites in Earth orbit, and an increasing number of private companies operate commercial satellite systems, both in the Geostationary Earth Orbit (GEO) (mostly telecommunications) and in Low Earth Orbits (LEO) (telecoms and earth observation). There are today about 1000 operational satellites, of which half are operating in the GEO ring, most of the rest either on the LEO orbits or the MEO orbits used by GPS and other global navigation satellite constellations.

### 1.3  THE MAIN CAUSES OF CONCERN

Increased crowding in Low Earth Orbit as well as in the region around the geostationary ring creates new challenges, among which managing the orbital and radio spectrum resources is becoming a real issue that will require more stringent rules and enforcement mechanisms by the International Telecommunications Union (ITU). In addition, the proliferation of space debris on and around the most widely used orbits is a major concern which calls for additional international mechanisms to limit debris creation and mitigate the negative impact of the present population, thereby ensuring a sustainable use of Outer Space.

Concerning the risk of outer space becoming a battlefield, one can note the positive fact that deployment of weapons in outer space has apparently not taken place so far. A less positive fact is that ground-based weapons can be used - and have been used - against spacecraft in Low Earth Orbit. If such weapons were activated during a conflict it is clear that the additional debris cloud resulting from such actions would jeopardize the secure use of near-Earth outer space for a very long time.

## 2.  INITIATIVES IN THE UNITED NATIONS CONTEXT

The issue of sustainability of Outer Space activities has already been addressed by many bodies set up by the international space community, for example the Inter-Agency Space Debris Coordination Committee (IADC) which focused on the proliferation of space debris and its impact on the safety of orbital systems, by the International Academy of Astronautics which published a report on space traffic management in 2006[1] and by the International Association for the Advancement of Space Safety (IAASS), which published a report called *An ICAO for Space?* in 2007.

The "space debris" work of the Inter Agency Debris Coordination Committee (IADC) and subsequent adoption in 2007 by the United Nations Committee on the Peaceful Uses of Outer Space (UN COPUOS) of its Space Debris Mitigation Guidelines[2] provides a good model of how the international community can make some progress towards a regime of sustainable space operations.

What is striking about the process for the development of these guidelines is that it was very much a bottom-up process, starting with a detailed assessment of the situation by technical experts from the IADC agencies,[3] complemented by many tests and simulations, continuing with technical discussions of possible mitigations measures and finally development of a consensus basis of orbital debris mitigation guidelines based on a very robust technical ground. The first version of the *IADC Debris Mitigation Guidelines* was published in 2002.[4] They formed the basis of the discussion within the COPUOS Scientific and Technical Sub-Committee (STSC) when it decided to take up this issue in 2003. The advantage of such a bottom-up, technically robust approach is that recommendations that emerge are agreed by all parties, thus making it difficult for the political level to reopen and thus disrupt the debate.

In addition, the continuous work of IADC provides a mean to constantly monitor the evolution of the orbital debris situation, to review regularly the adequacy of the mitigation guidelines and update them if needed. The IADC agencies are also exploring ways to go beyond debris mitigation and study possible "remediation" measures. The annual reporting of IADC activities at the February sessions of COPUOS/STSC in Vienna allows all Member States not represented in the IADC but participating in COPUOS to be regularly updated on its progress.

This positive assessment of the model of the IADC work and subsequent COPUOS/STSC debate on ways to mitigate the growing space

---

[1]  Cosmic Study on Space Traffic Management, IAA, 2006.
[2]  Endorsed by UNGA Resolution 62/217 of 21 Dec. 2007.
[3]  Member Agencies of IADC are ASI (Italy), CSA (Canada), CNES (France), CNSA (China), DLR (Germany), ESA, ISRO (India), JAXA (Japan), NASA (USA), Roskosmos (Russia), UK Space Agency (UK), the National Space Agency of Ukraine.
[4]  IADC-02-01, 12 April 2002, available on the IADC website : www.iadc-online.org

debris population and the threat they represent to safe operations in Earth orbit led the author of this paper to float the concept of a COPUOS/STSC Working Group dedicated to developing "rules of the roads for space traffic." This idea was first presented to the workshop held in Paris on May 15-16, 2006 on "Collective Security in Space, European Perspectives," organized by the Space Policy Institute of the George Washington University.[5]

During his term as chairman of UN COPUOS, the author subsequently raised the issue of long-term sustainability of Outer Space activities in the paper "Future role and activities of the UN COPUOS"[6] presented to member states' delegations at the plenary session of COPUOS in June 2007.

This proposal and the new impetus that it would give to COPUOS if it was accepted, led the United Nations Secretary General, via the Office for Disarmament Affairs, to invite the Chairman of COPUOS to brief the delegations at the First Committee of UN General Assembly. The First Committee oversees disarmament issues and is distinct from the Fourth Committee where COPUOS presents its annual report. This presentation took place on 22 October 2007 in New York, at a time when the space community worldwide was celebrating the 50th anniversary of the launch of the first artificial satellite, Sputnik. It reviewed past achievements of COPUOS and focused on the adoption of the COPUOS "Space Debris Mitigation Guidelines" at the plenary session in June. It stressed the usefulness of a bottom-up approach to achieve a consensus on possible "rules of behaviour" in outer space, based on sound technical assessment of issues and recommended solutions.

Following this first introduction of the issue in 2007, the initial intention was that the French delegation to COPUOS would table at the next plenary session of the Committee in June 2008 a formal proposal to include the topic "Long-Term Sustainability of Space Activities" as a new agenda item. If endorsed by consensus of all delegations, this topic would therefore become an official agenda item of COPUOS in 2009. However, consultations with other delegations who expressed strong support for this initiative, including the US delegation, showed that it would be premature to table this proposal in 2008. It was felt that it would be more realistic

---

[5]  Collective Security in Space, a Key Factor for Sustainable Long-Term Use of Space, Gérard Brachet, in « Collective Security in Space, European Perspectives. Workshop report, Space Policy Institute, The George Washington University, Washington DC, January 2007», pp. 1-16.

[6]  A/AC.105/L268 of 10 May 2007(section D).

and, in the end more efficient to start the process via a first phase of one year duration for international consultations involving government as well as and non-governmental space actors outside the formal COPUOS framework. The plan was to gather international support for the initiative, consult with actors not represented in COPUOS, and rely on a drafting group of experts contributing on a volunteer basis to produce a report describing what is at stake, before introducing a formal proposal to COPUOS for a new agenda item.

France agreed to take this up and invited representatives from 25 spacefaring nations, including some developing countries, as well as from the European Union and the European Space Agency, from commercial telecommunication satellite operators (Intelsat, Inmarsat, Eutelsat and SES) and from the UN Office for Outer Space Affairs (UNOOSA), the International Telecommunications Union (ITU), the International Space Environment Service (ISES) and the World Meteorological Organization (WMO) to participate on February 7 & 8, 2008 in Paris. This was an informal working meeting devoted to the topic of "Long-term sustainability of space activities." The purpose was to discuss the concept of long-term sustainability of space activities, exchange views on the assessment of the situation, debate on the way forward, and plan for the next steps.

The main conclusions of this informal meeting were:

- Brief the delegations at COPUOS on this initiative;
- Prepare a document presenting the various issues affecting Long-Term Sustainability of Space Activities;
- Organize communication channels with delegations involved in the discussions on Prevention of an Arms Race in Outer Space (PAROS) at the Conference on Disarmament.
- Brief the Non Governmental Organizations actively involved in "Space Security" topics and activities on "Safety of Space Systems" and obtain their support and contribution.

A presentation of the outcome of this informal meeting was made at the 44th session of the COPUOS Scientific and Technical Sub-Committee the following week.[7] The debate which followed during the STSC session and as well during the 51st plenary session of COPUOS in June 2008 confirmed that many delegations to COPUOS, perhaps a little wary of what was behind the concept of "Long-Term Sustainability of Space Activities,"

---

[7]   A/AC.105/C.1/2008/CRP 11, available at http://www.unoosa.org/.

were not ready to consider the inclusion of this topic as a new agenda item of the Committee. In particular some delegations from emerging space nations were suspicious that this initiative was a hidden attempt by established space-faring nations to slow down their access to Outer Space.

As a consequence it was felt preferable to postpone until 2009 the formal proposal for this new agenda item. However, despite this delay, the awareness of the importance of this issue had made significant progress and this was recognized by keen observers such as Theresa Hitchens, the current Director of UNIDIR, who published an insightful editorial in *The Bulletin* on the outcome of the 51st session of COPUOS entitled "COPUOS wades into the next great space debate."[8]

In parallel with the briefing of COPUOS delegations, the Canadian ambassador to the Conference on Disarmament in Geneva, Marius Grinius, coordinator of informal meetings on the Prevention of an Arms Race in Outer Space (PAROS) agenda item, invited the author in his capacity of current chairman of COPUOS to brief the CD delegations on COPUOS activities, which took place on 21 Feb. 2008. It highlighted the adoption in 2007 of the "Space Debris Mitigation Guidelines" and included an outline of the initiative on long-term sustainability of space activities, including the outcome of the informal working group meeting in Paris. Of interest here is to note that the Russia-China proposal for a treaty called "The Draft Treaty on the Prevention of the Placement of Weapons in Outer Space, and of the Threat or Use of Force Against Outer Space Objects," known as the "PPWT"[9] had been formally tabled at the CD by the Russian Minister of Foreign Affairs, Sergei Lavrov, on February 12. Also, coincidently, the COPUOS presentation took place one day after the United States destroyed their derelict USA 193 spacecraft, orbiting at a low altitude, by a ship-based missile. The low altitude of the spacecraft led the orbital debris generated by this destruction to re-enter the atmosphere within weeks after the event.

A presentation of the "Long-term sustainability of space activities" initiative was also made at the annual UNIDIR conference on space security, "Security in Space: The Next Generation" on 31 March and 1 April 2008 in Geneva.[10]

The "Paris" informal working meeting was followed by another informal meeting held in Glasgow, Scotland (United Kingdom) on October

---

[8]   www.thebulletin.org, 26 June 2008.
[9]   CD/1839
[10]   "Long-Term Sustainability of Space Activities," Gérard Brachet, Conference Report UNIDIR 2008/14, pp. 121-123.

3, 2008, in conjunction with the 59th International Astronautical Congress. Twenty three countries were represented, plus UNOOSA, ESA, the European Union, commercial satellite telecommunication operators Eutelsat and Inmarsat, as well as non-governmental organisations such as ISES, the International Association for the Advancement of Space Safety (IAASS) and the Secure World Foundation. This meeting reviewed the progress made since the Paris meeting in February, including preliminary reactions recorded in COPUOS and outcomes of the presentation of COPUOS activities at the Conference on Disarmament, and went into a detailed review of the structure of the basic document prepared by the drafting group on the topic of "Long-term sustainability of space activities."

A further meeting of the drafting group, enlarged at the request of many delegations attending the 46th session of COPUOS/STSC, took place in Vienna on 17 February 2009. Besides reviewing the draft document, this meeting served the purpose of informing these delegations of the progress made in preparing for a formal submission the proposed new agenda item to the plenary session of COPUOS in June 2009.

Following this, at the 52nd session of COPUOS in June 2009, the French delegation to UN COPUOS formally proposed the topic of "Long-Term Sustainability of Outer Space Activities" as a new agenda item of COPUOS in 2010.

COPUOS agreed to include this item as a new agenda item of its Scientific and Technical Sub-Committee (STSC) in 2010 and beyond. The COPUOS/STSC then decided to set up a formal Working Group to address this issue, as it had done in 2003 for the space debris issue. Dr. Peter Martinez (South Africa) was selected as chairman of this new, dedicated Working Group, and its first meeting took place in conjunction with the 53rd session of COPUOS in Vienna in June 2010. [Editor's note: Please see the following chapter describing this work, which was authored by Dr. Martinez.]

The UN COPUOS Working Group should conclude its work in 2015, based on discussions of its Chairman's report during the February 2014 session of the COPUOS Scientific and Technical Subcommittee, and further discussions in the margins of the plenary session of COPUOS in June.

## 3. THE EUROPEAN UNION PROPOSAL FOR AN INTERNATIONAL CODE OF CONDUCT FOR OUTER SPACE ACTIVITIES

In parallel to the COPUOS activities on Long-Term Sustainability of Outer Space Activities, the discussions on the Prevention of an Arms Race in Outer Space (PAROS) at the Conference on Disarmament (CD) in Geneva were not making any progress because the Conference could not agree on a work plan. As a consequence, the Council of the European Union had taken in 2007 an initiative to propose an "International Code of Conduct" for Outer Space Activities. A first version of the EU draft Code of Conduct was approved by the EU Council in December 2008 and widely circulated.

Bilateral consultations with many space-faring nations were conducted by the European Union in 2009 – 2010, leading to a new version of the EU-proposed International Code of Conduct that was circulated in September 2010. A first multilateral meeting took place on June 5, 2012 in Vienna, where an updated version of the draft Code was presented by the newly established European External Action Service (EEAS), and a second multilateral forum took place in Kiev, Ukraine, on 15th and 16th May 2013. This led to the issuing of yet another version of the draft international Code of Conduct in September 2013, presented and discussed at the third multilateral forum in Bangkok in November 2013. The EU European External Action Service is presently considering the comments it has received at this meeting in order to plans for the next stage of consultation.

## 4. THE UN GROUP OF GOVERNMENTAL EXPERTS

Every year, the first Committee of the UN General Assembly prepares and submit to the full General Assembly a draft Resolution on Transparency and Confidence Building Measures (TCBMs) in outer space activities. In 2010, Resolution 65/68 on Transparency and Confidence Building Measures (TCBMs) in outer space activities, the first such resolution adopted without a negative vote from the United States, requested the UN Secretary General to set up a Governmental Group of Experts (GGE) to conduct a study on outer space transparency and confidence measures and report to the UNGA by end of 2013.

The Group of Governmental Experts was formally set up at the beginning of 2012 and included representatives from 15 countries: Brazil, Chile, China, France, Italy, Kazakhstan, Nigeria, Romania, Russian

Federation (Chair), Rep. of Korea, South Africa, Sri Lanka, Ukraine, United Kingdom, United States.

It held its inaugural meeting in New York from July 23 to 27, 2012, and two working meetings in 2013, one from April 1 to 5 in Geneva and another from July 8 to 12 in New York. The GGE worked very efficiently under the chairmanship of Victor Vasiliev, at that time Deputy Chief of the Russian permanent mission to the United Nations and the Conference on Disarmament in Geneva, and was able to finalize its report, adopted by consensus, during its meeting in July in New York. It was submitted for endorsement to the first Committee of UN General Assembly in September 2013. The report was endorsed unanimously by the General Assembly in its Resolution 68/50 adopted on the 5th of December 2013 (reproduced in annex).

It is worth mentioning here some of the recommendations that appear in the GGE report. In section VI (Consultative mechanisms):

> *"Timely and routine consultations through bilateral and multilateral diplomatic exchanges and other government-to-government mechanisms including bilateral, military-to-military, scientific, and other channels can contribute to preventing mishaps, misperceptions and mistrust. They may also be useful in:*
>
> a.  *Clarifying information regarding exploration and use of space, including for national security purposes.*
> b.  *Clarifying information provided on space research and space applications programmes.*
> c.  *Clarifying ambiguous situations.*
> d.  *Discussing the implementation of agreed transparency and confidence-building measures in outer space activities.*
> e.  *Discussing the modalities and appropriate international mechanisms to address practical aspects of outer space uses.*
> f.  *Preventing or minimizing potential risks of physical damage or harmful interference.*
>
> *States are encouraged to consider using existing consultative mechanisms, for example as provided for in Article IX of the Outer Space Treaty of 1967 and the relevant provisions of the ITU Constitution and the Radio Regulations."*

Also, in its concluding section:

*"The GGE endorses efforts to pursue political commitments, for example, in the form of unilateral declarations, bilateral commitments or a multilateral code of conduct to encourage responsible actions in, and the peaceful use of, outer space. The GGE concludes that voluntary political measures can form the basis for considerations of concepts and proposals for legally binding obligations."*

Clearly, while not quoting directly the European Union proposed International Code of Conduct for Outer Space activities, the GGE report recognized the value of such an approach as a step towards more transparency and more international confidence in the conduct of space activities. The fact that the GGE report was adopted by consensus among its experts and was later unanimously endorsed by the 1st Committee and by the full General Assembly indicates a strong support by the international community for the directions it is pointing to.

## 5. CONCLUSIONS

The various initiatives described above of the UN Committee on the Peaceful Uses of Outer Space (COPUOS), of the European Union for an international Code of Conduct for Outer Space Activities, and the recent work of the UN Governmental Group of Experts on Outer Space TCBMs illustrate the serious concern that both spacefaring nations and non spacefaring nations hold for the future safety and sustainability of the uses of Outer Space for government-sponsored as well as commercial applications.

The recommendations and/or guidelines that these initiatives produce are for the most part converging, but will have to be incorporated into more binding rules and regulations, first at the national level and gradually at the international level.

An open question is how to convince the new actors in space activities, which are mostly from emerging nations, that it is in their best interest to abide with the recommendations and guidelines developed within the above mentioned bodies. The answer to this critical question lies in the ability of the historical spacefaring nations to include the new space actors and nations in the elaboration of such guidelines, and later, in the development of international conventions which might formalize the obligations resulting from them.

The complementary work done within the ITU to reinforce its ability to regulate the use of the radio-electric spectrum and the GEO orbital slots should eventually complete a whole new set of principles and guidelines which will provide for a better governance of the uses of Outer Space. Hopefully this new governance will lead to a safer and more sustainable outer space environment in the future.

•••

# ANNEX

Text of Resolution 68/50 adopted by the General Assembly of the United Nations 5 December 2013

## 68/50. Transparency and confidence-building measures in outer space activities

*The General Assembly,*

*Recalling its resolutions 60/66 of 8 December 2005, 61/75 of 6 December 2006, 62/43 of 5 December 2007, 63/68 of 2 December 2008, 64/49 of 2 December 2009 and 65/68 of 8 December 2010, as well as its decision 66/517 of 2 December 2011,*

*Recalling also the report of the Secretary-General of 15 October 1993 to the General Assembly at its forty-eighth session, the annex to which contains the study by governmental experts on the application of confidence-building measures in outer space,*

*Reaffirming the right of all countries to explore and use outer space in accordance with international law,*

*Reaffirming also that preventing an arms race in outer space is in the interest of maintaining international peace and security and is an essential condition for the promotion and strengthening of international cooperation in the exploration and use of outer space for peaceful purposes,*

*Recalling, in this context, its resolutions 45/55 B of 4 December 1990 and 48/74 B of 16 December 1993, in which, inter alia, the General Assembly recognized the need for increased transparency and confirmed the importance of confidence-building measures as a means of reinforcing the objective of preventing an arms race in outer space,*

*Noting the constructive debates that the Conference on Disarmament has held on this subject and the views expressed by Member States,*

*Noting also the introduction by China and the Russian Federation at the Conference on Disarmament of the draft treaty on the prevention of the placement of weapons in outer space and of the threat or use of force against outer space objects being the first State to place weapons in outer space,*

*Noting the presentation by the European Union of a draft of a non-legally binding international code of conduct for outer space activities,*

*Recognizing the work that takes place within the Committee on the Peaceful Uses of Outer Space, its Scientific and Technical Subcommittee and its Legal Subcommittee, which makes a significant contribution to the promotion of the long-term sustainability of outer space activities,*

*Noting the contribution of Member States that have submitted to the Secretary-General concrete proposals on international outer space transparency and confidence-building measures pursuant to paragraph 1 of resolution 61/75, paragraph 2 of resolution 62/43, paragraph 2 of resolution 63/68 and paragraph 2 of resolution 64/49,*

*Welcoming the work done in 2012 and 2013 by the group of governmental experts convened by the Secretary-General, on the basis of equitable geographical distribution, to conduct a study on outer space transparency and confidence-building measures,*

*1. Welcomes the note by the Secretary-General transmitting the report of the Group of Governmental Experts on Transparency and Confidence-building Measures in Outer Space Activities;*

*2. Encourages Member States to review and implement, to the greatest extent practicable, the proposed transparency and confidence-building measures contained in the report, through relevant national mechanisms, on a voluntary basis and in a manner consistent with the national interests of Member States;*

*3. Decides, in order to further advance transparency and confidence-building measures in outer space, to refer the recommendations contained in the report to the Committee on the Peaceful Uses of Outer Space, the Disarmament Commission and the Conference on Disarmament for consideration, as appropriate;*

*4. Requests the Secretary-General to circulate the report to all other relevant entities and organizations of the United Nations system in order that they may assist in effectively implementing the conclusions and recommendations contained therein, as appropriate;*

*5. Encourages relevant entities and organizations of the United Nations system to coordinate, as appropriate, on matters related to the recommendations contained in the report;*

*6. Decides to include in the provisional agenda of its sixty-ninth session, under the item entitled "General and complete disarmament", the sub-item entitled "Transparency and confidence-building measures in outer space activities."*

*60th plenary meeting*

*5 December 2013*

•••

## GÉRARD BRACHET

Gérard BRACHET is consultant in space policy, President of the Air and Space Academy (Académie de l'Air et de l'Espace) from 2009 to 2012, and Member of the International Academy of Astronautics.

- Engineering degree from the Ecole Nationale Supérieure d'Aéronautique (1967).
- Master of Sciences in Aeronautics and Astronautics from the University of Washington (1968).
- 1970 to 1982: various positions at Centre National d'Etudes Spatiales (CNES), the French Space Agency.
- Chairman of the European Space Agency's Programme Board for Earth observation from 1979 to 1981.
- 1982 to 1994: Chairman and Chief Executive Officer of SPOT IMAGE, the company set up by CNES to
- develop the market for remote sensing imagery from the SPOT series of satellites.
- 1995 to 1997, Director for programmes, planning and industrial policy at CNES.
- 1997 to 2002: Director General of CNES
- Chairman of the international "Committee on Earth Observation Satellites (CEOS)" in 1997.
- From 2003 onwards, consultant on space policy issues and space applications, advising industry and
- government institutions, in particular the European Commission and the European Space Agency.
- Research associate at Fondation pour la Recherche Stratégique (FRS), Paris.
- June 2006 to June 2008, chairman of the United Nations Committee for the Peaceful Uses of Outer Space (UN COPUOS).
- 2012 - 2013, expert from France in the Group of Governmental Experts (GGE) set up by the Secretary General of the United Nations to develop "Transparency and Confidence Building Measures" in outer space.

Honors and Awards:
- 1985: "1985 Laurels" of Aviation Week and Space Technology
- 1992: Brock Gold Medal Award of the International Society for Photogrammetry and Remote Sensing
- 1994: Gold Medal of the British Remote Sensing Society
- 2007: Social Sciences Award of the International Academy of Astronautics
- 2015: Distinguished Service Award of the International Astronautical Federation
- Officier de l'Ordre National du Mérite (France)
- Officier de la Légion d'Honneur (France)

CHAPTER 4

# SUSTAINABILITY, COPUOS AND THE LTSSA WORKING GROUP

## DR. PETER MARTINEZ
SPACELAB, DEPARTMENT OF ELECTRICAL ENGINEERING
UNIVERSITY OF CAPE TOWN

## I. INTRODUCTION

The term *space sustainability* has gained a certain currency in the past few years, and is now broadly used to refer to a set of concerns arising out of

---

[1] Adapted and extended from a presentation delivered at the 65th International Astronautical Congress in Vancouver, Canada, in October 2014.

the realization that near-Earth space and the electromagnetic spectrum are limited natural resources that are under increasing pressure from the steady growth in the number and diversity of space actors. However, the broad usage of the term causes problems in terms of understanding exactly what it means. This is an important consideration when discussing this subject in the context of the United Nations, where there are different understandings of the underlying concept of "sustainability."

As with many terms in wide use, there is no universally accepted definition of the term "space sustainability," and attempts to define it precisely have led to much discussion and debate nuanced by the political, cultural, developmental, linguistic and conceptual differences among different countries. A precise definition of the term remains an elusive goal – yet the hazards of the degrading space environment are very palpable to any operator of a spacecraft, so actually achieving sustainability is an important and globally-shared objective. It seems, then, the best one can do in terms of the definition is to come up with a sort of phenomenological characterization of the term, which describes what it encompasses, rather than to attempt to devise a dictionary definition of the term. This is the approach that the United Nations Committee on the Peaceful Uses of Outer Space (UN COPUOS) has adopted.

This chapter describes the evolution of the so-called "Long-Term Sustainability," or "LTS," discussions in COPUOS, and provides some indications of the content of the emerging guidelines for space sustainability. Extensive references to relevant UN documents and reports are provided for readers who wish to study this evolution in greater detail. We also consider the linkages between the LTS work in COPUOS and the report of the UN Group of Governmental Experts on Transparency and Confidence-Building Measures in Outer Space Activities.

## II. THE EMERGENCE OF SPACE SUSTAINABILITY IN THE AGENDA OF COPUOS

The UN Committee on the Peaceful Uses of Outer Space (COPUOS) and its two Subcommittees have addressed various aspects of space sustainability under a number of agenda items, such as "space debris" and "space weather" for many years. However, consideration of the topic from a more holistic standpoint began in 2005, as the Committee approached its fiftieth anniversary, with a reflection on the future role and activities of COPUOS. At its forty-eighth session in 2005, discussions on the future role and direction of the Committee[2] were spurred by an informal paper on "Planning for future roles and activities of the Committee," prepared by the Chair of the Committee for the period 2004-2005, Mr. Adigun Ade Abiodun (Nigeria), and by a special presentation made by Mr. Karl Doetsch (Canada), Chair of the Scientific and Technical Subcommittee for the period 2001-2003, on the scientific and technical aspects of the work of the Committee and the way ahead. (A/60/20, paras. 316-317). [Editor's note: Please see Chapter 2 by Dr. Doetsch.]

At the next session, its forty-ninth in 2006, the Committee considered a working paper titled "Future role and activities of the Committee on the Peaceful Uses of Outer Space" (A/AC.105/L.265), prepared by the Secretariat in response to a request by the Committee at its previous forty-eighth session in 2005. The Committee agreed to continue considering the issue at its fiftieth session, and further agreed that the Chair of the Committee for the period 2006-2007, Mr. Gerard Brachet (France), would conduct inter-sessional, open-ended informal consultations with the view of presenting a list of elements that could be taken into consideration in its future work (A/61/20, para. 297). [Editor's note: Please see Chapter 3 by Mr. Brachet.]

At the fiftieth session of the Committee in 2007 Mr. Brachet presented a working paper titled "Future role and activities of the Committee on the Peaceful Uses of Outer Space" (A/AC.105/L.268). The paper described the informal consultations carried out by the Chair and identified the long-term sustainability of outer space activities among the challenges facing the future peaceful uses of outer space (A/AC.105/L.268, paras. 26-29). It suggested that a working group could be set up within the Scientific and Technical Subcommittee to develop recommendations to deal with the new realities of space operations, and to suggest a way forward.

In 2008, the Scientific and Technical Subcommittee and the Committee discussed the idea of introducing the long-term sustainability of outer space activities as an agenda item of the Scientific and Technical Subcommittee and what such an agenda item might encompass. Subsequently, in 2009 at the forty-sixth session of the Scientific and Technical Subcommittee, a proposal was put forward by France to include the long-term sustainability of outer space activities as a new agenda item of the Scientific and Technical Subcommittee under a multi-year workplan (A/AC.105/L.274). The Working Group of the Whole agreed to submit the proposal for a decision by the Committee (A/AC.105/933, para. 170 and annex I, paras.

---

[2]  In this chapter "the Committee" refers to COPUOS. The subcommittees of COPUOS are referred to by their full names.

20-22).

At its fifty-second session in 2009 the Committee agreed that its Scientific and Technical Subcommittee should include, starting from its upcoming forty-seventh session in 2010, a new agenda item titled "Long-term sustainability of outer space activities," and it proposed a multi-year work plan that was to culminate in a report on the long-term sustainability of outer space activities and a set of best-practice guidelines for presentation to and review by the Committee (A/64/20, paras. 161-162).

Consequently, in 2010, the Scientific and Technical Subcommittee established the Working Group on the Long-term Sustainability of Outer Space Activities, and elected the author of this chapter, Peter Martinez (South Africa), as the Chair of the Working Group (A/AC.105/958, paras. 181-182). It is instructive to read the views of the member States as the Subcommittee began to frame this agenda topic (A/AC.105/958, paras. 186-203), bearing in mind the definitional challenges raised at the beginning of this paper.

The first order of business of this new Working Group was to establish its terms of reference and methods of work. These matters were discussed at the fifty-third session of the Committee in June 2010. The Committee agreed on the nomination of national focal points to coordinate the discussions within the member States, and also agreed to invite member States and the permanent observers of the Committee and other relevant entities to present information on their activities pertaining to the long-term sustainability of outer space activities, for consideration by the Working Group (A/65/20, paras. 156-158). The discussion of the draft terms of reference (A/AC.105/L.277) brought to the fore the difficulty of agreeing on what is encompassed by the phrase "long-term sustainability of outer space activities," with a wide range of views being expressed by the member States (A/65/20, paras. 159-168). Consideration of these matters resumed at the forty-eighth session of the Scientific and Technical Subcommittee in 2011. At its fifty-fourth session in 2011, the Committee adopted the terms of reference and methods of work of the Working Group (A/66/20, Annex II).

The Working Group was tasked to consider current practices, operating procedures, technical standards and policies associated with the long-term sustainability of outer space activities, throughout all the phases of a space mission life cycle. The Working Group was to take as its legal framework the existing United Nations treaties and principles governing the activities of States in the exploration and use of outer space; it was not to consider the development of new legally binding instruments.

The outputs of the Working Group will be a report on the long-term sustainability of outer space activities, and a consolidated set of voluntary guidelines that could be applied by States, international intergovernmental organizations, national non-governmental organizations and private sector entities to enhance the long-term sustainability of outer space activities for all space actors and for all beneficiaries of space activities. In order to cover the wide range topics within its purview, the Working Group established four expert groups as deliberative fora to consider and propose candidate guidelines for space sustainability.

The Working Group sought to avoid duplicating previous efforts of COPUOS and its subcommittees, and also the efforts of other relevant entities. For this reason, the Working Group took into consideration the inputs from member States, international intergovernmental organizations and other relevant entities. Inputs from non-State entities were submitted through the relevant member States.

In 2012 during the forty-ninth session of the Scientific and Technical Subcommittee, a workshop was held in which the expert groups presented the inputs collected from member States regarding their views on important issues and practices associated with promoting the long-term sustainability of outer space activities. Non-governmental and private sector entities had the opportunity to present their inputs to the Working Group at a similar workshop arranged during the fiftieth session of the Scientific and Technical Subcommittee in February 2013.[3] These inputs were very helpful to identify those areas in which the common understanding of scientific and technical issues and the accumulated experiences of successful practices were sufficient to allow the proposal of candidate guidelines for space sustainability. By mid-2013, the expert groups had identified 31 candidate guidelines, which were supplemented by a further two candidate guidelines proposed by the Chair. These 33 candidate guidelines were grouped under several related categories in document A/AC.105/C.1/L.339 for consideration by the Working Group at the fifty-first session of the Scientific and Technical Subcommittee in February 2014.

---

[3]   The presentations delivered at these two workshops may be viewed on the website of the UN Office for Outer Space Affairs at:
http://www.unoosa.org/oosa/en/COPUOS/stsc/2012/presentations.html#wglts
http://www.unoosa.org/oosa/en/COPUOS/stsc/2013/ltsworkshop.html

## III. EMERGING GUIDELINES FOR LONG-TERM SUSTAINABILITY OF SPACE ACTIVITIES

In developing its terms of reference, the Working Group identified a wide range of topics of relevance to the overall considerations of space sustainability, spanning from developmental issues to operational issues, space debris, space weather, and also regulatory issues. The topics were clustered to allow more efficient consideration of related matters, and four expert groups were established to consider these related sets of topics. The expert groups were deliberative fora that comprised experts nominated by COPUOS member States. Experts therefore served in an ad hominem capacity and did not necessarily represent their governments' positions in all matters.

The expert groups were tasked to contribute inputs to the report of the Working Group and to propose candidate guidelines for consideration by the Working Group. During their consideration of the various topics, the expert groups identified issues of concern for which the current state of knowledge is inadequate to propose best-practice guidelines. These issues were flagged for future consideration by COPUOS and/or its appropriate Subcommittee.

The mandates, compositions and outcomes of the four expert groups are as follows:

### EXPERT GROUP A: SUSTAINABLE SPACE UTILIZATION SUPPORTING SUSTAINABLE DEVELOPMENT ON EARTH

Co-chaired by Mr. Filipe Duarte Santos (Portugal) and Mr. Enrique Pacheco Cabrera (Mexico)

Experts from 23 States and 5 international intergovernmental organizations took part in the deliberations of Expert Group A, which addressed the societal benefits of space activities and their contribution to sustainable development on Earth. It considered space as a shared natural resource, which raises issues of equitable access to outer space and to the resources and benefits associated with it. This expert group also considered the role of international cooperation in ensuring that outer space continues to be used for peaceful purposes for the benefit of all nations, proposed seven candidate guidelines, and identified four topics for future consideration.

## Expert Group B: Space debris, space operations and tools to support collaborative space situational awareness

Co-chaired by Mr. Richard Buenneke (United States of America) and Claudio Portelli (Italy)

Experts from 23 States and 4 international intergovernmental organizations took part in the deliberations of Expert Group B, which addressed the issues that make the space environment unpredictable and unsafe for space actors. These included an analysis of risks from space debris and measures to reduce the creation and proliferation of space debris. The implementation of such measures requires strengthened cooperative space situational awareness, which in turn requires the collection, sharing and dissemination of data on space objects. This expert group also considered tools to support collaborative space situational awareness, such as registries of operators and contact information and procedures for sharing relevant operational information among space actors. The group proposed eight candidate guidelines, and identified three topics for future consideration.

## Expert Group C: Space Weather

Co-chaired by Takahiro Obara (Japan) and Ian Mann (Canada)

Experts from 27 States and 5 international intergovernmental organizations took part in the deliberations of Expert Group C, which focused on ways to reduce the risks of detrimental effects of space weather phenomena on operational space systems. Such risks may be reduced through the sharing and dissemination of key data on phenomena related to space weather in real or near-real time, as well as sharing of models and forecasts. This group proposed five candidate guidelines and identified two topics for future consideration.

## Expert Group D: Regulatory regimes and guidance for actors in the space arena

Co-chaired by Anthony Wicht (Australia), Michael Nelson (Australia) and Sergio Marchisio (Italy)

Experts from 25 States and 6 international intergovernmental organizations took part in the deliberations of Expert Group D, which addressed the contribution of international and national legal instruments and regulatory practices to promote the long-term sustainability of outer space activities. This included considerations of how the existing Treaties and Principles that define the international legal framework for space activities are being implemented at national level through legal and regulatory regimes, and how such national regulatory frameworks for space activities can be developed or further strengthened to support the long-term sustainability of space activities. The importance of ensuring appropriate participation of all actors affected by new regulatory measures during the development of such measures was also considered. This expert group proposed eleven candidate guidelines and identified five topics for future consideration.

In total, thirty-three guidelines emerged from the Expert Group phase of the Working Group activities. The Expert Groups proved to be a productive way in which to consider a wide range of topics within the time allowed by the work plan for the Working Group, and it is worth noting that even though the experts serving in the expert groups were all nominated by States and international intergovernmental organisations, and even if the consensus rules of COPUOS were followed by the Expert Groups, the Expert Groups were deliberative fora that produced recommendations for consideration by the Working Group as a negotiating forum at the inter-governmental level.

## IV. PROGRESS DURING THE 2014 SESSIONS OF COPOUS AND ITS SUBCOMMITTEES

At the fifty-first session of the Scientific and Technical Subcommittee in February 2014, Expert Groups A, C and D presented their working reports, containing their recommended guidelines and topics for further consideration, to the Working Group for its consideration. The working reports of expert groups A, C and D are contained in documents A/AC.105/C.1/2014/CRP.13, A/AC.105/C.1/ 2014/CRP.15 and A/AC.105/C.1/2014/CRP.16, respectively. These working reports are instructive to read in the sense that they provide the background context and rationale for the proposed candidate guidelines.

Clearly, there are several cross-cutting issues that do not fit neatly within the purview of any one of the four expert groups. This is reflected in a certain amount of duplication and overlap in the candidate guidelines proposed by the expert groups, and for this reason the Working Group began to combine related guidelines with the view to producing a shorter set of consolidated guidelines.

During the fifty-first session of the Scientific and Technical

Subcommittee in February 2014, the Russian Federation also presented three working papers addressing a number of considerations in the long-term sustainability of outer space activities. One of these working papers (A/AC.105/L.290) also contained a proposal for three additional guidelines.

The first of these addressed the building of an international information and data sharing system, and proposed the establishment of a unified Centre for information sharing on near-Earth space monitoring, to be established under the auspices of the United Nations. The second proposed guideline addressed the observance of certain criteria for operations on active space debris removal, and the third proposed guideline addressed the security of foreign space-related ground and information infrastructures.

At the fifty-seventh session of the Committee in June 2014, Expert Group B presented its working report and candidate guidelines (A/AC.105/2014/CRP.14) and the Working Group commenced its consideration of a proposal by the Chair of the Working Group for the consolidation of the set of thirty-three draft guidelines proposed by the expert groups into a more streamlined set of sixteen guidelines (A/AC.105/2014/CRP.5).

Member States gave their comments on the proposals by the Chair and proposed some alternative groupings of the original draft guidelines. During the fifty-seventh session of COPUOS, the Swiss delegation proposed an additional guideline on the investigation and development of new measures to promote space sustainability in the medium and long term. This was intended to address the issue that the measures proposed by the Expert Groups would make a contribution to promoting space sustainability in the short term, and some measures (such as those addressing collision avoidance) are only applicable to a very small fraction of the number of space objects in orbit that have the capability to change their trajectory. Other States indicated their intention to consider and perhaps propose further guidelines in 2015.

During the development of the draft guidelines a number of difficulties were identified regarding the usage of certain terms and their translation into the official UN languages. The Working Group agreed to establish an informal translation and terminology reference group to assist in the Chair in the consideration and resolution on these issues.

One of the most important considerations that took place during the fifty-seventh session of the Committee in June 2014 was the extension of the work plan for the Working Group. The work plan adopted in June 2011 came to an end in June 2014, and a mandate for an extension was required. The duration of the extension was discussed at length, with some States arguing for a one-year extension, while others argued that the work should not be rushed, and that it should be completed by 2017, in time for the 50th anniversary of the Outer Space Treaty. In the end, a compromise was reached whereby States agreed to an extension up until 2016, but with a possibility to prolong the work by a further year if the circumstances so dictate. Part of the work plan extension compromise reached in June 2014 was the agreement that delegations would have until the fifty-eighth session of the Committee in June 2015 to propose new guidelines and/or substantive changes to the current draft guidelines.

## V. LINKAGES WITH THE REPORT OF THE UN GROUP OF GOVERNMENTAL EXPERTS ON TRANSPARENCY AND CONFIDENCE BUILDING MEASURES IN OUTER SPACE ACTIVITIES

The COPUOS work on Long-Term Sustainability is not the only initiative in the UN system that is addressing the topic of space sustainability. In 2010, the UN General Assembly adopted Resolution A/Res/65/68, which called for the establishment of a "Group of Governmental Experts on Transparency and Confidence-Building Measures in Outer Space Activities," the so-called "GGE."

The mandate of the GGE was to conduct a study on outer space transparency and confidence-building measures (TCBMs), making use of the relevant reports of the UN Secretary-General, and without prejudice to the substantive discussions on the prevention of an arms race in outer space within the framework of the Conference on Disarmament, and to submit to the General Assembly at its sixty-eighth session a report with an annex containing the study of the governmental experts.

The GGE, which comprised 15 experts selected on the basis of their knowledge and geographical representation, held its first meeting in New York in July 2012 and concluded its programme of work in July 2013 with the adoption of a consensus-based report (A/68/189). In December 2013 the UN General Assembly adopted Resolution 68/50, which welcomed the report of the GGE and requested the Secretary-General to circulate the report to all other relevant entities and organisations in the UN system. It further encouraged relevant entities and organisations of the UN system to coordinate as appropriate on matters related to the recommendations contained in the report.

In this context, and as also mandated in its terms of reference, the UN

COPUOS Working Group on LTS considered the linkages of the report of the GGE with the emerging COPUOS LTS guidelines. A number of linkages were identified in issues relating to information exchange on orbital parameters of space objects and potential orbital conjunctions; the provision of public access to national registries of space objects; exchange of information on forecast natural hazards in outer space; notifications on scheduled maneuvers that may affect spaceflight safety; notifications and monitoring of uncontrolled re-entries; and notifications of intentional breakups. A number of potential linkages in the domain of international cooperation could also be identified, and indeed, the LTS work in COPUOS complements the objectives of the GGE in that the emerging COPUOS LTS guidelines could provide the technical basis for the implementation of a number of TCBMs proposed by the GGE.

That these two UN processes have identified a number of issues in common is not surprising. However, what is important is that the issue of space sustainability is being seen as a cross-cutting issue that requires interaction between the UN's First and Fourth Committees, which have not interacted in the past on space issues. Indeed, the report of the GGE envisages that the First and Fourth Committees of the General Assembly may also decide to hold a joint ad hoc meeting to address possible challenges to space security and sustainability. This would be a very positive step towards promoting a more integrated global governance of space issues in the UN system.


## VI. Concluding Remarks

Of the seventy-seven member States in COPUOS, roughly half (mainly the countries with established or emerging space capabilities) participated in the expert group discussions. As the LTS process in COPUOS enters into the negotiation stage, more countries are beginning to engage in the discussions, and some of the more political considerations are coming to the fore. It is to be expected that additional guidelines and/or amendments to the current draft guidelines will be proposed. However, in the interests of delivering its guidelines in a timely fashion, the Working Group set itself a cut-off date of June 2015 for the introduction of substantive changes to the existing draft guidelines and/or new proposed guidelines.

A number of developing countries have expressed the view that the LTS guidelines should not raise the barriers to entry for countries aspiring to develop their national space capabilities. Others have expressed the view that voluntary non-binding instruments are inherently fragile and would not prove effective in promoting the long-term sustainability of outer space activities. However, although such instruments may be legally non-binding, they are in a sense politically binding.

Another important point to appreciate is that non-binding does not mean non-legal, in the sense that States can choose to domesticate their politically binding agreements to such voluntary frameworks in their national regulatory practices. Given the current lack of consensus on the need to develop new legally binding instruments, the development of voluntary frameworks provides a pragmatic alternative to address the pressing issue of space sustainability and provides technically based guidance that may be implemented by all space actors in their own interest.

•••

## DR. PETER MARTINEZ

Dr. Peter Martinez is the Chairman of the South African Council for Space Affairs, the national regulatory authority for space activities in South Africa, and Professor of Space Studies at the University of Cape Town. He has served on the South African delegation in several multilateral fora dealing with space affairs and is currently the Chairman of the United Nation Committee on the Peaceful Uses of Outer Space (UN COPUOS) Scientific and Technical Subcommittee's Working Group on the Long-Term Sustainability of Outer Space Activities. He was South Africa's representative on the United Nations Group of Government Experts on transparency and confidence building measures for space activities.

Dr. Martinez holds a PhD in astrophysics from the University of Cape Town. He is a member of the International Academy of Astronautics, the International Astronomical Union and the International Institute of Space Law. He has authored, co-authored or edited over 190 publications in journals and books and is an Associate Editor of the COSPAR journal *Advances in Space Research*.

CHAPTER 5

# STRENGTHENING STABILITY IN OUTER SPACE FROM THE BOTTOM UP:
## THE ROLE OF GUIDELINES FOR THE LONG-TERM SUSTAINABILITY OF OUTER SPACE ACTIVITIES

KENNETH D. HODGKINS,
RICHARD H. BUENNEKE,
AND
AMBER CHARLESWORTH
UNITED STATES DEPARTMENT OF STATE

During the past 50 years, the international community has made substantive progress in cooperative efforts to strengthen stability and ensure the long-term sustainability of space activities. Recently, in July 2013, the United

Nations Group of Governmental Experts (GGE) on Transparency and Confidence-Building Measures in Outer Space Activities issued a consensus report. This study's recommendations, which are now a topic for further consideration in several UN bodies, provide a unique opportunity to advance consensus on the importance and priority of pragmatic measures that can be implemented by States. The GGE report specifically supports efforts of the UN Committee on the Peaceful Uses of Outer Space and its Working Group on Long-Term Sustainability (LTS) of Outer Space Activities. Established in June 2010, the LTS Working Group is taking a "bottom-up" approach to develop guidelines based on the best practices of government and commercial space entities. This paper will discuss ongoing efforts and new opportunities for the implementation of specific transparency and confidence-building measures in bilateral and multilateral contexts, with a particular focus on the role of voluntary guidelines for sustainability that can be implemented by States and private sector entities.


## I. INTRODUCTION

The outer space environment has changed significantly over the past 50 years. As more and more States are realizing the benefits space has to offer, the number of space actors continues to rise. These benefits include communications, weather forecasting, environmental monitoring, navigation, surveillance and treaty monitoring.

Space assets and the benefits they provide were originally available only to a few nations, but today nearly 60 nations, international organizations, commercial entities, and academic and government consortia operate satellites, and many more benefit from the information derived from space systems. This evolution in the use of outer space has greatly benefited society and has brought people around the world closer together, but it also presents challenges.

As more nations and people benefit from space applications and space-derived information and the demand for satellite use has grown, the orbital environment has become increasingly congested, contested, and competitive. To overcome these challenges, dangers, and risks requires, among other things, building confidence among nations, which can be achieved with transparency, openness, and predictability.

Transparency and confidence-building measures (TCBMs) are an important means by which governments can address challenges and share information with the aim of creating mutual understanding and reducing tensions. TCBMs can reduce or even eliminate misunderstandings, mistrust, and miscalculations with regard to the activities and intentions of States in outer space. Building on existing national and multilateral TCBMS, in July 2013, the United Nations Group of Governmental Experts (GGE) on Transparency and Confidence-Building Measures in Outer Space Activities issued a consensus report (Report) that highlighted the importance of pragmatic and voluntary TCBMs that can be implemented readily by States.[1] In particular, the GGE noted the importance of ongoing efforts that build on existing actions taken by States as well as a growing set of "best practices guidelines" for information sharing and coordination of operations. The GGE further recommended that Member States take measures to implement, to the greatest extent practicable, these principles and guidelines to contribute to the development of "top down" TCBMs such as a multilateral Code of Conduct for Outer Space Activities.

## II. IMPLEMENTATION AT THE NATIONAL LEVEL

In General Assembly Resolution 68/50, Member States are encouraged to review the proposed transparency and confidence-building measures contained in the Report, and implement them to the greatest extent practicable, through relevant national mechanisms, on a voluntary basis and in a manner consistent with the national interests of Member States.[2]

A number of States have well-established unilateral, bilateral, and multilateral TCBMs already in place. Indeed, the GGE drew heavily on practices that were already in existence. For example, the United States and many other spacefaring nations have published information on their national space policies and strategies. Additionally, the United States has published information related to its national space activities, including budgets and research and development initiatives.

This exchange of information provides regular and routine opportunities for States to describe their current and planned space activities, thus leading to more transparency among nations. Additionally, many States are currently involved in exchanging information on the orbital parameters of outer space objects and potential orbital

[1] Report of the Group of Governmental Experts on Transparency and Confidence-Building Measures in Outer Space Activities, Summary, UN document A/68/189.

[2] Resolution adopted by the General Assembly on 5 December 2013 [on the report of the First Committee (A/68/411)] 68/50. Transparency and confidence-building measures in outer space activities.

conjunctions.    Space situational awareness (SSA) is foundational to understanding what is going on in space and to managing the increasing amount of space debris.  Finally, many States have voluntarily opened their national registries of space objects to the public, which has provided for insight into specific space activities, and has also perhaps contributed to the avoidance of misperceptions and mistrust.

While substantial progress has been made to promote and implement TCBMs at the national level, there is still work to be done in this area. States which have ongoing TCBM practices might find possible areas in which to improve, while others might utilize the proposed TCBM measures contained within the Report and identified as current State practice to identify potential areas where TCBMs might be initiated in the first instance.

## III.  IMPLEMENTATION AT THE MULTILATERAL LEVEL

In addition to encouraging Member States to review and implement the proposed TCBMs contained in the Report, the General Assembly also referred the recommendations contained in the Report to the Committee on the Peaceful Uses of Outer Space (COPUOS) for consideration, as appropriate.[3]

Over the past 50 years, COPUOS has developed a governance framework that has been instrumental in promoting international cooperation in the peaceful uses of outer space.   Under this framework, outer space activities by States, intergovernmental organizations, and private actors have flourished. This success is due, in part, to the existence of key elements of the framework, described below, that focus on specific areas of cooperation and information exchanges. These important elements have also contributed to the emerging TCBM regime and have led to work within COPUOS on developing guidelines for the long-term sustainability of outer space activities (LTS Guidelines).

Beginning with General Assembly (GA) Resolutions 1472 (XIV) of 1959 and 1721 (XVI) of 1961, the international community recognized the need for international cooperation in the peaceful uses of outer space. Expanding on the principle of cooperation, GA Resolution 1721 B calls upon all States launching objects into orbit or beyond to furnish

---

[3]    Resolution adopted by the General Assembly on 5 December 2013 [on the report of the First Committee (A/68/411)] 68/50. Transparency and confidence- building measures in outer space activities.

information to COPUOS, through the Secretary-General, for the registration of launchings, and requests the Secretary-General to maintain a public registry. This established, for the first time, information exchanges on outer space activities and provided a baseline for the development of additional instruments governing the use of outer space.

## A. DECLARATION OF LEGAL PRINCIPLES GOVERNING THE ACTIVITIES OF STATES AND THE EXPLORATION AND USE OF OUTER SPACE AND THE FOUR CORE OUTER SPACE TREATIES

General Resolution 1962 (XVIII), the Declaration of Legal Principles Governing the Activities of States and the Exploration and Use of Outer Space of 1963 (Declaration of 1963), expanded on GA Resolutions 1472 and 1721, establishing the principle that "the exploration and use of outer space shall be carried on for the benefit and in the interest of all mankind."

Additionally, it inferred, for the first time, that States should establish a national registry stating that: "[t]he State on whose registry an object launched into outer space is carried shall retain jurisdiction and control over such object...." This is important, as it expands on the principles associated with information sharing of GA Resolution 1721, inferring that beyond submitting information to the Secretary-General, launching States might have a national registry which could be utilized to exchange information on their space objects.

Ultimately the Declaration of 1963 led to the drafting and adoption of four core outer space treaties, all of which further reinforced the fundamental principles of international cooperation in the peaceful uses of outer space. Additionally, they have continued to establish key elements which have served as the basis for developing long-term sustainability guidelines and TCBMs.

The first of the four core outer space treaties, the Treaty on Principles Governing the Activities of States in the Exploration and Use of Outer Space, including the Moon and Other Celestial Bodies (Outer Space Treaty), is the overarching treaty that sets out obligations for the use of outer space. The Outer Space Treaty reaffirms and expands on key elements of cooperation contained in the Declaration of 1963. In this regard, the Outer Space Treaty states that the exploration and use of outer space shall be carried out for the benefit and in the interests of all countries, and shall be the province of all mankind. Further, it states that the exploration and use of outer space shall be guided by the principle of cooperation and that States shall conduct activities in outer space with due regard to the corresponding interests of all other States that are Parties to the Treaty.

The Outer Space Treaty also expands on the Declaration of 1963 associated with information exchanges, stating that, "if a State has reason to believe that an activity or experiment planned by it or its nationals in outer space would cause potentially harmful interference with activities of other States Parties in the peaceful exploration and use of outer space, it shall undertake appropriate international consultations." It is worth noting that this element might have served as the basis for some of the TCBM measures contained in the GGE Report to include "notifications on scheduled maneuvers that may result in risk to the flight safety of other space objects" and "notification of intentional orbital break-ups."

Finally, the Outer Space Treaty establishes the obligation to inform other States and the Secretary-General of any phenomena they discover in outer space which could constitute a danger to the life or health of astronauts. The GGE Report recommendation entitled "exchanges of information on forecast natural hazards in outer space" reaffirms this obligation, and further expands on it stating that "States should also consider providing, on a voluntary basis, timely information to other governmental and non-governmental spacecraft operators of natural phenomena that may cause potentially harmful interference to spacecraft engaged in the peaceful exploration and use of outer space."

These elements of cooperation and information exchanges in outer space activities are further reinforced in the Agreement on the Rescue of Astronauts, the Return of Astronauts, and the Return of Objects Launched into Outer Space (Rescue and Return Agreement), the Convention on International Liability for Damage Caused by Space Objects (Liability Convention), and the Convention on Registration of Objects Launched into Outer Space (Registration Convention).

Most notably, although not originally drafted to serve as a TCBM or as a mechanism to sustain the outer space environment, the Registration Convention has increasingly served these functions, obliging States, under certain situations, to provide information regarding space objects both by means of entries in their national registries and also to the Secretary-General. Indeed, the Registration Convention is explicitly cited and expanded upon in both the TCBM Report and in the draft LTS Guidelines.

Although the Committee enjoyed much success in drafting the four core treaties, in the 1980s it realized there was much to be gained by utilizing non-binding mechanisms to address emerging outer space challenges and to further reinforce the key principles contained within the outer space treaties. To this end, over the following 20 years the Committee adopted three sets of non-binding principles and one set of guidelines. Each of these non-binding mechanisms has served an important role in furthering cooperation in the peaceful uses of outer space. The elements described within these mechanisms have equally served as the basis for the development of TCBMs and for ongoing efforts to sustain the outer space environment.

## B. REMOTE SENSING PRINCIPLES, 1986

The Principles Relating to the Remote Sensing of the Earth from Outer Space introduced the concept of non-discriminatory data availability. The Principles have enabled the use of timely, high-quality space-derived geospatial data for sustainable development in areas such as agriculture, deforestation assessment, disaster monitoring, and land management, yielding significant societal benefits.

Global open data access policies provide geospatial data free of charge or for only a nominal cost. For example, the United States Geological Survey (USGS) provides, without cost, electronic access to all Landsat scenes. Additionally, several other satellite missions have similar data distribution policies. The principle of open and non-discriminatory data availability has promoted transparency and confidence building among nations, and is of vital importance to the sustainable development on Earth.

## C. PRINCIPLES RELEVANT TO THE USE OF NUCLEAR POWER SOURCES IN OUTER SPACE, 1992

The 1978 uncontrolled re-entry of a Russian military reconnaissance satellite and its nuclear reactor payload power source highlighted the importance of exchanges of data on potentially hazardous re-entry objects.[4] In 1992, the United Nations adopted principles that call on States launching space objects with nuclear power sources on board to protect individuals, populations, and the biosphere against radiological hazards.[5] These principles call for the design and use of space objects with on-board nuclear power sources to ensure that potential hazards, in foreseeable operational or accidental circumstances, are kept at acceptable levels. Specifically, the Principles call on the launching states to conduct a thorough and comprehensive safety assessment prior to launch, and in the event of an unplanned reentry, to provide advance notification to all concerned States.

---

[4] Gus W. Weiss, "The Life and Death of Cosmos 954," Studies in Intelligence, Vol 22 (Spring 1978), http://www.loyola.edu/departments/academics/politic al-science/strategic-intelligence/intel/cosmos954.pdf accessed on 17 August 2014.

[5] United Nations General Assembly, "'Principles Relevant to the Use of Nuclear Power Sources in Outer Space," A/RES/47/68, 14 December 1992.

Additionally, these voluntary Principles provide for States operating spacecraft with nuclear power sources to "inform in a timely fashion" other States concerned in the event the spacecraft is malfunctioning with a risk of re-entry of radioactive materials to the Earth. Over the following decade and a half these Principles served as a key cooperation tool among nations to provide transparency and predictability in outer space activities. As cooperation in this area expanded, it led to the establishment in 1997 of procedures for data exchanges between the members of the Inter-Agency Space Debris Coordination Committee for both nuclear and non-nuclear "high risk reentry events."[6]

Recent experience involving the monitoring of several uncontrolled spacecraft re-entries that did not pose a "high risk" has highlighted the value of the IADC data exchanges, as well as the need for considering the development of processes and procedures to ensure technical assessments can better support informed decision making by national policymakers and emergency management agencies.

## D. DECLARATION ON INTERNATIONAL COOPERATION IN THE EXPLORATION AND USE OF OUTER SPACE FOR THE BENEFIT AND IN THE INTEREST OF ALL STATES, TAKING PARTICULAR ACCOUNT THE NEEDS OF DEVELOPING COUNTRIES, 1996.

In 1996, the Committee adopted the Declaration on International Cooperation in the Exploration and Use of Outer Space for the Benefit and in the Interest of all States, Taking into Particular Account the Needs of Developing Countries. This instrument has been important as it further promotes international cooperation in an effort to maximize the benefits of the use of space applications for all States. To this end, the Declaration states that all States, particularly those with relevant space capabilities and with programs for the exploration and use of outer space, should contribute to promoting and fostering international cooperation on an equitable and mutually acceptable basis. It also states that particular attention should be given to the benefit for and the interests of developing countries and that States are free to determine all aspects of their participation (including with regard to such concerns as appropriate technology safeguard arrangements, multilateral non-proliferation commitments, and relevant standards and practices). Finally, in terms of information exchanges and promoting

---

6  Interagency Space Debris Coordination Committee, "Annex VI: Risk Object Reentry Data Exchange," Terms of Reference, 14 October 2008, http://www.iadc-online.org/Documents/IADC_ToR_October%202009.pdf, accessed on 17 August 2014.

transparency, the Declaration states that COPUOS should be strengthened in its role as a forum for the exchange of information on national and international activities in the field of international cooperation in the exploration and use of outer space. This is important in that it underscores the important role of COPUOS in promoting cooperation by providing a platform for openness and transparency.

## E. UN SPACE DEBRIS MITIGATION GUIDELINES

In 2007, COPUOS achieved a major success by drafting its own Space Debris Mitigation Guidelines. Through the adoption of these guidelines, there is a general agreement among States that the voluntary mitigation of space debris at the national level would help to ensure the sustainability of the outer space environment – thus, allowing the continued use of space applications for sustainable development on Earth.

Although the Debris Mitigation Guidelines have not played a major role in the development of TCBMs, COPUOS work on space debris mitigation guidelines has served as key catalyst for further efforts to develop a more extensive set of guidelines for the long-term sustainability of space activities.[7]

## F. COPUOS – CURRENT WORK

COPUOS continues to promote cooperation and transparency in the peaceful use of outer space. Its Legal Subcommittee has two standing agenda items intended to foster the exchange of information among Member States. One of these agenda items invites States to exchange information on national legislation, working to ensure outer space is used for peaceful purposes, and that the obligations under international law are implemented. Another item invites States to exchange information on national mechanisms relating to space debris mitigation measures. This exchange of information is an important transparency mechanism as it allows countries to gain insight and "lessons learned" from their neighbors and partners, and to potentially implement similar mechanisms and processes.

The work of the Scientific and Technical Subcommittee (STSC) has also contributed to the key principles of cooperation and transparency in the peaceful uses of outer space. Many emerging issues that require urgent

---

[7]  Martinez, P. "Assuring the Long-Term Sustainability of Outer Space Activities: The Role of UN COPOUS," *Proceedings of the 64th International Astronautical Congress*, Beijing, China, IAC-13- E3.4.3, 2013 and Brachet, G. "The origins of the Long-term Sustainability of Outer Space Activities initiative at UN COPUOS," Space Policy, 28 (2012), 161-165.

international action are first identified in STSC. For example, States are increasingly concerned about the potential impacts of increased and more intense solar radiation on satellites, humans in space, and terrestrial infrastructure such as power grids and communications systems. STSC has been very active in recent years in promoting international cooperation and the exchange of information on space weather as well as methods to mitigate its effects. Additionally, STSC has done important work to develop a set of guidelines to enhance the long-term sustainability of the outer space environment.

As Dr. Yasushi Horikawa, former COPUOS Chair, has said,

> "The work currently being carried by the Scientific and Technical Subcommittee through its Working Group on the Long-term Sustainability of Outer Space Activities is of critical importance; its goal is to ensure the safe and sustainable use of outer space over many years by future generations."

The objective of this Working Group on Long-Term Sustainability (LTS) under the Scientific and Technical Subcommittee is to develop a set of guidelines for the sustainability of outer space activities. To that end, four expert groups were established to recommend guidelines in areas including space situational awareness, space operations, space weather, and the use of space for sustainable development. The recommendations of these four expert groups have served as important inputs for deliberations by the full LTS Working Group, which is scheduled to complete its work in 2016. [Editor's note: Please see Chapter 4, "Sustainability, COPUOS and the LTSSA Working Group" by Dr. Peter Martinez for a more detailed discussion of the four Expert Groups.]

## IV. ENHANCING STABILITY IN SPACE

As the COPUOS LTS Working Group has progressed, it has become increasingly apparent that it could also contribute to the development of TCBMs that would enhance stability. This linkage was directly noted by the GGE on space TCBMs, which included several experts with extensive personal experience with the work of COPUOS and its subcommittees. In its report, the GGE "noted the importance" of the development and implementation of LTS working group guidelines by all States and intergovernmental organizations. "These guidelines will have characteristics similar to those of transparency and confidence-building measures; some of them could be considered as potential transparency and confidence-building measures, while others could provide the technical basis for the implementation of certain transparency and confidence-building measures proposed by [the GGE]."

The interrelationship between sustainability guidelines and TCBMs can be seen in ongoing efforts to increase international cooperation and information sharing for satellite collision avoidance. Many have noted the need to establish improved measures to ensure timely notification of potential satellite conjunctions and exchanges of information between satellite operations centers – military as well as civilian and commercial – and space situational awareness centers. In response, the United States Strategic Command's Joint Space Operations Center has sought to obtain and maintain a roster of contact information for the full range of satellite operations centers. To help obtain this information, the United States has used its bilateral diplomatic dialogues on space security as well as participation in COPUOS to build support for the principle that orbital collision avoidance serves the common interest of the exploration and use of outer space for peaceful purposes. During these exchanges the United States highlights the importance of exchanges of contact information for spacecraft operations and conjunction assessment.[8] In addition to the obvious value for information exchanges for collision avoidance, such contacts also could support information exchanges in the event of space weather events that result in satellite outages or other unexpected interference with satellite operations – thus reducing the chances of misperceptions and mistrust during a crisis.

At the multilateral level, the guidelines that are being developed in the COPUOS LTS Working Group will help inform discussions on space TCBMs at the 58th session of the Committee scheduled for 2016. Results from these discussions could also be an important input for a potential joint ad hoc session of the United Nations General Assembly's two committees responsible for space security, the First and Fourth Committees. Such a joint meeting could also provide an opportunity for UN Member States to identify opportunities for improved coordination on the implementation of space TCBMs across the United Nations system. In particular, this meeting could explore the role of "whole of government" cooperation that spans all sectors of national space activity in the peaceful exploration and use of

---

[8]   U.S, Department of State, Office of the Spokesperson, "U.S.-China Strategic and Economic Dialogue Outcomes of the Strategic Track" July 14, 2014, http://www.state.gov/r/pa/prs/ps/2014/07/229239.htm, accessed on 17 August 2014.

outer space. It could consider how a cross-UN coordination mechanism could provide a useful platform for the promotion and effective implementation of TCBMs involving various entities of the United Nations Secretariat and other institutions involved in outer space activities.

In conclusion, "bottom-up" efforts such as the development of TCBMs and guidelines for space debris and space operations have played key roles in advancing space cooperation. In particular, COPUOS continues to provide a unique forum for the exchange of information among developed and developing countries on the latest developments in the use and exploration of outer space. There are tangible opportunities to advance this cooperation so that, in the words of the 2010 U.S. National Space Policy, "all nations and peoples—space-faring and space-benefiting—will find their horizons broadened, their knowledge enhanced, and their lives greatly improved."

•••

# KENNETH D. HODGKINS

Ken Hodgkins has been with the Department of State since 1987 and presently is the Director for the Office of Space and Advanced Technology in the Bureau of Oceans, Environment and Science. The office is responsible for bilateral and multilateral cooperation in civil and commercial space and high technology activities, including the International Space Station, collaboration in global navigation satellite systems, the International Thermonuclear Experimental Reactor (ITER), and nanotechnology, and represents the Department in national space policy review and development. Mr. Hodgkins serves as the U.S. Representative to the UN Committee on the Peaceful Uses of Outer Space (COPUOS). He has been the State representative for major Presidential policy reviews on remote sensing, the Global Positioning Satellite (GPS) system, orbital debris, and the use of space nuclear power sources in space. Before coming to the State Department, he was the Director for International Affairs at the National Environmental Satellite Data and Information Service (NESDIS) of the Department of Commerce.

# RICHARD H. BUENNEKE

Richard H. Buenneke is senior advisor for national security space policy in the Office of Emerging Security Challenges, Bureau of Arms Control, Verification and Compliance, at the United States Department of State in Washington, D.C. In his current position, Mr. Buenneke participates in planning and implementation of diplomatic and public diplomacy activities relating to U.S. national security space policy.

From 2011-2014, Mr. Buenneke served as the co-chairman of an international expert group on space debris, space operations and tools for collaborative space situational awareness, which is an element of a working group on the long-term sustainability of space activities of the United Nations Committee on the Peaceful Uses of Outer Space.

Before joining the State Department as a Foreign Affairs Officer in March 2007, Mr. Buenneke was a senior policy analyst at The Aerospace Corporation, a federally-funded research and development center. In this position, he led Aerospace's support to the planning and strategy directorate of the National Security Space Office in U.S. Department of Defense's Executive Agent for Space. In this assignment, he served as lead analyst for NSSO's work on commercial satellite protection.

From 2001 to 2003, he led Aerospace's Center for Space Policy and Strategy. Prior to joining Aerospace, Mr. Buenneke was a space policy analyst at Booz Allen Hamilton in McLean, Va., and the RAND Corporation in Santa Monica, Calif. In these positions, he participated in studies on the health of the U.S. space industrial base, military use of commercial space capabilities, and operational requirements for satellite early warning systems.

Mr. Buenneke holds bachelors degrees in economics and systems engineering from the Wharton and Engineering schools of the University of Pennsylvania. He also holds masters' degrees in policy analysis from George Washington University's Elliott School of International Affairs and the Pardee RAND Graduate School.

## AMBER CHARLESWORTH

Amber Charlesworth has been with the U.S. Department of State's Office of Space and Advanced Technology since 2011. The office is responsible for bilateral and multilateral cooperation in civil and commercial space and high technology activities. Amber serves as a U.S. representative to the United Nations Committee on the Peaceful Uses of Outer Space (UNCOPUOS), where a wide range of space policy issues are addressed. Most recently, UNCOPUOS has been a vital forum for U.S. efforts to develop new international guidelines on emerging issues such as minimizing the generation of orbital debris and ensuring safe space operations and sustainable access to space. Prior to working at the U.S. Department of State, Amber served as an air traffic controller in the United States Air Force from 1998-2006. Amber holds a J.D. from Creighton University School of Law and an LL.M. in Space, Cyber and Telecommunications Law from the University of Nebraska College of Law.

CHAPTER 6

# FORWARDING MULTILATERAL GOVERNANCE OF OUTER SPACE ACTIVITIES: NEXT STEPS FOR THE INTERNATIONAL COMMUNITY

THERESA HITCHENS
SENIOR RESEARCH SCHOLAR AT THE CENTER OF INTERNATIONAL AND
SECURITY STUDIES AT THE UNIVERSITY OF MARYLAND

INTRODUCTION

Over the past decade, concerns about ensuring stability and security in outer space have led the international community to pursue initiatives to improve the governance of outer space activities. This reflects the increasing importance of satellites and space services to everyday life on Earth. At a global level, banking, communications, transportation, and the Internet are all underpinned by access to space, as is weather prediction, natural disaster mitigation and sustainable farming. Militaries around the world are also increasingly reliant on space operations. There are now some 1,200[1] active spacecraft on orbit, and more than 70 states and commercial/civil entities own and/or operate satellites. Unfortunately, increasing activity in space has engendered increasing congestion of the space environment, and increasing competition for access.

Governance issues regarding the use of space are complicated both because of the physical realities of the space environment, and the legal status of space as a global resource. In particular, as the number of space users grows and the types of activities in space expand, competition is growing for access to the most productive orbital bands in near-Earth orbit and to the most useful parts of the radio-frequency spectrum for satellite operations.

In addition, as more and more militaries around the world turn to space assets to improve military and national security operations, tensions are rising regarding the potential development of counterspace weapons, including debris-creating anti-satellite systems. Given the somewhat opaque nature of national security space operations, one of the critical foundations for better multilateral governance will be improved transparency – a fact widely recognized by the international community. The multilateral work has concentrated on voluntary measures, reflecting the widespread recognition that obtaining a legally binding treaty for space security likely cannot be achieved at this time. While the Russian Federation and the People's Republic of China long have been advocating a treaty approach to prevent an arms race in outer space, and ban the placement of weapons in space, the United States, in particular, has been loathe to move forward on legally binding measures that might restrict its space activities – with particular concerns about spill over to missile defense. Indeed, there is an argument to be made that voluntary approaches, such as codes of behavior and transparency and confidence building measures, are a necessary first step toward any legal resolutions. There are currently three major multilateral initiatives in the field, related to each other but each aimed at a slightly different aspect of the governance problem.

---

[1]  http://www.ucsusa.org/nuclear_weapons_and_global_security/space_weapons/ technical_issues/ucs-satellite-database.html

In 2008, the European Union (now comprising 28 members) released a draft code of conduct for outer space activities. The draft code, designed as a voluntary but politically binding instrument, is primarily a norm-setting exercise that looks to distinguish between responsible and irresponsible behavior in space. The code thus straddles the (somewhat blurry) line between space security issues and space safety/stability, addressing both norms for behavior in peacetime and times of conflict. The latest EU revised draft is dated March 31, 2014.[2] Meanwhile, the EU and likeminded States have been seeking to obtain international support for the code, and intend to begin formal negotiations later this year. However, the effort faces an uphill battle as the BRICS countries, in particular, are leery about the code proposal for a variety of political reasons, including the fact that it is being negotiated outside of the United Nations. Indeed, Russia has directly charged that the code effort is undermining the work of UN on space security.

In 2010, the 77-member United Nations Committee on the Peaceful Uses of Outer Space (COPUOS) established the Working Group on the Long-term Sustainability of Outer Space Activities (LTS) under the COPUOS Scientific and Technical Subcommittee (STSC) to develop a set of voluntary "best-practices" for space activities. The draft guidelines, which are of a more technical nature than a political one and focus on protecting the space environment, are now being negotiated.[3] (It should be noted that the COPUOS also includes non-governmental organizations and private industry as observers.) While the long-term sustainability exercise is not officially aimed at dealing with space security or military uses of space, the guidelines if accepted by States would by their nature have an effect on the conduct of national security space activities. Many members had hoped to finalize the draft guidelines by 2016, after which they would be referred to the UN General Assembly for adoption. However, the negotiations have become bogged down – in particular due to a

---

[2] "Draft International Code of Conduct for Outer Space Activities," (here after European Draft Code) European Union External Action Service, http://www.eeas.europa.eu/non-proliferation-and-disarmament/pdf/space_code_conduct_draft_vers_31-march-2014_en.pdf

[3] While there is no comprehensive document reflecting all of the proposed guidelines and language, a good compilation by the chairman of the Working Group is available: "Draft Report of the Working Group on the Long-Term Sustainability of Outer Space Activities: Working paper by the Chair of the Working Group," (hereafter "COPUOS Draft Working Group Report), Committee on the Peaceful Uses of Outer Space, Scientific and Technical Subcommittee, Fifty-second session, December 10, 2014, A/AC.105/C.1/L.343, http://www.unoosa.org/pdf/limited/c1/AC105_C1_L343E.pdf

complicated set of amendments and working papers proposed by Russia during the February 2-13, 2015 STSC session in Vienna. Particularly controversial is the Russian proposal to incorporate an official interpretation of "self defense" in space.[4]

Finally, in 2011, the General Assembly's First Committee (which is the UN body responsible for international security affairs) called upon the Secretary-General to establish a Group of Governmental Experts (GGE) on transparency and confidence-building measures (TCBMs) in space. The 15-member GGE began deliberations in 2012 and issued a report in July 2013, which was adopted by the General Assembly at its 68th session.[5] The work of the GGE is most directly related to space security, seeking to create mutual understanding and build trust among nations in order to reduce risks of misperceptions, miscalculations and conflict. The report lays out basic TCBMs that could be undertaken by states unilaterally, bilaterally or multilaterally. The GGE report is important, since it is the first UN agreement in many years directly to focus on improving space security. The question remains how, or even if, this agreement will be taken up UN Member States.

This chapter reviews these initiatives, and seeks to elucidate ways to forward their progress. In addition, the paper looks to identify further steps beyond current activities at the multilateral level for establishing a foundational space governance framework. The goal is to help illuminate a pathway that capitalizes on gains already made, and continues to forward momentum in what is arguably a deteriorating geopolitical context. In particular, there are actions that could be taken by space-faring States unilaterally, and in the near- to medium-term, that could help to shore up multilateral approaches. It is imperative that the gains already made toward defining a space governance framework, as limited as they may be, are not lost to increasing tensions among space actors.

---

[4]   "Achievement of a uniform interpretation of the right of self-defence in conformity with the United Nations Charter as applied to outer space as a factor in maintaining outer space a safe and conflict-free environment and promoting the long-term sustainability of outer space activities: Working paper submitted by the Russian Federation," Committee on the Peaceful Uses of Outer Space, Scientific and Technical Subcommittee, Fifty-second session, February 2, 2015, A/AC.105/C.1/2015/CRP.22, http://www.unoosa.org/pdf/limited/c1/AC105_C1_2015_CRP22ER.pdf

[5]   "Report of the Group of Governmental Experts on Transparency and Confidence-Building Measures in Outer-Space Activities," United Nations General Assembly, Sixty-eighth session, A/68/189*, July 29, 2013, (hereafter, GGE Report) http://www.unidir.org/files/medias/pdfs/outer-space-2013-doc-2-a-68-189-eng-0-580.pdf

## EU CODE OF CONDUCT

The UN First Committee in 2009 endorsed the effort by the European Union to draft a "Code of Conduct on Outer Space Activities" – which was adopted by the EU Council of Ministers in 2008. The proposed code, which was presented to the Conference on Disarmament in 2009 and which has been revised several times since, most recently on March 31, 2014, is aimed at reinforcing and expanding  norms of behavior that define acceptable and unacceptable actions in space. Rather than a legally binding treaty, the EU has shaped the proposed code as a politically binding set of commitments. Key principles enshrined in the proposed Code include:

- *"The freedom of all States "to access, to explore and to use outer space for peaceful purposes, without harmful interference, fully respecting the security, safety and integrity of space objects, and consistent with internationally accepted practices, operating procedures, technical standards and policies associated with the long term sustainability of outer space activities ... "*[6]
- *"The responsibility of States to refrain from the threat or use of force against the territorial integrity or political independence of any State, or in any manner inconsistent with the purposes of the Charter of the United Nations, and the inherent right of States to individual or collective self-defense as recognized in the Charter of the United Nations. "*[7]
- *"The responsibility of States, in the conduct of scientific, civil, commercial and military activities to promote the peaceful exploration and use of outer space for the benefit, and in the interest, of humankind and to take all appropriate measures to prevent outer space from becoming an arena of conflict. "*[8]

In particular, the draft code would pledge signatories to:

*"Refrain from any action which brings about, directly or indirectly, damage, or destruction, of space objects unless such action is justified:*
- *by imperative safety considerations, in particular if human life or health is at risk; or*

---

[6] European Draft Code, http://www.eeas.europa.eu/non-proliferation-and-disarmament/pdf/space_code_conduct_draft_vers_31-march-2014_en.pdf
[7] Ibid.
[8] Ibid.

- *in order to reduce the creation of space debris; or*
- *by the Charter of the United Nations, including the inherent right of individual or collective self-defense.*

*And where such exceptional action is necessary, that it be undertaken in a manner so as to minimize, to the greatest extent practicable, the creation of space debris."*[9]

While this does not clearly stake out a prohibition against the testing and use of debris-creating antisatellite weapons (ASATs), it does represent a call for restraint.

The draft code would also commit States to a number of notification measures, including when scheduled maneuvers might result in "dangerous proximity to space objects", as well as to adhere to the existing legal framework governing space.

In order to dampen complaints from non-EU governments that the development of the draft code had not been inclusive, the EU held three rounds of multilateral Open-ended Consultations, in Kiev (May 2013), Bangkok (November 2013), and Luxembourg (May 2014). Overall, 95 UN Member States participated in the consultation process, and 61 countries were present in each round of consultations. In addition, the United Nations Institute for Disarmament Research (UNIDIR) UNIDIR organized, on behalf of the EU, regional seminars in Malaysia, Ethiopia, Mexico, and Kazakhstan, and helped with the preparation of the Open-ended Consultations.[10]

Those consultations, however, have so far failed to convince the BRICS States (Brazil, Russia, India, China and South Africa) of the worthiness of the proposed code. Russia has taken the position that support for the proposed code would undermine the ongoing COPUOS efforts – reflecting a general discomfort among the BRICS States about negotiating an international agreement outside of the UN structure. Russia, in particular, has noted that many precepts in the draft code fall within COPUOS's purview and should be a part of the latter's ongoing discussions. China also continues to express concerns that acceptance of a code of conduct will undermine the goal of achieving a legally binding treaty to prevent an arms race in outer space, a text of which China and

---

[9]  Ibid.
[10]  See: http://www.unidir.org/programmes/emerging-security-threats/facilitating-the-process-for-the-development-of-an-international-code-of-conduct-for-outer-space-activities

Russia first proposed to the Conference on Disarmament in 2008.[11] In addition, many African and Latin American countries object to the draft's references to a right to self-defense, and worry that the code's precepts would restrict their future space activities. On the other hand, the United States, Australia, Canada and Japan have vocally supported the draft code.

The EU currently has a goal of opening the draft code for international negotiations in July 2015, under the chairmanship of Australia (as a non-EU, but supportive, State). If this plan proceeds, it is unlikely that the BRICS nations will sign on. It is debatable as to whether the lack of participation by the BRICS, all of whom are substantial space actors, would render the new code politically inert. There is precedent (such as the Ottawa Treaty on landmines and the Hague Code of Conduct on ballistic missiles) for multilateral agreements by like-minded States to set norms that eventually are accepted by others, whether formally or informally.

## COPUOS LTS WORKING GROUP

The COPUOS LTS Working Group is in some ways an outgrowth of the EU code initiative, in that it seeks to improve space transparency and establish principles of safe operations in space – thereby reducing risks. There are 69 member states in the Vienna-based COPUOS and a large number of non-governmental and intergovernmental organizations are observers. Technically, COPUOS is the only formal UN body empowered to negotiate new international space treaties; however, COPUOS's mandate does not include military space activities, which has meant that discussions of space weapons have been ceded to the Conference on Disarmament n Geneva. COPUOS activities are divided between two subcommittees, the Legal Subcommittee and the Scientific and Technical Subcommittee. While no new space treaties have emerged from the Legal Subcommittee since the mid-1980s, COPUOS has made progress in addressing space safety and security within the Scientific and Technical Subcommittee.

In 2007, for example, COPUOS adopted a set of voluntary guidelines for space debris mitigation based on technical guidelines developed by the Inter-Agency Debris Coordinating Committee (IADC) and subsequently endorsed by the General Assembly in January 2008.[12] The accord is a

---

[11]  See: http://www.acronym.org.uk/official-and-govt-documents/draft-text-placement-weapons-in-outer-space-submitted-russia-and-china

[12]  See: http://orbitaldebris.jsc.nasa.gov/library/Space%20Debris%20Mitigation%20Guidelines_COPUOS.pdf

significant achievement for space safety and also helps to underpin space security, especially regarding Guideline 4, which pledges nations to avoid the intentional break-up of space objects and the creation of long-lived debris.

Building on the success of the debris mitigation effort, COPUOS in February 2010 initiated a new working group under the Scientific and Technical Subcommittee on the "long-term sustainability of outer space."

The group was empowered to:

> "[E]xamine the long-term sustainability of outer space activities in all its aspects, consistent with the peaceful uses of outer space, and avail itself of the progress made within existing entities, including but not limited to the other working groups of the Subcommittee, the Conference on Disarmament, the International Telecommunication Union, the Inter-Agency Space Debris Coordination Committee, the International Organization for Standardization, the World Meteorological Organization and the International Space Environment Service."[13]

The Subcommittee agreed that the Working Group should avoid duplicating the work being done within those bodies and instead identify areas of concern for the long-term sustainability of outer space activities that are not covered by them. [The Subcommittee also agreed that the Working Group should consider organizing an exchange of information with the commercial space industry to understand the views of that community.]

The Working Group has been charged to consider new measures to enhance the sustainability of space activities and a possible set of "best practice guidelines."[14] Some of these eventual guidelines in effect fall under the rubric of "space traffic management" – i.e., processes, procedures, and new regulations for how spacecraft are launched, operated and disposed of at the end of their working lifetimes.

According to the group's terms of reference established by General Assembly Resolution A/AC.105/C.1/L.307/Rev.1, published Feb. 28, 2011, the objective of the working group is the production of "a set of guidelines that could be applied on a voluntary basis by international organizations,

---

[13]  "Report of the Scientific and Technical Subcommittee on its forty-seventh session, held in Vienna 8-19 February 2010," Committee on the Peaceful Uses of Outer Space, fifty-third session, March 11, 2010, A/AC.105/958, http://www.unoosa.org/pdf/reports/ac105/AC105_958E.pdf

[14]  Ibid.

non-governmental entities, individual States and States acting jointly to reduce collectively the risk to space activities for all space actors and to ensure that all countries are able to have equitable access to the limited natural resources of outer space."[15]

The Working Group's deliberations are divided among four expert groups:

A. Sustainable space utilization supporting sustainable development on Earth
B. Space debris, space operations and tools to support space situational awareness sharing
C. Space weather
D. Regulatory regimes and guidance for new actors in the space arena.

The scope section notes that topics to be studied include several items that overlap or expand upon the UN GGE's recommendations:

1. Collection, sharing and dissemination of data on functional and non-functional space objects, and the creation of contact points responsible for timely communications;
2. Re-entry notifications regarding substantial space objects, and also on the re-entry of space objects with hazardous substances on board;
3. Capabilities to provide a comprehensive and sustainable network of key data in order to observe and measure space weather phenomena adequately in real or near-real time;
4. Pre-launch and maneuver notifications; and,
5. Adherence to existing treaties and principles on the peaceful uses of outer space, including the Registration Convention and consideration of supplying enhanced information to the registry. [16]

The working group's work plan is multiyear, initially stretching from 2011 through 2014. All four of the Expert Groups submitted their reports to

---

[15] "Terms of reference and the methods of work of the Working Group on the Long-Term Sustainability of Outer Space Activities of the Scientific and Technical Subcommittee, Working Paper submitted by the Chair," Committee on the Peaceful Uses of Outer Space, February 28, 2011, A/AC.105/C.1/L.307/Rev.1, http://www.unoosa.org/pdf/limited/c1/AC105_C1_L307Rev01E.pdf

[16] Ibid.

the STSC at its 51<sup>st</sup> session February 10-21, 2014.[17] In total, the Expert Groups recommended 31 guidelines.[18] At the STSC 52nd session, held February 2-13, 2015, a draft Working Group report was submitted by the Chair based on the Expert Group findings, which also looked at issues for possible future consideration.[19]

The Working Group's deliberations have been complicated and the hope now is that a consensus on a set of draft guidelines can be formed in 2016. However, at the most recent meeting of the Science and Technical Subcommittee was held Feb. 2-13, 2015 in Vienna, proposals for either new guidelines or amendments were made by Belgium, Brazil, Germany, Iran, Russia and the United States. Indeed, a host of proposals from the Russian Federation – including several that are controversial – and a less than productive attitude on the part of Moscow has dampened optimism about a 2016 wrap up. The next meeting of the Working Group will be in June, during the 58<sup>th</sup> Session of COPUOS, June 10-19, 2015. At that time, the Chair is expected to present a summary of all the suggested guidelines for review by the group. One criticism that has arisen is that there are too many of the guidelines, and that many are too "in the weeds." While there currently is no reason for optimism that the Working Group will be able to conclude in 2016, the issues causing difficulties in the discussions are not primarily of a fundamental nature and there is a good baseline for eventual consensus.

---

[17]  See: Expert Group A,
     http://www.unoosa.org/pdf/limited/c1/AC105_C1_2014_CRP13E.pdf;
     Expert Group B, http://www.unoosa.org/pdf/limited/l/AC105_2014_CRP14E.pdf;
     Expert Group C,
     http://www.unoosa.org/pdf/limited/c1/AC105_C1_2014_CRP15E.pdf;
     Expert Group D,
     http://www.unoosa.org/pdf/limited/c1/AC105_C1_2014_CRP16E.pdf
[18]  "Compilation of proposed draft guidelines of expert groups A to D on the Long-Term Sustainability of Outer Space Activities, as at the fiftieth session of the Scientific and Technical Subcommittee, held in February 2013," Committee on the Peaceful Uses of Outers Space, Fifty-sixth session, March 26, 2013, A/AC.105/1041, http://www.unoosa.org/pdf/reports/ac105/AC105_1041E.pdf
[19]  "Draft Report of the Working Group on the Long-Term Sustainability of Outer Space Activities: Working paper by the Chair of the Working Group," Committee on the Peaceful Uses of Outer Space Scientific and Technical Subcommittee, Fifty-second session, December 10, 2014, A/AC.105/C.1/343, http://www.unoosa.org/pdf/limited/c1/AC105_C1_L343E.pdf

## Group of Governmental Experts (GGE)

The work of the GGE, unlike that of the EU and COPUOS, is highly focused on issues of space security, and took place under the auspices of the UN First Committee which deals with disarmament and global challenges to peace and security. The First Committee also oversees the work of the Conference on Disarmament.

Transparency and Confidence-Building Measures (TCBMs) have long been a part of multilateral statecraft, enshrined in United Nations resolutions as potentially useful for improving mutual understanding, reducing misunderstandings and tensions, and promoting a more favorable climate for arms control and non-proliferation. In the case of space, TCBMs can be seen as a foundation stone to future multilateral approaches to governance of space as a global resource.

The GGE on Space Activities met in three sessions: July 23-27, 2012 in New York; April 1-5, 2013, in Geneva; and July 8-12, 2013 in New York. The GGE was chaired by Russia and comprised representatives of 15 member UN member states.[20] GGEs work by consensus; and the group's consensus final report was transmitted by the Secretary-General to the First Committee in October 2013.

The GGE focused on TCBMs that "could be adopted voluntarily by States on a unilateral, bilateral, regional or multilateral basis." The group laid out its recommendations in five broad categories: Enhancing the transparency of outer space activities; international cooperation; consultative measures; outreach and coordination.

Recommended transparency measures include information exchanges on: space policies; military space spending and national security space activities; orbital parameters of satellites and conjunction potentials; and forecasts of natural hazards.

Specifically, the report calls for "Exchanges of Information on the orbital elements of space objects and the provision, to the extent practicable, of notifications of potential orbital conjunctions involving spacecraft to affected government and private sector spacecraft operators."[21] Currently only the United States (through its Space-Track program) routinely shares notifications of potential collisions with other

---

[20]  GGE Members were: Brazil, Chile, China, France, Italy, Kazakhstan, Nigeria, Republic of Korea, Romania, Russian Federation, South Africa, Sri Lanka, Ukraine, United Kingdom of Great Britain and Northern Ireland, and United States of America. The GGE was chaired by Victor L. Vasiliev of the Russian Federation.

[21]  GGE Report, paragraph 39(a), http://www.unidir.org/files/medias/pdfs/outer-space-2013-doc-2-a-68-189-eng-0-580.pdf

governments and the private sector. Russia, China, France and a number of other European countries have space tracking capabilities, but do not routinely share that information. In addition, the European Union has for a number of years been seeking to establish a European space surveillance capability. Since September 2014, the European Satellite Center has been working on a project, "Preparation for the Establishment of A European SST Service provision function (PASS), that is due to result in a proposal for how to integrate current European capabilities, securely share data and perform collision analysis. The study is due to be completed August 31, 2016.[22]

The GGE noted the need for improved compliance with current agreements. Compliance with the Registration Convention has been uneven. There is a lack of standardization regarding what data is actually registered; confusion over the term "launching State," and administrative issues. According to Dr. Jonathan McDowell, an astrophysicist at the Harvard-Smithsonian Center for Astrophysics, about 95% of the satellites launched between 1957 and 2013 have been registered (although sometimes with faulty data sets), and about 21% of satellites launched between 2013 and 2015 so far. The states with the most unregistered satellites launched prior to end-December 2013 are the United States, the People's Republic of China, Israel and Saudi Arabia.[23] McDowell said that U.S. registrations are only up to November 2013; Russian registration is up to date to August 2014.

Thus, the GGE report calls for improved compliance with the 1976 Convention on Registration of Objects Launched into Outer Space. Fifty-six UN Member States have ratified the Convention and another four have signed it. The Convention provides that "launching States should furnish to the United Nations, as soon as practicable, the following information concerning each space object:

- Name of launching State;
- An appropriate designator of the space object or its registration number;
- Date and territory or location of launch;

22   "Preparation for the Establishment of A European SST Service provision function," European Union Satellite Center, http://www.satcen.europa.eu/ index.php?option=com_content&task=view&id=70&Itemid=92
23   "Adherence to the 1976 Convention on Registration of Objects Launched into Outer Space," Jonathan McDowell, *Jonathan's Space Page*, http://planet4589.org/space/un/un_paper1.html

- Basic orbital parameters, including:
  - Nodal period (the time between two successive northbound crossings of the equator - usually in minutes);
  - Inclination (inclination of the orbit - polar orbit is 90 degrees and equatorial orbit is 0 degrees);
  - Apogee (highest altitude above the Earths surface - in km);
  - Perigee; (lowest altitude above the Earths surface - in km);
- General function of the space object."[24]

Several types of notifications by States regarding space activities also are recommended, including notification of planned launches; scheduled maneuvers that might result in risk to other space objects; uncontrolled "high risk" re-entries; emergency situations and orbital breakups. Finally, in pursuit of transparency, the report recommends that States make opportunities for site visits and technology demonstrations.

The recommendation on notification of maneuvers is especially important, as one of the lacunae in the Registration Convention is that maneuvers are not required to be registered. The recommendation reads: "States should notify, in a timely manner and to the greatest extent practicable, potentially affected States of scheduled maneuvers that may result in risk to the flight safety of the space objects of other States."[25]

Regarding international cooperation, the GGE report notes that "the disparity in the space capabilities of States, the inability of most States to participate in space activities without the assistance of others, uncertainty concerning sufficient transfer of space technologies between States and the inability of many States to acquire significant space-based information are factors contributing to a lack of confidence." The report thus notes that technical and capacity building cooperation should be undertaken; and endorses "an open satellite data collection and dissemination policy for sustainable economic and social development."

Timely and routine consultations through bilateral and multilateral government-to- government exchanges further are endorsed; as well as outreach to international organizations and NGOs. Coordination among States and multilateral organizations also is encouraged, including the development of national focal points and stronger coordination among UN entities. Regarding focal points, the GGE report recommends that this

---

[24] "Registration of Objects Launched into Outer Space," United Nations Office of Outer Space Affairs, http://www.unoosa.org/oosa/SORegister/regist.html

[25] GGE Report, paragraph 42, http://www.unidir.org/files/medias/pdfs/outer-space-2013-doc-2-a-68-189-eng-0-580.pdf

practice should be put into place not only by States, but also by international organizations and private sector actors.

Specifically, the GGE report recommends a joint meeting of the UN First Committee (which deals with security and the Geneva-based Conference on Disarmament) and the Fourth Committee (which deals with scientific issues and the Vienna-based COPUOS) on challenges to space security and sustainability– a meeting that is expected to take place in October this year during the next session of the General Assembly. Indeed, the GGE specifically mentions the importance of the COPUOS work on long-term sustainability and the fact that the guidelines will have "characteristics similar to transparency and confidence-building measures. ..."[26] This meeting will provide an opportunity for States to present proposals on how to forward the work of the GGE, if they so desire.

## MOVING FORWARD

The establishment of the EU, GGE and COPUOS initiatives underscore the general consensus that multilateral solutions to the challenges in maintaining stability, sustainability and security in space are necessary. This expanding consensus is underscored by the fact that there is a great deal of overlap between their work. However, actual progress remains achingly slow. For example, up to now, no State, regional group or international organization has moved to substantively work on implementation of the GGE recommendations. And as noted above, the COPUOS and EU efforts are somewhat politically fraught; although the LTS initiative has made substantial steps toward its goal.

It must be said that part of the current problem is the chill between Russia and the West regarding the situation in Ukraine, which has leaked over to almost every international engagement. It is an undeniable fact that Russia-U.S. cooperation on space governance has been one of the drivers moving the multilateral ball down the field as far as it has gone. As noted, the tensions over Ukraine have affected Moscow's attitude and willingness to cooperate in almost every international forum. The U.S.-Russia-China dynamic also has been negatively impacted in recent years by increased U.S. concern regarding Russian and Chinese efforts to develop and deploy offensive counterspace capabilities. Given the U.S. reliance on space assets in its military operations, the concerns are understandable. Up to now the

---

[26] GGE Report, paragraph 13, http://www.unidir.org/files/medias/pdfs/outer-space-2013-doc-2-a-68-189-eng-0-580.pdf

U.S. strategy has been to focus on the need for international restraint in the space regarding military activities -- including strong support for all three of multilateral processes at hand – although (according to U.S. sources involved in the policy process) this focus may be starting to shift in favor of a more aggressive approach. This dynamic is worrisome, particularly as the United States moves into its presidential election season, when domestic political considerations trump all else.

Another obstacle to progress continues to be the growing diversity in priorities, rationales and perspectives on the use of space as the number of space actors grows. Differences exist on issues such as potential constraints on space activities, especially constraints that increase the cost of entry, such as a requirement for specific technical measures to mitigate debris creation. In addition, there are deep differences regarding what might constitute legitimate military or self-defense activities in space. The space arena is not immune from long-standing North-South political issues, or from the economic issues that divide developed and developing nations. Resolving these tensions and developing mutual understanding about the threats and solutions to space security will require much good will and concerted diplomatic engagement from all parties to avoid the creation of political "blocks" that can only impede progress.

However, it is imperative that the progress made so far is not allowed to stall. And there are still steps that can be taken, by individual States, coalitions of the willing, regional groups, multilateral fora and by NGOs.

## IMPLEMENTING THE GGE, A FIRST STEP

Given that the GGE has completed its work and made a number of recommendations, a logical place to start is with the implementation of those recommendations. Importantly, the fact that the GGE calls on States to take up the recommendations on a voluntary, unilateral, bilateral, regional or multilateral basis opens the path for leadership by States, especially the United States, as the leading space-faring power, as well as Russia and China. In addition, the GGE report lobbies for more government interaction and cooperation with private sector and NGO actors. This provides opportunities for private sector and NGO-led initiatives that could be supported by individual States or groups of likeminded States. No longer are nation-states the only space actors: private commercial activities have eclipsed government activities in economic value, and further, barriers of entry have fallen to the point that even universities and NGOs can undertake space operations. Thus it is important that these non-governmental space entities are engaged and educated about the value of a multilateral space governance framework. In particular, commercial operators and space entrepreneurs, who have a natural skittishness regarding regulation, must be brought into the large space security conversation.

## REVIEW OF SPACE REGIME IMPLEMENTATION

The planned joint meeting of the UNGA First and Fourth Committee in October 2015 will provide an initial platform for discussion of possible first steps toward implementing the GGE recommendations. One idea that has been circulating as a possible outcome of this meeting is a resolution to call a meeting of States Parties to the 1967 Treaty on Principles Governing the Activities of States in the Exploration and Use of Outer Space, including the Moon and Other Celestial Bodies (OST) in the year of its 50 anniversary (2017) to review implementation of the treaty, as well as other parts of the multilateral space regime. The GGE report recommends "universal participation in, implementation of and full adherence to the existing legal framework relating to outer space activities."[27] Such a UNGA resolution could be proposed by one or more States during the joint meeting. For example, the meeting could address the problems highlighted by the GGE report, including compliance with the Registration Convention, as well as the Debris Mitigation Guidelines. One caveat, however, is that the resolution should make clear that the debate would not in any way represent a re-opening of the OST to amendments, nor seek to reinterpret the treaty's provisions.

## CONTACTS AND FOCAL POINTS

One of the first, and relatively easy steps that could be taken unilaterally by any number of States would be to take up the GGE recommendation regarding the creation of focal points and contacts for data exchange, particularly in the case of potential collisions. Similar recommendations are included in the draft COPUOS guidelines, which call for exchanges of information on spacecraft operators and entities performing conjunction analysis.[28] Likewise, the EU draft code calls for each subscribing State to establish a "central point of contact" responsible for reporting to the code's

---

[27]  GGE Report, paragraph 71, http://www.unidir.org/files/medias/pdfs/outer-space-2013-doc-2-a-68-189-eng-0-580.pdf

[28]  COPUOS Draft Working Group Report, section E.1 (paragraphs 60-62), http://www.unoosa.org/pdf/limited/c1/AC105_C1_L343E.pdf

management organization of subscribing States. That State contact would not only provide notifications to the group, but also be responsible for serving as a conduit for consultations.[29]

The value of setting up contact nodes within organizations responsible for spacecraft management, including the private sector, is manifold. First, it forces national governments to identify and create linkages to all stakeholders and sets up channels for both internal and external communications. The identification of individuals as contact points creates ownership within stakeholder organizations. And finally, the creation of an international "space phonebook" lays the foundation for the development of easily accessible dispute resolution methods. Once a network of contacts is developed at the national level, the information could be reported to various UN bodies, including the First and Fourth Committees, COPUOS, the Conference on Disarmament, the Office of Outer Space Affairs and the International Telecommunication Union.

Indeed, the UN Office of Outer Space Affairs (OOSA) already has established a Registration Information Submissions Form (as part of its responsibility for managing the UN Registry of Outer Space Objects), which allows States to provide additional information when registering launched objects, including contacts for satellite operators.[30] Further, General Assembly Resolution 62/101, "Recommendations on enhancing the practice of States and international intergovernmental organizations in registering space objects," adopted in December 2007, calls on OOSA to make public on its website known focal points – although this has not yet occurred.[31]

## IMPROVING COMPLIANCE AND ENHANCEMENT OF THE UN REGISTRY

States could also undertake unilateral and multilateral initiatives to improve compliance with, and enhancement of, the UN Registry of Space Objects. Again, the GGE calls for universal adherence to and implementation of the Registration Convention; and both the EU draft code and the COPUOS draft guidelines address the issues of inadequate compliance, problems

---

[29] European Draft Code, section 9, http://www.eeas.europa.eu/non-proliferation-and-disarmament/pdf/space_code_conduct_draft_vers_31-march-2014_en.pdf

[30] See: http://www.unoosa.org/pdf/misc/reg/regformE.pdf

[31] "Resolution adopted by the General Assembly on 17 December 2007, *on the Report of the Special Political and Decolonization Subcommittee (Fourth Committee) (A/62/403),* Recommendations on enhancing the practice of States and international intergovernmental organizations in registering space objects," General Assembly Sixty-second session, agenda item 31, January 10, 2008, A/RES/62/101, http://www.un.org/en/ga/search/view_doc.asp?symbol=A/RES/62/101

with harmonization of reporting data, and the need for more information (such as notification that a space object is no longer functional). The COPUOS chair's draft working group report notes: "The lack of comprehensive information on objects launched into orbit results in a patchy and incomplete picture of what is in orbit and where. This affects space situational awareness, and ultimately safety too, if a potentially hazardous situation arises and inadequate information is available to identify a space object and/or its operators, or it is unclear under whose control or jurisdiction the object falls."[32]

In particular, the final report of Expert Group B, made to the Scientific and Technical Subcommittee in June 2014, explains in some detail that there are a number of gaps in satellite registration – including the fact that a number of nations are not parties to the Registration Convention.[33] Individual or groups of states parties should begin awareness-raising and lobbying non-parties to join – through bilateral or multilateral consultations, Track 1 or Track 1.5 conferences, etc. This is an area where advocacy by NGOs could be of assistance.

Compliance is the responsibility of individual States, and is an issue on which the United States could lead, but also Russia and China as major space actors and fellow laggards in registration compliance could become more active on. In the case of the United States, the key stumbling block seems to be administrative lack of oversight and accountability rather than a policy choice – thus a relatively simple problem to fix. All states parties should unilaterally review their standing vis-a-vis the Registration Convention's obligations and fix any shortcomings with registry as soon as possible.

One of the thorny issues at hand is agreement regarding establishing the responsible launching State, and thus responsibility for registering a space object – particularly for payloads launched by one state on behalf of another, an increasingly common practice. According to McDowell, "the tradition that 'launching state' is to be interpreted as 'owner state' is now being challenged by some states."[34] Indeed, Switzerland has refused to register any of the satellites that it operates, arguing that this is the job of the state in which the launch vehicle operator is located. This situation too

---

[32]  Ibid, section D.1, paragraphs 50-53.
[33]  "Working report of expert group B: Space Debris Space Operations and Tools to Support Collaborative Space Situational Awareness," (hereafter Expert Group B report) Committee on the Peaceful Uses of Outer Space," Fifty-seventh session, 11-20 June 2014, June 16, 2014, A/AC.105/2014/CRP.14, p. 26, http://www.unoosa.org/pdf/limited/l/AC105_2014_CRP14E.pdf
[34]  McDowell, "Adherence to the 1976 Convention … ."

must be addressed by states parties to the Convention and resolved as soon as possible. It is certainly possible for the states parties to call a meeting to discuss issues of implementation; this could also be an agenda item for review at the joint meeting of the Fourth and First Committees.

States could also unilaterally decide to include enhanced information in their registrations, such as: reporting the final destination orbit of a satellite as well as its initial insertion orbit, loss of functionality, maneuvers including movement to a disposal orbit, expected re-entry, and a change of ownership/operation authority. Indeed, Resolution 62/101 calls upon States to provide such additional information, specifically: the date of change in supervision; the identification of the new owner or operator; any change in orbital position; and any change of function of the space object.[35] Most of this data is easily available from space operators, whether they be governmental bodies or private sector actors, due to ongoing operational requirements.

Harmonization of reporting data to the Registry would require multilateral consultations. One critical problem to be solved is establishing a fixed set of metadata on space objects. Currently, countries provide differing data sets (sometimes not even completing those required by the Convention,) which makes it hard to undertake comparative analysis. The COPUOS Expert Group B notes with some dismay "the absence of internationally established and maintained system of registration of orbital launches … ."[36] Such consultations could be established via a working group at COPUOS under the Scientific and Technical Subcommittee; this could easily be initiated by an individual State or a group of like-minded States. Another route would be a proposed UNGA resolution to establish a working group under the Fourth and First Committees to recommend solutions.

## OTHER NOTIFICATIONS

While the UN Registry is not set up to provide the other specific notifications recommended by the GGE, it could be modified to do so via expansion of the current OOSA-developed submission form. States also have plenty of other options for doing so – both to UN bodies such as the Secretariat and OOSA, as well as to other States. Developing a network of

---

[35] A/RES/62/101,
http://www.un.org/en/ga/search/view_doc.asp?symbol=A/RES/62/101
[36] Expert Group B report, p. 26,
http://www.unoosa.org/pdf/limited/l/AC105_2014_CRP14E.pdf

focal points, of course, would ease this process. These notifications include:

- Pre-launch notifications
- Notifications and monitoring of predicted high-risk re-entry
- Notifications of emergency situations such as an out-of-control object on a collision course with another space object
- Notifications of any intentional orbital break-ups.

The GGE report noted that with regard to pre-launch notifications there is a model for doing so: The Hague Code of Conduct Against Ballistic Missile Proliferation (HCOC). The 2002 HCOC, which was a U.S. initiative, has 137 signatories and includes Russia but not China and India, both of whom are major space-launching States.[37] While the HCOC is not universally considered to be a success, the actual reporting methodology could be copied – either by expanding the Registration Convention or via an ad hoc arrangement.

As for intentional orbital break-ups, the GGE recommends that States inform others of their plans for any destruction of an on-orbit object, including "measures that will be taken to ensure that intentional destruction is conducted at sufficiently low altitudes to limit the orbital lifetime of resulting fragments."[38] In addition, the report stresses that all such actions should conform to the UN Space Debris Mitigation Guidelines endorsed by the General Assembly in 2008.[39]

Obviously, developing a notifications process would require multilateral consultations and agreement – although such discussions could be kick-started by one or more States. The joint First and Fourth Committee meeting scheduled for October 2015 could serve as a venue to present a proposal or UN resolution to begin consultations. Another possibility would be for regional organizations, such as the European Union, NATO or ASEAN, to begin sharing such information among their memberships. However, it would not be optimal if different notification regimes were conceived at the regional level – thus some coordination would be required. In any event, the United States, by virtue of its

---

[37] "Hague Code of Conduct Against Ballistic Missile Proliferation (HCOC)," U.S. Department of State, http://www.state.gov/t/isn/trty/101466.htm

[38] GGE Report, paragraph 45, http://www.unidir.org/files/medias/pdfs/outer-space-2013-doc-2-a-68-189-eng-0-580.pdf

[39] "Space Debris Mitigation Guidelines of the Committee for the Peaceful Uses of Outer Space," United Nations Office for Outer Space Affairs, 2010, http://orbitaldebris.jsc.nasa.gov/library/Space%20Debris%20Mitigation%20Guideli nes_COPUOS.pdf

leadership role in HCOC, would be a logical choice to galvanize this process.

## Potential New Multilateral Initiatives

Going beyond implementation of current treaties, resolutions and the GGE recommendations, new multilateral initiatives and approaches could be launched – even simultaneously with current unfinished work such as the COPUOS LTS guideline development. One fairly obvious action would be the *convening of a follow-on GGE*, with a slightly differing geographical makeup in order to expand "ownership" of the process, to discuss issues of implementation including those elaborated above. This would have the advantage of working off of an already-agreed agenda of recommendations, and provide the opportunity to elaborate specific implementation actions. In addition, the COPUOS LTS Working Group has identified a handful of important issues of space governance that the Group believes require more in-depth study before any guidelines could be developed, including active debris removal.

### Institutionalization and Implementation of the Debris Mitigation Guidelines

A concerted effort to encourage space-faring States to institutionalize the UN Space Debris Mitigation Guidelines – through national legislation and licensing procedures for satellite operators – could also be discussed at the joint Fourth/First Committee meeting; or taken up by COPUOS. This could include capacity building in emerging space powers; something that in and of itself would be useful for stabilizing the space environment and creating trust. The COPUOS draft guidelines discuss debris mitigation at length, and Expert Group B specifically recommends as Guideline B.2: "In accordance with the Space Debris Mitigation Guidelines of the Committee on the Peaceful Uses of Outer Space, States and intergovernmental organizations should address, establish and implement space debris mitigation measures through applicable mechanisms."[40] The EU draft code explicitly would commit subscribing States to implementing the UN Guidelines.

Unfortunately, not all States have moved to incorporate the Debris Mitigation Guidelines, and just as importantly, the IADC's technical

---

[40] Expert Group B report, p. 8,
http://www.unoosa.org/pdf/limited/l/AC105_2014_CRP14E.pdf

guidelines that serve to flesh out the UN principles, into either their practices or their regulatory systems for space launch and satellite operations.   ESA, in its 2015 annual report on "Classification of Geosynchronous Objects," notes that while 17 years after adoption of the IADC guidelines for geosynchronous (GEO) operations "there is widespread compliance," there are also problems. In 2014, 18 satellites in GEO reached the end of their lives but only 13 were moved to the graveyard orbit set by the IADC technical guidelines.[41] In addition, at least four rocket bodies were left adrift in orbits close to or crossing through GEO. From reviewing the past several years of the ESOC report, it seems that a common practice among even major spacefaring states is re-orbiting too low, and disconcertingly satellites continue to simply be abandoned in GEO.[42] In addition, a study by the French space agency CNES found that 40% of all satellites and rocket bodies launched into Low Earth Orbit (LEO) between 2000-2012 were abandoned in orbits too high for them to re-enter the Earth's atmosphere within the 25-year window specified in the Space Debris Mitigation Guidelines.[43] While governments and UN organizations have been loath to "name and shame," it is also clear that some space actors are less concerned with abiding by the guidelines. It might behoove, therefore, and NGO or group thereof to undertake a study of compliance over the past five to 10 years (based on the ESA data as well as observational data from amateurs) in order to put political pressure on outliers.

If the COPUOS LTS guidelines are adopted, this will provide another tool for motivating States to comply with the debris mitigation guidelines. Still, more work will need to be done to reach agreement in detail how to implement those guidelines – something that COPUOS could continue to address through extending the mandate of Expert Group B. However, it is also clear that awareness-raising and capacity building among emerging space actors is necessary. A group of likeminded States could, perhaps through OOSA or in tandem with select NGOs and research universities,

---

[41]   "CLASSIFICATION OF GEOSYNCHRONOUS OBJECTS, Produced with the DISCOS Database," ESA European Space Operations Center (ESOC), 28 March 2015, Issue 17, http://www.astronomer.ru/data/0128/ESAclassification_Issue17.pdf
[42]   For access to the ESOC reports going back to 2005, see: http://www.astronomer.ru/reference.php
[43]   "Cubesat Revolution, Spotty Compliance with Debris Rules Fuel Dangerous Congestion in Low Earth Orbit," Peter B. deSelding, Space News, Oct. 3, 2014, http://spacenews.com/4207665th-international-astronautical-congress-cubesat-revolution-spotty/

put together a campaign for debris mitigation education that includes presentations at regional fora.

## Public SSA Database

As noted above, the GGE report cites lack of access by many States to "space-based information" as contributing to a lack of trust and confidence at the international level. The recommendation regarding sharing of orbital data suggests a need for an internationally accessible database of space objects. The draft EU code also calls for the better sharing of orbital data[44], and the COPUOS Working Paper notes the need for improved debris monitoring as well as internationally agreed methodologies for observations and verification of data from various sources[45]. The report of Expert Group B recommends in Guideline B.1: "Promote the collection, sharing and dissemination of space debris monitoring information;" and in Guideline B.4, "Promote techniques and investigation of new methods to improve the accuracy of orbital data for spaceflight safety."[46]

In addition, Russia made a proposal at the most recent COPUOS LTS Working Group meeting that a UN-managed space object database be seriously studied. The proposal states: "The Russian Federation, having proposed for consideration the basic elements of the concept of establishing a unified Centre for Information on Near-Earth Space Monitoring under the auspices of the United Nations ... is of the view that an in-depth examination of the feasibility of a UN-centric information hub gathering information on objects and events in outer space from different sources as a tentatively promising means of meeting general needs and aspirations, in particular, the needs of emerging space-faring States, would be reasonable."[47] Russia foresees the database being managed by UN OOSA, initially gathering data from States. A later stage might see OOSA providing conjunction analysis.

---

[44]  EU Draft Code, section 3.5, http://www.eeas.europa.eu/non-proliferation-and-disarmament/pdf/space_code_conduct_draft_vers_31-march-2014_en.pdf

[45]  COPOUS Draft Working Group report, http://www.unoosa.org/pdf/limited/c1/AC105_C1_L343E.pdf

[46]  Expert Group B report, pp. 3, 14, http://www.unoosa.org/pdf/limited/l/AC105_2014_CRP14E.pdf

[47]  "Proposal on the review and consideration of the concept of a United Nations information platform serving common needs in collecting and sharing information on Near-Earth space monitoring in the interests of safety of space operations, and its architectural and programmatic aspects: Working paper submitted by the Russian Federation," Committee on the Peaceful Uses of Outer Space Scientific and Technical Subcommittee, February 9, 2015, A/AC.105/C.1/2015/CRP.32, http://www.unoosa.org/pdf/limited/c1/AC105_C1_2015_CRP32E.pdf

While the proposal has met with a chilly reception from Western States (which always are leery of creating new UN bodies that require budgetary resources,) there is widespread acknowledgement that better access for all space operators to SSA data is a crucial transparency and safety measure. The Expert B report emphasizes this requirement, as well as the need for more accurate data, and uniform standards of data collection and processing. In addition, the United States is considering a counter-proposal to create a new informal international group, along the lines of the International Committee on Global Navigation Satellite Systems (ICG), created in 2005 under the auspices of the UN to promote voluntary cooperation on commercial satellite navigation issues,[48] to discuss the challenges to space object data sharing. The creation of such a body would be an excellent first step, and again increase "buy in" by a wide range of space actors. Any such body should include commercial operators in its membership.

In the meantime, however, there remains the problem of lack of a widely accessible space object catalog. While the United States should be commended for its efforts to share Space Situational Awareness (SSA) data and conjunction warnings with other States and private sector actors, because the U.S. space object catalog and subsequent data analyses are the province of the U.S. military (U.S. Strategic Command's Joint Space Operations Center -- JSpOC), there is some concern on part of non-U.S. actors regarding reliability and availability of the data. For example, the European Union decision in 2008 to pursue SSA capability was predicated on the perceived requirement for independence from the U.S. network.

Further, the U.S. Space Surveillance Network (SSN) faces both technical and budgetary challenges to providing 24/7 space tracking capacity for even U.S. national security requirements, much less for use by the international community and global space operators. For example, most of the radar facilities used to track space objects are located in the Northern Hemisphere, which makes continuous coverage of an orbital object impossible.[49] In addition, the conjunction warnings provided by the JSpOC are advisory in nature, and according to satellite operators often inaccurate and thus not reliable for decision-making about collision avoidance. "We discovered that the majority of conjunctions, or close approaches, were missed by the Joint Space Operations Center, and the majority of

[48] See: http://www.unoosa.org/oosa/SAP/gnss/icg.html
[49] "USSTRATCOM Space Control and Space Surveillance," U.S. Strategic Command, http://www.stratcom.mil/factsheets/11/Space_Control_and_Space_Surveillance/

conjunction summary messages[50] that went out advising us of close approaches were wrong," Richard DalBello, then vice-president of Intelsat General for legal and governmental affairs, told a Feb. 23, 2012, conference in Washington organized by the Techamerica Space Enterprise Council and the George C. Marshall Institute.[51]

It is for this reason that industry decided to take matters into their own hands with the creation of the Space Data Association (SDA) in 2009. Industry operators, after all, have the best telemetry on their own satellites. The non-profit SDA collects telemetry data from space operators who join, and provides conjunction analysis to members.[52] Of course, SDA conjunction analysis is limited by its need for JSpOC tracking data on debris (under an August 2014 agreement[53]) and while any space operator is welcome to join, data is not shared outside of the group. In April 2015, SDA and the JSpOC reached an agreement to stand up a "commercial prototype cell" whereby chosen space industry officials would sit inside the JSpOC to better share SSA information, importantly by taking steps to allow industry data to be used by the JSpOC.[54]

In addition, U.S. firm Analytical Graphics, Inc. (AGI) in March 2014 established the Commercial Space Operations Center (ComSpOC), which is providing operators with highly accurate space object detection, tracking and characterization. The ComSpOC uses its own sensor network of 28 (so far) optical telescopes and one passive radar, with observations updated hourly.[55]) The goal is providing better quality conjunction analysis than can currently be provided by the JSpOC.

While industry efforts are to be welcomed, they too are inadequate to provide truly internationally accessible data. For one thing, pay-to-play may be out of reach for developing nation space operators. Secondly, initiatives such as SDA and ComSpOC may be constrained in sharing data

---

[50] Note bene: the conjunction summary messages were replaced in 2014 by conjunction data messages based on an improved and internationally agreed data format.

[51] "USAF Satellite-Conjunction Advisories Called Inaccurate," Frank Morring Jr., *Aviation Week*, February 24, 2012, http://aviationweek.com/space/usaf-satellite-conjunction-advisories-called-inaccurate

[52] See the Space Data Association website, http://www.space-data.org/sda/

[53] "Space Data Association and USSTRATCOM Reach Data Sharing Agreement," Marcia S. Smith, *SPACEPOLICYONLINE.com*, August 11, 2014, http://www.space-data.org/sda/

[54] "US JPoC to Create 'Commercial Cell Prototype' In Coming Months," Caleb Henry, *ViaSatellite*, April 10, 2015, http://www.satellitetoday.com/regional/2015/04/10/us-jspoc-to-start-commercial-cell-prototype-in-coming-months/

[55] See http://comspoc.com/

with some entities (such as Chinese space operators) due to their close, contractual connections to the JSpOC.

Therefore, the creation of an independent, publicly accessible space data catalog could fill a gap. In addition to industry SSA data, observations of space objects are also made and provided publicly by amateur space watchers and space scientists.

The astronomical research arm of AGI, the Center for Space Standards & Innovation, provides the website CelesTrak, which monitors active satellites, as a public database for space scientists and astronomers. CelesTrak, managed by T.S. Kelso, features an online satellite catalog of active and decayed satellites, based on data pulled from the public JSpOC catalog.[56] Using CelesTrak requires a certain amount of technical expertise, however. McDowell also maintains an online list of satellites, as well as a separate list of geostationary satellites. Like CelesTrak, McDowell's data also requires technical competency to use.[57] The Union of Concerned Scientists (UCS), a U.S.-based NGO, maintains a more lay-person friendly satellite database of operational satellites.[58] While all of these sources provide information on civil and military satellites globally, and are updated relatively frequently, they do not provide information on debris and some do not provide information on the orbital positions of space objects

Amateur sky-watchers also share satellite observation and tracking data through their informal networks. One such network is Heavens Above, which also maintains a satellite database on its website based on input from the JSpOC[59] Another network is SeeSat-L, which is a mailing list for satellite observers.[60] Observer networks have proven important in tracking phenomena such as failed and drifting satellites, including the undisclosed failure of U.S. DSP-23 reconnaissance satellite in 2008,[61] as well as keeping tabs on the U.S. Air Force's secret space plane, the X-37B.[62] Indeed, the U.S. Defense Advanced Research Projects Agency (DARPA) is

---

[56]  See https://celestrak.com/
[57]  See Jonathan's Space Home Page, http://www.planet4589.org/space/
[58]  UCS Satellite Database, Union of Concerned Scientists, http://www.ucsusa.org/nuclear_weapons_and_global_security/solutions/space-weapons/ucs-satellite-database.html#.VTArNWTBzGc
[59]  See http://www.heavens-above.com/
[60]  See http://www.satobs.org/seesat/seesatindex.html
[61]  "The On-Going Saga of DSP Flight 23," Brian Weeden, The Space Review, January 19, 2009, http://www.thespacereview.com/article/1290/1
[62]  "Secret X-37B Space Plane Spotted Again by Amateur Skywatchers," Joe Rao, Space.com, March 28, 2011, http://www.space.com/11240-video-secret-x37b-space-plane-skywatchers.html

running a small program called Orbit Outlook, designed to find a methodology for integrating amateur observations into the JSpOC catalog.[63]

Research on space debris is carried out by many States around the world, primarily using optical telescopes that are optimized for astronomical observations. While, as noted, the United States has the most comprehensive space object catalog including debris, the Russian Space Surveillance System (which, like the U.S. SSN, doubles as an early-warning radar network) also tracks both satellites and debris.[64] The European Space Agency as well keeps a space object catalog, called the Database and Information System Characterizing Objects in Space (DISCOS).[65] One of the most important organizations for multilateral work on the debris problem is the Inter-Agency Debris Coordinating Committee (IADC), which is an intergovernmental body comprising 13 space agencies from around the world, including China, Russia and the United States.[66] The IADC functions as a platform to share debris observations and research, and promote collaborative efforts. The UN Debris Mitigation Guidelines came out of an IADC exercise, and the IADC continues to refine its own, more technical guidelines that can be voluntarily taken up by member agencies and other States. The IADC reports on its activities ever year at the annual meeting of COPUOS STSC.[67]

Another informal body, the International Scientific Optical Network (ISON), based at the Russian Academy of Sciences Keldysh Institute of Applied Mathematics, specializes in observations of both satellites and debris in Geosynchronous Orbit[68] and highly elliptical orbits.[69] The ISON network is made up of 35 facilities in 15 countries (primarily in the Northern Hemisphere) utilizing more than 80 telescopes of varies sizes and

---

[63]   See http://www.darpa.mil/Our_Work/TTO/Programs/OrbitOutlook.aspx
[64]   "Early Warning," *Russian strategic nuclear forces*, http://russianforces.org/sprn/
[65]   See http://www.esa.int/Our_Activities/Operations/Space_Debris/Analysis_and _prediction
[66]   See IADC, http://www.iadc-online.org/
[67]   The reports can be found at: http://www.iadc-online.org/index.cgi?item=docs_pub
[68]   Geosynchronous Orbit is roughly 36,000 kilometers above the Earth's equator, where satellites move in such a way as to remain over the same spot on Earth. This orbit is the key band for large communications satellites.
[69]   Highly Elliptical Orbits have a perigee at very low altitudes (often under 1,000 kilometers) and an apogee at very high altitudes (often in the GEO band).

types.[70] ISON is well known for its work discovering very faint objects in GEO.

Each of these systems and networks use different methodologies for collecting, reporting and analyzing orbital data. There is no "one stop shop" that integrates these various sources of information. Thus, an initiative that would seek to do so, and provide a "master" catalog in the process would be doing a public service, and arguably enhance ongoing government efforts. Such a "crowd-sourced" database, run by an independent organization, could serve as a kind of "halfway house" between what is publically available today and the higher quality data kept by the United States and some other nation states for national security purposes. In particular, an independently operated space object catalog could provide much needed transparency to those States and space operators that have little or no SSA capacity.

## ACTIVE DEBRIS REMOVAL AND ON-ORBIT SERVICING

The Expert Group B report identifies active debris removal (ADR) and on-orbit servicing of satellites (i.e. for refueling or repair) as issues that should be raised to the Scientific and Technical Subcommittee for consideration. Both represent serious technical challenges, as well as political and legal challenges. In addition, there is concern – with reason – that these technologies could be applied to create space-based ASAT weapons. Thus, developmental efforts by individual States could raise suspicions about intentions and contribute to instability in space.

There is a scientific consensus that even if strong mitigation measures are adopted by all space actors, the debris population is very likely to grow – raising serious concerns about the safety and security of future space activities. ADR concepts have been studied for three decades, but there has been little impetus to begin to seriously explore and develop the technology. The technical challenges differ for removal of small debris versus large debris, and for debris in LEO versus GEO. The Expert Group B report states: "Numerous independent robotic concepts, ranging from classical space-based propulsive tugs to momentum and electrodynamic tethers, drag augmentation devices, solar and magnetic sails, and ground- and space-based lasers have also been considered."[71] In addition, a DARPA study, called "Catcher's Mitt," in 2011 found that while medium-sized

---

[70]  "Current status of the ISON optical network," Igor Molotov, Vladimir Agapov, et al, abstract presented to the 40th COSPAR Scientific Assembly, August 2-10, 2014, Moscow, http://adsabs.harvard.edu/abs/2014cosp...40E2157M

[71]  Expert Group B report, p. 27. http://www.unoosa.org/pdf/limited/l/AC105_2014_CRP14E.pdf

debris (5mm to 10cm) is the greatest threat to satellites, there is no feasible way to remove it at this time. Thus, the study recommended that development focus on removing large debris from LEO and GEO as a preventive measure (i.e., to prevent this large debris from fracturing into smaller, more lethal pieces.)[72]

In addition, there are legal questions surrounding how to ensure that ADR methods are compliant with the Outer Space Treaty and the Liability Convention, among other international legal instruments. For example, there is no legal definition of space debris. Further, under the Outer Space Treaty, it would be illegal for a State or commercial entity to remove a piece of debris without obtaining permission from the "owner" of that object – a fact that is complicated by the problem of identifying who owns what pieces of debris.

On-orbit servicing concepts rely on so-called rendezvous and proximity operations (RPO) – the ability to maneuver into an orbit that is closely matched with another satellite and/or "dock" with that satellite to perform refueling or repairs. The ability to re-fuel and repair satellites would extend their lifetimes, and potentially slow the growth of debris. NASA's Goddard Space Flight Center in 2010 did a comprehensive study of the value of on-orbit servicing and necessary technologies, and called for a dedicated U.S. program.[73] This study, as have various others, found that technologies to conduct on-orbit services exist, although as of today there are no active systems. The Expert Group B report notes: "However, other than NASA's Hubble Space Telescope, experience with on-orbit servicing (OOS) is very limited, and if done incorrectly OOS could create debris instead of reducing it."[74] In addition, the ability to maneuver close to and/or physically connect with a non-cooperative satellite would enable disabling or destroying it, meaning that development projects to do so are politically fraught.

While the Expert Group B report stops short of making a recommendation that these issues be the subject of formal study by COPUOS, this only makes sense. COPUOS should either expand the remit

---

[72] "Threats to U.S. National Security Interests in Space: Orbital Debris Mitigation and Removal," Stephen A. Hildreth and Allison Arnold, Congressional Research Service, January 8, 2014, p. 11. https://www.fas.org/sgp/crs/natsec/R43353.pdf

[73] "On-Orbit Satellite Servicing Study: Project Report," National Aeronautics and Space Administration Goddard Space Flight Center, October 2010, http://ssco.gsfc.nasa.gov/images/NASA_Satellite%20Servicing_Project_Report_0511.pdf

[74] Expert Group B report, p. 28. http://www.unoosa.org/pdf/limited/l/AC105_2014_CRP14E.pdf

of Expert Group B or create a special working group under the Scientific and Technical Subcommittee to study ADR and on-orbit servicing technologies, determine the most feasible of them, and make recommendations for development programs. COPUOS should at the same time create a working group under the Legal Subcommittee to examine the legal and political challenges to implementing ADR and on-orbit servicing systems, with an eye toward enabling and encouraging international cooperative development (or at a minimum establishing processes to ensure transparency regarding national developments.)

## REVIEW OF LEGAL OBLIGATIONS REGARDING MILITARY ACTIVITIES

One of the foundational issues for multilateral space governance is the lack of clarity surrounding both space law, and, in particular, the application of the laws of armed conflict (LOAC) to military space activities. The COPUOS Legal Subcommittee has for years been debating foundational issues of space law, such as a definition of where outer space begins, with little progress. In addition, it is apparent that States have differing interpretations of how LOAC is to be applied in space, for example, how does one define the principle of "proportionality" in any response to an attack, especially a less than destructive attack, on a satellite? What restraints, if any, are inherent in LOAC on "space war?" Is radio frequency jamming a use of force, or an "armed attack" as legally defined? How do you distinguish "legitimate" military targets when most of the world's militaries use commercial satellites for communications, mapping and other key functions?

Some legal scholars have been discussing the creation of an independent legal review body to look at these questions, with an eye towards establishing an informal, but scholarly, understanding of current obligations for space actors regarding military operations. Additionally, the application of environmental law to the "means and methods" of conflict involving satellites could be included in such a review. Indeed, the Australian government submitted a paper in 2012 to the GGE Chair that sought to identify some key principles of international law that might be applicable to space activities, in particular with an eye to debris, including LOAC , environmental law and international telecommunications law. The paper noted that: "Acknowledgement of and commitment to these obligations by States could be expected to generally discourage the development and use of kinetic counter-space weapons."[75]

---

[75] "Submission by the Government of Australia on building confidence through transparency on international law applicable to international security issues in outer

The Australian paper suggested that the GGE include in its report:

- "A reaffirmation of the applicability of international law to outer space, recognition of the contribution that working toward a common understanding of how international law applies to outer space can make to confidence-building and international security, and,
- "A recommendation that States consider potentially applicable international law obligations and principles when planning their space activities."[76]

Unfortunately, the GGE did not take up the Australian suggestions, as legal issues were, and remain, contentious. While discussions of constraints on military space operations are uncomfortable (indeed, unwelcome) for many States, the current situation where military space powers espouse differing interpretations of legitimate activities is a cause of misunderstanding and tension. Thus, an effort to develop a common understanding of the legal framework that applies to space activities by independent legal scholars would serve as a less-threatening starting point for multilateral discussion.

As explained by Dr. Cassandra Steer, Executive Director of the Center for Research in Air and Space Law at McGill University, at a March 23, 2015 meeting co-sponsored by Secure World Foundation and the George Washington Military Law Society, there are many precedents of the development of such legal "manuals" by legal and military scholars that have subsequently been used by the militaries in many States to guide their actions. These include the Harvard Manual of International Law Applicable to Air and Missile Warfare, and the Tallinn Manual on International Law Applicable to Cyber Warfare.[77] Such an exercise could serve as a confidence-building measure as more and more States integrated space operations into their militaries. Further, such an initiative would not necessarily require State endorsement to be launched, although obviously participation from national governments/militaries would improve the

---

space, submitted to the Chairman of the Group of Governmental Experts during the Plenary Session of the Conference on Disarmament in August 2012, found in "Disarmament Series 34, Transparency and Confidence Building in Outer Space Activities," UN Office of Disarmament Affairs, 2013, p. 79, http://www.un.org/disarmament/publications/studyseries/en/SS-34.pdf

[76] Ibid p. 80

[77] See http://swfound.org/events/2015/international-law-and-military-activities-in-space/

chances of successful update of the legal findings. Indeed, the Center for Research in Air and Space Law intends to hold an initial seminar to discuss this concept in September 2015.

## SPACE TRAFFIC MANAGEMENT AND SMALL SATELLITES

The COPUOS Legal Subcommittee over many decades has found consensus on issues, including the question of what issues should be on its agenda, to be elusive. However, at its most recent meeting, April 13-24, 2015, the subcommittee agreed to two new agenda items for discussion with relevance to future space governance. These are:

- "General exchange of views on the legal aspects of space traffic management," and,
- "General exchange of views on the application of international law to small satellite activities."[78]

The proposal to discuss space traffic management (STM) was made by Germany. It states that the purpose of the discussions "would be to reflect on the concept of STM, on what it entails and on what consequences it would have for the organization and governance of space activities. In particular, the contribution of STM to the safety of space operations benefitting all users of outer space (whether they are established users or recent and future users) could be investigated. The item would also provide the opportunity to discuss the status of academic research in that field and to possibly invite presentations on the technical as well as legal background of this issue."[79]

The German proposal jumped off of a seminar held in the margins of the meeting (on April 13) by the International Institute of Space Law (IISL) and the European Centre for Space Law (ECSL) on STM.[80] Concept for STM have been mooted for a number of years, with perhaps the seminal study on the issues being the International Academy of Astronautics

---

[78]  "Draft Report (Addendum 4)," Legal Subcommittee, Committee on the Peaceful Uses of Outer Space, 23 April 2015, A/AC.105/C.2/L.296/Add.4, http://www.unoosa.org/pdf/limited/c2/AC105_C2_L296Add04E.pdf

[79]  "Proposal for a Single Issue/Item for discussion at the fifty-fifth session of the Legal Subcommittee in 2016 on: Exchange of views on the concept of Space Traffic Management," Proposal by Germany, Legal Subcommittee, Committee on the Peaceful Uses of Outer Space, Fifty-fourth session, 14 April 2015, A/AC.105/C.2/2015.CRP13, http://www.unoosa.org/pdf/limited/c2/AC105_C2_2015_CRP13E.pdf

[80]  See: http://www.unoosa.org/oosa/en/COPUOS/lsc/2015/symposium.html

"Cosmic Study on Space Traffic Management" published in 2006.[81]  While there is no accepted definition of STM or its scope, that study used the following definition: "Space traffic management means the set of technical and regulatory provisions for promoting safe access into outer space, operations in outer space and return from outer space to Earth free from physical or radio-frequency interference." The study, while asserting a need for STM, also recognized that: "Space traffic management however, will limit the freedom of use of outer space. Therefore an international consensus on internationally binding regulations will only be achieved, if States identify certain urgency and expect a specific as well as collective benefit – including an economic benefit - from this."

The Cosmic Study essentially laid out a framework that covered all phases of space operations from launch to end-of-life disposal, including: a process for provision of orbital data; a notification system including pre-launch and maneuver notifications; "zoning," "right of way" rules for maneuvering; safety provisions for launches, human spaceflight including tourism, and re-entries; debris mitigation and environmental pollution measures; and liability laws. Many studies since – including the 2007 report of the International Association for the Advancement of Space Safety (IAASS) that called for a new international agency to manage space traffic similar to the International Civil Aviation Organization – have elaborated on these ideas. Further, IAA intends to publish a follow-on study in 2016. However, despite 10 years of debate, there is no international consensus that a STM regime is necessary, much less what it would entail. Obviously, however, some of the foundational concepts of STM dovetail with recommendations from the GGE, the COPUOS LTS work and the EU draft code, especially with regard to orbital object notifications. Therefore, despite the fact that it will take much time, a COPUOS Legal Subcommittee debate is to be welcomed. While the current mandate is only for a year long discussion beginning in 2016, it would be wise for the Legal Subcommittee to make this issue a standing item on the agenda. In addition, national governments should begin to consider their views on STM, how STM principles might be incorporated into national legislation and licensing regimes, and whether and how national government structures are appropriate to handle any new responsibilities that this might entail.

---

[81] "Cosmic Study on Space Traffic Management," International Academy of Astronautics, 2006, https://iaaweb.org/iaa/Studies/spacetraffic.pdf

Concern among major satellite operators over the proliferation of small and very small satellites has grown over recent years. Four key issues are at hand:

1. Very small satellites are difficult to track, especially when multiples are released into orbit at the same time.
2. Many of these satellites are working at radio frequencies that do not require registration and coordination with the ITU, which in turn means that they fall between the regulatory cracks (including at the national level.)
3. Most of these satellites do not have capabilities for de-orbiting in line with the COPUOS and IADC Debris Mitigation Guidelines, ergo at the end of their lives become de facto debris.
4. Many operators of small satellites are new to space activities, and are not necessarily well versed in their obligations under international law and regulatory regimes – where those actually apply.

According to UNOOSA and the ITU, in a newly developed set of guidelines for small satellite operators:

*"Satellites may be grouped into different categories based on their mass (for example, mini satellites <1000 kg, micro satellites < 100 kg, nano satellites < 10 kg, pico satellites < 1 kg, femto satellites < 0,1 kg). However, as of today, there is no consensus or universally accepted standard on the definition of a small or very small satellite. A small satellite is not necessarily physically small as it may have deployable structures, it is not necessarily low-weight and neither does it have to be less complex or less capable compared to a satellite that is not considered to be small. Typical characteristics of small satellite missions include: a) reasonably short development times; b) relatively small development teams; c) modest development and testing infrastructure requirements; and d) affordable development and operation costs for the developers, in other terms "faster, cheaper and smaller". Some other characteristics often seen in small satellite missions are: a) they often involve actors new to space activities mainly non-governmental actors (academic institutions, private companies etc.); b) for various reasons, very often due to inexperience or unfamiliarity with the national and international*

> *regulatory framework, they are not always conducted in full compliance with international obligations, regulations and relevant voluntary guidelines (authorization, supervision, registration, ITU radio regulations, space debris mitigation guidelines etc.); and c) they have raised concerns to worsening the space debris situation.*"[82]

The COPUOS agenda item was proposed by the Group of Latin American and Caribbean states (GRULAC). The proposal explained that the purpose of the discussion is "to allow Member States and organizations that participate in the COPUOS Legal Subcommittee (LSC) to exchange views on and discuss the implications of the growth in small satellite activities in order to ensure these devices can fulfill their role in supporting the development of space programs. The discussion will illuminate whether Member States wish to undertake further work on this topic."[83] In addition, the GRULAC submission noted that "While regulations for the small satellite segment seems to be required, this activity should also be protected from over regulation" because of their role in fostering emerging space actors.

Again, the discussion of this extremely important and immediate issue within the COPUOS Legal Subcommittee is to be welcomed. It will be important that there is multilateral agreement on new recommendations for best practices and/or regulatory regimes in order to create a level playing field for operators.

At the same time, individual States must move out to find ways of incorporating these operators into their national regulatory and licensing regimes. For one thing, even if such satellites are not being currently registered with the UN or ITU, under the Outer Space Treaty, States are responsible for national activities in space, and are required to supervise activities by non-governmental actors. Further, under the Liability

---

[82] "Guidance on Small and Very Small Satellite Registration and Frequency Management," United Nations Office of Outer Space Affairs and the International Telecommunication Union, presented to the Legal Subcommittee of the Committee on the Peaceful Uses of Outer Space, Fifty-fourth session, April 13-24, 2015, A/AC.102/C.2/2015/CRP.17, 13 April 2015, http://www.unoosa.org/pdf/limited/c2/AC105_C2_2015_CRP17E.pdf

[83] "Proposal for a Single Issue/Item for discussion at the fifty-fifth session of the Legal Subcommittee in 2016 on: 'Exchange of views on the application of international law to small satellite activities,' Submission by GRULAC, Legal Subcommittee, Committee on the Peaceful Uses of Outer Space, Fifty-fourth session, 13-2 April 2015, A/AC.105/C.2/2015/CRP.23/Rev.1, http://www.unoosa.org/pdf/limited/c2/AC105_C2_2015_CRP23Rev01E.pdf

Convention, States are responsible for damage caused by national actors. At the March 3, 2015 meeting in Prague, participants at the ITU Symposium on Small Satellite Regulation and Communication Systems, a declaration was issued that confirmed "the importance of implementing national legal and regulatory frameworks in conformity with the above international instruments, clearing defining rights and obligations of every stakeholder participating in small satellite initiatives."[84]

Indeed, the United States government is already working at the interagency level to assess how this might be done, according to officials involved in the discussions. One option under consideration is creating legal jurisdiction for the Federal Aviation Administration (FAA) to license small operators and their payloads. Conclusion of this discussion and rapid movement to implement a pathway forward should be strongly urged by all stakeholders.

The Prague Declaration, while urging the small satellite community itself to figure out how to comply with applicable laws, also recommended capacity-building for small satellite operators – both in the area of technological capabilities related to, for example, space debris mitigation, and in the area of legal obligations. This is an area where the NGO and academic research community could contribute to progress on space governance by organizing and holding seminars, workshops and training activities. Major space actors should support such independent activities, because it is in the interest of both governmental and commercial space operators to ensure that the small satellite sector is both encouraged and prevented from causing harm to the space environment.

## Commercial Property Rights on Asteroids, the Moon and Celestial Bodies

The question of property rights on, and exploitation of, celestial bodies has long been a highly contentious arena of international space law. But only in the past few years, as the activities of several commercial ventures have become more viable, has the issue gained immediacy, particularly in the United States. In particular, the plans by Bigelow Aerospace to develop lunar habitats and the goal of commercial companies such as Planetary Resources and Deep Space Industries to develop technology to mine asteroids have pushed the U.S. government and Congress to focus on the legal and regulatory issues surrounding commercial access to the Moon and

---

[84]  "Prague Declaration on Small Satellite Regulation and Communication Systems," http://www.itu.int/en/ITU-R/space/workshops/2015-prague-small-sat/Documents/Prague%20Declaration.pdf

asteroids.

Bigelow Aerospace, founded by billionaire Robert Bigelow in 1999, is under contract by NASA to test its expandable space habitat technology on the International Space Station in 2015, but its ultimate game plan is to deploy these habitats on the Moon beginning in 2025.[85] In late 2013, Bigelow made a formal request to the FAA's Office of Commercial Space Transportation asking for a so-called "payload review request" and making a case that the company has a right, under the Outer Space Treaty, to base its habitats on the Moon and should be protected by the U.S. government from interference.[86]

Bigelow's legal team argues in essence that, while Article II of the 1967 Outer Space Treaty prohibits "national appropriation" of property on the Moon and other celestial bodies,[87] it does not specifically prohibit commercial companies doing so. Other space law scholars, however, beg to differ. As long ago as 2004, the International Institute of Space Law (IISL) issued a statement asserting that there can be no private ownership of celestial bodies, due to Article VI of the Outer Space Treaty, which makes nations responsible for the activities of private sector actors and ensuring that those actors are in compliance with the treaty.[88] The IISL stated:

*"Article VI of the Outer Space Treaty provides that "States bear international responsibility for national activities in outer space, including the Moon and other celestial bodies, whether such activities are carried on by governmental agencies or by non-governmental entities," the is, private parties, and for "assuring that national activities are carried out in conformity with the provisions set forth in the present Treaty." Article VI further provides that "the activities of non-governmental entities in outer space, including the Moon and*

[85] "The FAA: Regulating business on the moon," Irene Klotz, *Reuters*, Feb. 3, 2015, http://www.reuters.com/article/2015/02/03/us-usa-moon-business-idUSKBN0L715F20150203

[86] "Hard Cheese," by K.R., *The Economist*, Feb. 16, 2014, http://www.economist.com/blogs/babbage/2014/02/lunar-property-rights

[87] Article II of the Treaty on the Principles Governing the Activities of States in the Exploration and User of Outer Space, including the Moon and Other Celestial Bodies (the Outer Space Treaty) states that: "Outer space, including the Moon and other celestial bodies, is not subject to national appropriation by claim of sovereignty, by means of use or occupation, nor by any other means." See: http://www.unoosa.org/pdf/publications/ST_SPACE_061Rev01E.pdf

[88] "Statement of the Board of Directors of the International Institute of Space Law (IISL) On Claims to Property Rights Regarding The Moon and Other Celestial Bodies," 2004, http://www.iislweb.org/docs/IISL_Outer_Space_Treaty_Statement.pdf

*other celestial bodies, shall require authorization and continuing*
*supervision by the appropriate State Party to the Treaty. "*
*Therefor, according to international law, and pursuant to Article VI,*
*the activities of non-governmental entities (private parties) are*
*national activities. The prohibition of national appropriation by Article*
*II thus includes appropriation by non-governmental entities ...."*

The FAA, in a December 2014 letter of response to Bigelow, agreed to the review, stating that the agency will "leverage the FAA's existing launch licensing authority to encourage private sector investments in space systems by ensuring that commercial activities can be conducted on a non-interference basis."[89] The FAA was quick to clarify that this did not represent an endorsement of Bigelow's claims to commercial property rights on the Moon; but instead a first step toward developing a U.S. government framework to clarify regulatory and international legal obligations regarding such proposed activities.[90] Such a review requires input from the plethora of U.S. agencies involved in space governance, including the White House Office of Science and Technology Policy, the State Department, the Defense Department, the Federal Communications Commission and NASA. Indeed, the State Department has already signaled concern about the Article VI obligations involved, according to officials. It is therefore unclear how, and when, Bigelow's petition will be handled.

Meanwhile, the U.S. Congress – at the urging of space exploitation activists and Planetary Resources – has seriously taken up the issue of asteroid mining for resources such as water, oxygen, platinum and other rare Earth minerals. Planetary Resources, the firm founded by space entrepreneurs Peter Diamandis and Eric Anderson in April 2015, launched its first technology demonstrator aimed at asteroid mining. A second test spacecraft is slated to be launched in December.[91] On April 13, 2015, the House Committee on Science, Space and Technology passed H.R. 1508, "The Space Resource Exploitation and Utilization Act of 2015," which seeks to "establish and protect property rights for commercial space exploration and utilization of asteroid resources," according to its sponsor,

---

[89]  Klotz, op cit.

[90]  "FAA Moves to Establish Framework for Commercial Lunar Operations," Douglas Messier, *Parabolic Arc*, Feb 7, 2015, http://www.parabolicarc.com/2015/02/07/faa-moves-establish-framework-commercial-lunar-operations/

[91]  "Akryd 3 Reflight (A3R) Launches from Cape Canaveral on Space X CRS-6," Planetary Resources website, April 14, 2015, http://www.planetaryresources.com/2015/04/arkyd-3-reflight-a3r-launches-from-cape-canaveral-on-spacex-crs-6/

Rep. Bill Posey of Florida.[92] The legislation asserts that asteroid resources are the property of the entity that obtains them, and calls on the Obama Administration to within 180 days issue recommendations on the responsibilities of various federal agencies with regard to space exploitation.[93] A similar bill, S.976 – Space Resource Exploitation and Utilization Act, proposed by Sen. Patty Murray of Washington, has been referred to the Senate Committee on Commerce, Science and Transportation.

Any congressional action on asteroid mining faces the same legal issues surrounding lunar property rights. In addition, implementation of any legislation would be reliant on a determination by the Obama Administration that the legislation was in accordance with U.S. obligations under the Outer Space Treaty and international law. Even if the U.S. government decides that such activities do not violate the Outer Space Treaty, that interpretation will be subject to challenge by other Member States – many of whom may be concerned with the possibility of a "gold rush" on space resources in which most States cannot currently compete.

While there are potentially many benefits to the exploitation of space resources, including to economic and social development on Earth, as well as the enabling of space exploration and even the advent of space-based solar power, there are also many risks to international stability and security. Unregulated commercial competition in space raises could lead to interstate rivalries and even conflicts over access. Given the enthusiasm with which commercial space entrepreneurs are pushing the edges of the envelope on these issues, it behooves the international community to begin a serious discussion. The COPUOS Legal Subcommittee could, and should, add this item to its agenda – as resolving these questions will be critical to future space governance. In the meantime, the U.S. government needs to act quickly to establish its own legal and regulatory policies, because of its responsibilities and potential liability under Article VI of the Outer Space Treaty. The academic community could contribute to this debate by convening an international panel of legal scholars to study the issues and make recommendations.

---

[92] "House Science Committee Approves Posey's Bipartisan Legislation to Promote Commercial Space Ventures," Office of U.S. Congressman Bill Posey press release, April 13, 2015, http://posey.house.gov/news/documentsingle.aspx?DocumentID=394237

[93] The text of H.R. 1508 is found here: https://www.congress.gov/bill/114th-congress/house-bill/1508/text

## CONCLUSION

As the number and diversity of space actors grows, the challenges to multilateral approaches to space governance are increasing. Established space powers have different priorities than do emerging space powers; military space powers have fundamentally conflicting goals (i.e. to do harm to each other if considered necessary) and different understandings of their legal constraints; and capacity to uphold international legal and political commitments varies widely. Thus, while there is a widespread (if not universal) consensus on the problems facing the space domain, there is not consensus on what should be done, and what should be done first.

In addition, the explosion of commercial actors – who have less stake in (indeed even some antipathy to) current or any future governance regimes – has complicated interactions between States. Growing commercial presence will require States to put in place either new, or modified, national laws, policies and regulatory regimes to ensure against chaos. Already, the priorities of commercial actors – such as would-be asteroid mining ventures and micro-sat operators – are forcing States to re-examine existing regulatory and legal regimes (both national and multilateral) with an eye to potential adaptation.

Finally, current multilateral fora for establishing space governance are stovepiped, and sometimes at odds regarding priorities and needs. This creates a problem in addressing the challenges, which are by and large cross-cutting. Certain issues fall through the cracks, and other issues are being debated through a narrow lens that fails to take into account potential second-order or cross-sectoral consequences. This is apparent, for example, in the Cold War created mandates of COPUOS and the Conference on Disarmament, which seek to create a separation from issues of peaceful uses of space and military ones – despite the fact that those uses and functions have largely converged. In addition, there is little room in current fora (with the exception of the ITU) for the participation of the private and civil sectors. This makes little sense in that private sector actors and NGOs are seeking, and are gaining, influence with their national governments regarding legal and regulatory regimes to support their own priorities, which may not coincide with international security needs.

These factors in part account for the current focus in the international community on voluntary measures, norms of behavior and transparency and confidence-building measures. The current environment is simply not ripe for the pursuit of legally binding treaties, nor, given the uptick in tensions among the three major space powers – China, Russia and the United States – is it likely to be any time soon.

But despite these growing tensions, there are steps that could – and indeed should -- be taken or supported by individual or groups of States that would forward progress toward improved multilateral governance of outer space activities, starting with the already agreed recommendations of the UN GGE on transparency and confidence-building. Transparency and confidence-building measures are necessary precursors to eventual multilateral accord. Even unilateral actions by States would create forward momentum and establish foundation stones for future multilateral agreements. The October 2015-slated joint meeting of the UNGA's First and Fourth Committees provides the first opportunity for States to propose follow-on initiatives.

In particular, a review of implementation of the current space regime could, and should, be initiated. States could develop lists of contacts and focal points for space governance issues, in order to underpin transparency and provide avenues for discussion when worrisome situations arise. In particular with the Registry Convention can, and needs to be, improved – especially by major space-faring States that have little excuse for failing in their compliance obligations. In addition, the Registry could be expanded, or an ad hoc process could be developed, to expand notification of specific space activities such as pre-launch notifications and notification of satellite maneuvers.

In addition, there are other opportunities – for example, based on the work of the COPUOS LTS Working Group, as well as in the legal arena, to put flesh on the emerging skeleton framework of space governance. As a first step, States could move to institutionalize and implement the UN Debris Mitigation Guidelines to their fullest extent. This is an action that can be taken unilaterally by any State. Active debris removal and on-orbit servicing are both issues that could be put onto the agenda of COPUOS in the near-term, as both will require multilateral accord to move forward. Other issues that could be elevated to COPUOS include: space traffic management, with a particular emphasis on establishing controls on small satellite activities; and commercial property rights on the Moon and other celestial bodies, including resource extraction.

Further, there are activities that could be taken by NGOs and the private sector to forward progress on multilateral space governance in a manner that encourages the expanding utilization of space while protecting against risks to a safe and stable environment. Near-term initiatives could include the creation by a group of NGOs and private sector actors of a public space situational awareness database, and a review by independent legal scholars of the legal obligations regarding military activities in space.

These are no doubt only a few of the potential avenues for future work on the vital establishment of multilateral space governance. What is needed at this time is political will and leadership. It would be a pity if the progress made so far is allowed to stagnate, or worse, to be eroded.

•••

## THERESA HITCHINS

Theresa Hitchens is a Senior Research Scholar at the Center of International and Security Studies at the University of Maryland, where she focuses on space security, cyber security, and governance issues surrounding disruptive technologies. Prior to joining CISSM, Hitchens was the director of the United Nations Institute for Disarmament Research (UNIDIR) in Geneva from 2009 through 2014. Among her activities and accomplishments at UNIDIR, Hitchens served as a consultant to the U.N. Group of Governmental Experts on Transparency and Confidence Building Measures in Outer Space Activities, provided expert advice to the Conference on Disarmament regarding the prevention of an arms race in outer space (PAROS), and launched UNIDIR's annual conference on cyber security.

From 2001 to 2008, Hitchens worked at the Center for Defense Information, where she served as Director, and headed the center's Space Security Project, setting the strategic direction of the center and conducting research on space policy and other international security issues. She was also previously Research Director of the Washington affiliate of the British American Security Information Council (BASIC), where she managed the organization's program of research and advocacy in nuclear and conventional arms control, European security and North Atlantic Treaty Organization (NATO) affairs.

Hitchens's latest publications include, "Space Security-Relevant International Organizations: UN, ITU, ISO," 2014, which was penned for the *Handbook of Space Security*; "Preserving Freedom of Action in Space: Realizing the Potential and Limits of U.S. Spacepower," 2011, which was coauthored with Michael Krepon and Michael Katz-Hyman; and "Saving Space: Threat Proliferation and Mitigation," 2009. Hitchens holds a Bachelor of Science in journalism from Ohio University in Athens, Ohio.

Research support for this chapter was provided by the Secure World Foundation.

# PART II

# ECONOMIC AND LEGAL FOUNDATIONS OF SPACE SUSTAINABILITY

## INTRODUCTION TO PART II

In these chapters we examine some of the most critical issues pertaining to the short and long term future of space operations from the perspectives of economics and law, hearing from practicing experts and analysts.

CHAPTER 7

# THE ECONOMICS
# OF SPACE SUSTAINABILITY

BRIAN WEEDEN
SECURE WORLD FOUNDATION

AND

TIFFANY CHOW
SECURE WORLD FOUNDATION

During the last several years, the long-term sustainable use of outer space, specifically Earth orbit, has emerged as a significant public policy issue in the United States and among many spacefaring nations. A crucial element of space sustainability is the problem of human-generated space debris in orbit around the Earth, and the risks it poses to active satellites. Recent events such as the 2007 Chinese antisatellite test (Weeden, 2012a) and the 2009 collision between an American Iridium and Russian Cosmos satellite (Weeden, 2012b) are only the most publicized instances of the issue. The changing geopolitical landscape since the end of the Cold War and the lowering of barriers to entry to space are spurring an increase in the number of states and non-state actors conducting activities in space. This increase in the number of space actors, many of whom are new to the domain, and the resulting increase in congestion in key orbital regimes means that the challenges of maintaining space sustainability will only become more acute as time passes.

This chapter examines the economic principles behind the sustainability challenges in the most congested regions of Earth orbit. It explains why microeconomic policy interventions have so far failed to find traction, largely due to confusion over the type of economic good that outer space actually is, and the lack of private actors in outer space that would be responsive to such economic incentives. It concludes with a discussion of progress that is being made on several initiatives to enhance space sustainability, and traces their roots to the relatively new economic fields of information theory and sustainable management of common-pool resources.

## SATELLITES AND SPACE DEBRIS

More than 1,200 active satellites are currently in orbit around Earth (Union of Concerned Scientists, 2014) providing a wide range of social and private benefits, including enhanced national and international security, more efficient use and management of natural resources, improved disaster warning and response, and reliable global communications and navigation. The vast majority of these active satellites exist in two distinct regions. Approximately 650 are in Low Earth Orbit (LEO) orbiting between 200 and 2,000 km in altitude, and another 460 or so active satellites are in the Geosynchronous Earth Orbit (GEO) region approximately 36,000 km above the equator.

In addition to the 1,200 active satellites, the U.S. military currently tracks close to 22,000 pieces of human-generated debris in Earth orbit that are bigger than 10 cm in size, each of which could destroy an active satellite in a collision, consisting of dead satellites, spent rocket stages, and fragments associated with humanity's six decades of activity in space. In economic terms, such space debris is a producer-to-producer negative externality, and research by scientists and space agencies indicates there is a population of another 500,000 pieces of space debris between 1 cm and 10 cm in size, any of which could severely damage an active satellite if they were to collide.

[Editor's note: Please see Figure 1, Chapter 15, page 307.]

As a product of human activity in space, space debris is concentrated in the most actively used regions (LEO and GEO), resulting in increased risks of collisions with active spacecraft. This increased risk raises the private costs of satellite operations due to greater expenditure of fuel and interruption of mission from avoidance maneuvers, and in the future, through increased production costs in designing and maintaining satellites. Taken together, these rising costs may make it financially unviable to perform certain types of space missions in the future, leading to a loss of the social benefits attained through those missions.

## Space as a Public Good and Global Commons

Outer space as a whole is traditionally considered to be non-excludable and non-rivalrous, thus making it a public good. Non-excludability arises from the Treaty on Principles Governing the Activities of States in the Exploration and Use of Outer Space of 1967 (more commonly known as the Outer Space Treaty) which states that outer space is free for exploration and access by all countries, and is not subject to national appropriation by claim of sovereignty, by means of use or occupation, or by any other means.

The non-rivalrous aspect is a function of the physical characteristics of the space environment and the physics of orbital mechanics. A satellite placed into orbit does not occupy a fixed location: it moves in an elliptical path (an orbit) whose size, shape, altitude, and exact dimensions are a result of the forward velocity of the object and the gravitational pull exerted by the Earth. The immense volume of space around the earth (trillions of cubic kilometers) means that the placement of a satellite into orbit by one country does not impede in a significant way the placement of a satellite into orbit by another country.

As a result, outer space is often referred to as being a "global commons," and closely linked to other perceived global commons of the Earth's atmosphere, oceans and the Internet. This view of outer space has led the historical economic policy discussions to focus on using environmental economics to price and regulate space debris as a pollutant to achieve the target amount released. Molly Macauley,[1] one of the few economists to focus on space debris, presented the broad outlines of some

---

[1]   http://www.rff.org/people/profile/molly-k-macauley

of these economic approaches to the space community as far back as 1994. She stated that at the time space debris experts had not considered economic incentive-based approaches stemming from classic microeconomic theory, and were instead focused on implementing traditional command-and-control approaches to regulation (Macauley, 1994). These regulations included the mandatory de-orbiting of spent rockets and satellites at the end of their operational lifespan. Macauley points out the drawbacks to the command and control approach, notably the costs of compliance and monitoring, and the lack of flexibility (Macauley, 1994).

As an alternative, Macauley discusses some of the traditional incentive-based microeconomic approaches for dealing with an externality. One suggestion she made was to require a deposit on launch of a spacecraft, which would be returned in the event the spacecraft is de-orbited (Macauley, 1994). If unclaimed, the deposit could go to a fund for cleaning up legacy debris or compensating owners of spacecraft hit by debris. Another approach is the assessment of taxes or fees with creation of space debris, and perhaps even a cap-and-trade regime to limit the amount of space debris produced (Macauley, 2003).

In addition to these brief mentions by Macauley, work by Bradley and Wein (2009) provides estimates of various fees and costs for regulating space debris. Their model of space debris dynamics over the medium (centuries) and long (millennium) term estimated damage to operational spacecraft from debris, and by choosing an average cost of a destroyed satellite and a discount rate, the model calculates the fee a policy mechanism needs to implement to ensure a tolerable level of risk. For a discount rate of 5% and a satellite replacement cost of $500 million, Bradley and Wein calculate a launch fee of $980 to offset future expected damage caused by a satellite that will be deorbited at the end of its mission.


## THE SHORTCOMINGS OF MICROECONOMIC POLICY MECHANISMS FOR SPACE DEBRIS

Despite this early recognition of the problem, microeconomic policy approaches to space debris have not been widely explored, and the topic is currently absent from any of the serious national or international policy discussions. At the core of this shortcoming are two root causes: (1) a misdiagnosis of the economic nature of the most commonly used regions of Earth orbit, and (2) the lack of private actors using these regions that would be responsive to market incentives.

## LEO AND GEO AS COMMON-POOL RESOURCES

While space as a whole is indeed non-rivalrous, the heavily-used regions of LEO and GEO are both rivalrous and congestible, and as a result are more appropriately classified as common-pool resources (CPRs) within the larger global commons of outer space, in the same way that fisheries and undersea oil fields are CPRs within the global commons of the oceans.

The orbital mechanics of the specific regions of LEO and GEO provide unique benefits, and the common engineering solutions used by almost all space actors result in a clustering of satellites at certain altitudes in LEO and at a more or less single altitude in GEO. Thus, as an increasing number of countries place satellites into LEO and GEO, those regions have begun to exhibit efficiency problems, such as congestion and crowding. These problems stem from each satellite operator taking into consideration only their own private costs of building and operating new satellites (what economists call marginal private costs) instead of factoring in the costs to everyone as a whole of adding more satellites (what are known as marginal social costs).

The congestion in GEO is particularly acute due to GEO's small size, high demand, and the need for all satellites in the region to use the same or similar portions of the radiofrequency spectrum. GEO is still an economically useful orbit, largely because the international community has created exclusion mechanisms in the form of international and national legal mechanisms, primarily through the International Telecommunication Union (ITU), to regulate and allocate the spectrum used by GEO satellites. Although the principles of non-appropriation from the Outer Space Treaty are still in effect, and only partial property rights are afforded to satellite operations in GEO, the exclusion mechanisms established by the ITU have led to a fairly efficient use of GEO, and have also likely contributed to GEO having a less acute space debris problem relative to LEO.

Although there have been general discussions of developing similar regulatory exclusion mechanisms for the most congested parts of LEO (Weeden & Shortt, 2008; Bilimoria & Krieger, 2011), these discussions have yet to gain wide acceptance. This is largely due to the high cost of putting such mechanisms in place, the thousands of pieces of legacy space debris that would be unaffected by a new zoning scheme, and the lack of measureable economic benefits from LEO that would justify the effort.

## LACK OF PRIVATE ACTORS DERIVING PRIVATE BENEFITS

Although it is true that space as a whole is very valuable and humanity derives much benefit from use of space, an economic analysis reveals very little measureable private benefits from our current use of LEO. The Space Foundation's 2014 "Space Report" estimates the value of the total space economy at $314 billion in 2013 (Space Foundation, 2014). However, very little of this figure comes from private benefits in LEO: $117 billion is from commercial ground infrastructure and support industries, and another $73 billion from government space budgets. Of the $122 billion in commercial space products and services, almost all of it is provided by satellites operating in GEO.

In LEO, the single biggest source of private benefit is the Earth observation sector, with an estimated total value in 2011 of $2.24 billion. However, this revenue stems largely from civil government and military customers, and governments are expected to fund development of most Earth observation satellites during the next 10 years. Aside from Earth observation, the only other meaningful revenue from satellites in LEO comes from three communications satellite constellations: Globalstar (46 satellites), Orbcomm (26 satellites), and Iridium (71 satellites). These firms had combined annual net income in 2014 of less than $100 million (Iridium, 2015). Data from the insurance industry supports this lack of private value in LEO as well – only 24 satellites in LEO are currently carrying commercial insurance, for a total insured value of about $2 billion, out of a total satellite insurance market of $20 billion.

The February 2009 collision between the U.S. Iridium 33 and Russian Cosmos 2251 satellites also demonstrates this lack of private value in LEO. This was the first collision between two satellites in orbit, and it destroyed both parent objects and created nearly 2,000 pieces of new debris greater than 10 cm in size (Weeden, 2012b) in what was already the most congested region of LEO.

Prior to the collision, an executive with Iridium had stated publicly that the company believed in the "big sky" theory – that space was big enough to not have to worry about collisions between objects. In addition, Iridium determined that the positional data available to them before the collision was not precise enough to be actionable. As a result, at some point before the collision Iridium stopped performing assessments to determine if there was a possible conjunction between one of its 65 satellites and other space objects (Wolf, 2009).

Neither the U.S. nor Russia decided to pursue damages over the incident, because the private loss for both parties due to the collision was near-zero and the liability regime in space is based on actual damages (Jakhu, 2010). Cosmos 2251 had ceased functioning years earlier, and Iridium 33 had been fully depreciated. Although Iridium-33 was being actively used to provide telecommunications services, it was part of a large constellation and the loss of a single satellite had minimal operational impact, and an on-orbit spare was maneuvered into Iridium 33's slot in a matter of weeks. Most observers would agree that the debris field created by the collision did impose a social cost on other satellites in that region, but future liability claims for satellites hit by a piece of debris from the Cosmos-Iridium collision must prove fault by either the U.S. or Russia to recover damages, which is a very unlikely scenario (Jakhu 2010).

The vast majority of economic value currently derived from human activity in LEO is either from satellites owned and operated by governments, or governments providing the bulk of demand for services provided by privately owned satellites. The value of LEO is almost entirely in the form of social benefits such as national security, science, climate and weather monitoring, management of natural resources, disaster response, and space exploration, and as the public sector is traditionally much less responsive to prices and markets than the private sector, LEO is thus a poor candidate for microeconomic policy mechanisms aimed at using economic incentives to change behavior on the generation of space debris.

At some point in the future, however, the situation may change, and LEO may see the degree of private benefits that would change the economics. Ideas such as orbital manufacturing, tether-based launch systems, space elevators, propellant depots, and space habitats have all been postulated as potential future private space sectors, and while some of these ideas are promising and pioneering firms such as Bigelow Aerospace are actively developing new markets, during the near- to mid-term the chances are extremely slim of developing a thriving space economy in LEO that will lead to a free market solution to the region's space debris problems.

## A New Approach: Information Economics and Common-Pool Resources (CPR)

Despite the lack of a private space economy and lack of progress in putting into place classical microeconomic policy mechanisms to deal with space debris, there has been progress on other initiatives. In 2007, the Inter-

Agency Space Debris Coordination Committee (IADC), a body formed by several national space agencies, published a set of voluntary guidelines for minimizing the creation of space debris. These guidelines were endorsed by the United Nations in 2007, and many countries around the world are in the process of implementing them into national regulations.

In 2011, the United Nations Committee on the Peaceful Uses of Outer Space (UN COPUOS) initiated a Working Group on the long-term sustainability of space activities, with the goal of developing a set of voluntary best practice guidelines to promote the sustainable use of space. And in 2012, the United States declared it would work with the European Union and other spacefaring states to develop an International Code of Conduct for Space Activities.

## GAME THEORY AND INFORMATION ECONOMICS

At first glance, these initiatives all seem to be purely political and unrelated to economics. They even appear to violate a core theory in economics, first postulated by Mancur Olson, which suggests rational, self-interested individuals will not act to achieve their common or group interests (Ostrom, 2000). However, these initiatives do take on new meaning when viewed through the lens of recent developments in economics such as information and game theory.

Although purely self-interested behavior and rampant free-ridership does hold true for a single-round game such as the traditional Prisoner's Dilemma, a large amount of evidence shows it does not hold true for games involving multiple rounds (iterations) and multiple types of players (Ostrom, 2000). In particular, the development of norms of behavior, and mechanisms for players to reward those who follow norms or punish those who break norms can lead to situations in which players act against what would be considered their private self-interest. The same principles underpin much of the strategic bargaining concepts developed by Thomas Schelling, and successfully implemented during the Cold War as part of the nuclear standoff between the U.S. and Soviet Union.

In the space context, these principles were behind the development and implementation of the voluntary debris mitigation guidelines, and the diplomatic pressure on the Chinese over their 2007 antisatellite test, which violated the principles at least in spirit. The diplomatic pressure and the accompanying international criticism, combined with the United States' "responsible" destruction of its own USA 193 satellite the following year, arguably resulted in a change in Chinese behavior. Subsequent tests of the same SC-19 system in 2010 and 2013, characterized by China as missile defense tests, targeted suborbital targets and did not result in the creation of any long-lived space debris.

Information economics also plays a role in the current space sustainability and security initiatives, and in particular the effort underway to increase the space situational awareness (SSA) available to space actors. SSA is information about the space environment, human activities there, and the relationship between the two. The United States and Russia currently have the most significant SSA capabilities, via networks of optical and radar sensors operated by their militaries. Despite recent efforts to improve SSA, the limits of current technical capabilities, and the unwillingness by many to share certain types of information with other space actors, have created a situation where knowledge of the current state of the resource environment is not uniform across all actors. Efforts to increase SSA among all space actors can have collective benefits by both providing missing information to all actors and eliminating information asymmetry that could lead to mistrust, misperceptions, or divergence of private values from group norms.

## AVOIDING THE TRAGEDY OF THE COMMONS

Despite the progress shown by ongoing political efforts, there is still the central challenge of how to overcome the tragedy many assume to be unavoidable in the use of a commons. The concept originated from Garrett Hardin's influential 1968 paper "The Tragedy of the Commons." In Hardin's view, continuing development of human civilization requires extracting and exploiting more natural resources (Hardin, 1968), but the increasing demand and use of natural resources leads to a pattern of over-exploitation and environmental degradation resulting in "tragedy," especially in commons scenarios. The open access and subtractable nature of commons incentivizes users to consume as much as possible before others do, leading to a vicious cycle of mismanagement and over-consumption that ultimately causes environmental degeneration beyond repair. Hardin argued that the only way to avoid this tragedy was either to install a Leviathan authority, which would manage and oversee the resource management, or to privatize the commons.

In 2008, political scientist Elinor Ostrom won the Nobel Prize in economics for her work on sustainable management of Common Pool Resources (CPRs), which showed that the tragedy of the commons could be avoided. Ostrom argues that many CPRs have been successfully governed without resorting either to a centralized government or a system of private property (Ostrom, 1998), and cites numerous cases where resource users have effectively self-organized and sustainably managed a CPR in spite of centralized authorities and without instituting any form of private property. Ostrom also points out that both the "Leviathan" and private property management schemes are just as likely to fail as other efforts.

Ostrom developed an eight-principle framework from her extensive research on successful and unsuccessful attempts to govern CPR, which outlines the conditions necessary to sustainably manage commons resources without a centralized government or private property regime. Her research shows that in every otherwise dissimilar case where CPRs were successfully managed, these eight elements were present:

- Clearly defined boundaries of the CPR (effective exclusion of external unentitled parties);
- Congruence between governance structure or rules and the resource context;
- Collective-choice arrangements that allow most resource appropriators to participate in the decision making process;
- Effective monitoring by monitors who are part of or accountable to the appropriators;
- Graduated sanctions for resource appropriators who violate community rules;
- Low-cost and easy-to-access conflict resolution mechanisms;
- Self-determination of the community, recognized by higher level authorities;
- In the case of larger common-pool resources, organization in the form of multiple layers of nested enterprises.

Over the last few years, Secure World Foundation has engaged in research aimed at examining Ostrom's principles of sustainable governance of a CPR in the context of space sustainability. The goal of this research is to use Ostrom's theoretical framework for sustainable polycentric governance of CPRs to identify gaps in the current space governance structures, and inform both current initiatives and potential future initiatives that foster the long-term sustainable use of outer space. The particular usefulness of Ostrom's approach is that it is developed for situations where neither of the two traditional solutions to the tragedy of the commons, complete privatization or a Leviathan to impose rule of law, are feasible, as is the case for Earth orbit.

The initial results from this research indicate several important focus areas for space governance (Weeden & Chow, 2012). Historically, nearly all of the discussion and negotiation of formal governance mechanisms has been through United Nations structures such as the UNCOPUOS and the Conference on Disarmament (CD). While the UNCOPUOS and the CD give voice to nation states, they exclude private entities from formal membership and restrict the role of non-governmental entities. Ostrom's principles indicate that for there to be useful progress on space governance, resource appropriators affected by rules governing that resource should have a voice in both making and modifying the applicable rules. Thus, a discussion on potential new fora for true multi-stakeholder discussions on space governance is an important issue.

Additional research has delineated the growing diversity of space actors, and to define a more granular assessment of the gaps between space actors and the existing governance fora (Weeden & Chow, 2013). According to this research, space actors can be categorized using several dimensions, including the level of engagement in space activities (direct or indirect), the spectrum of engagement in space activities (full spectrum or limited to only specific types or sectors), the level of dependence on space activities, and the level of an actor's prioritization of space activities relative to its peers.

Governance fora can similarly be characterized using dimensions of the type of decision-making process they employ, who is allowed to be a member, the scope of space activities that fall under a forum's purview, and whether decisions made by a forum are binding or enforceable.

The research done by SWF on applying Ostrom's principles for sustainable governance of CPRs in the space domain also provides guidance for efforts to improve SSA. Ostrom's principles indicate that for maximum effectiveness, the monitoring of the CPR should be done by the resource appropriators themselves. As a result, Ostrom's principles suggest that rather than pushing for a single nation or international entity to provide SSA to all space actors, efforts should instead focus on increasing SSA capabilities among and data sharing between many space actors. If many States and private sector entities each have at least some independent SSA capability, it would allow them to verify at least some of the activities in space using their own capabilities, leading to greater overall trust.

The most recent aspect of SWF's research on Ostrom's principles for sustainable governance of space CPRs concerns dispute resolution mechanisms for resolving disagreements between space actors. Ostrom's principles state that low-cost and easy to access dispute resolution mechanisms contribute significantly to sustainable governance of a CPR because they give resource appropriators ways to voice and resolve disagreements over the rules.

However, existing dispute resolution mechanisms in the space world are neither low-cost nor easy to access. The Convention on International Liability for Damage Caused by Space Objects of 1971, referred to as the Liability Convention, established the legal basis for damage liability resulting from space activities. The Convention requires that all claims for damage compensation be presented by a claimant state to a launching state through formal diplomatic channels or the UN Secretary-General. This has the disadvantage of raising the issue at hand to formal diplomatic status, probably bringing into play the broader political relationship between the states even for cases that might involve only commercial actors. It is probably not a coincidence, then, that the only case to date that has invoked the Liability Convention was the 1978 crash of the Soviet Cosmos 954 satellite in northern Canada. At the end of its life the satellite's nuclear reactor failed to separate, and re-entered the Earth's atmosphere along with the satellite, which resulted in a radioactive debris field over a large swath of northern Canada. The issue was eventually settled by a formal agreement between the USSR and Canada, with the former paying CAN $3 million in compensation. However, the settlement was accomplished without actually invoking the Claims Commission mechanism outlined in the Convention, and in fact to this day the Claims Commission mechanism has never been formally invoked.

## CONCLUSION

Given the recent prominence of space debris and space sustainability in official policy statements, and the near-term potential for policy action, it is extremely important for those in the field to have a solid understanding of the economic principles involved.

Although the problem of space debris in LEO appears on the surface to be related to pollution in a global commons, this is an incomplete understanding of the situation. While outer space as a whole may indeed be a global commons, depending on which definition of the term is used, the most highly used areas of Earth orbit such as LEO and GEO are neither public goods nor global commons. They are more appropriately viewed as CPRs within the global commons of outer space.

The tendency to treat all of outer space as a global commons obscures some of the underlying characteristics of the problems faced in use of Earth orbit, and can lead to misapplication of policy solutions. Moreover, a narrow-minded focus on just classical microeconomics ignores recent developments in game theory, information economics, and behavioral economics that could yield valuable insights for dealing with space debris and improving space sustainability.

The measureable private benefits from LEO and the costs due to space debris are small, and the vast majority of the social benefits are derived by publicly-funded activities. These two factors prevent standard microeconomic policy mechanisms from being a feasible solution to space debris and space sustainability. Until a private space economy is developed in LEO, it appears that policy approaches and initiatives that utilize principles from information theory, game theory, and polycentric approaches to governance of CPRs are more relevant for dealing with space debris and the long-term sustainability of space activities as a whole.

•••

## BRIAN WEEDEN

Brian Weeden is the Technical Advisor for Secure World Foundation and has 15 years of professional experience in the national and international space security arena. Mr. Weeden conducts research and facilitates discussions on space debris, global space situational awareness, space traffic management, protection of space assets, and space governance. He is formerly an officer in the U.S. Air Force in space and missile operations.

Mr. Weeden holds a Bachelor's Degree in Electrical Engineering from Clarkson University and a Master's Degree in Space Studies from the University of North Dakota. He is currently a Doctoral Candidate in Public Policy and Public Administration at George Washington University in the field of Science and Technology Policy.

## TIFFANY CHOW

Tiffany Chow is currently Partnership Strategist at Roadtrip Nation. She works with new and existing partners to brainstorm collaboration opportunities, ultimately growing the impact of Roadtrip Nation. She also serves as commissioning editor for *ROOM: The Space Journal* in the United States. Previously, Tiffany worked as a Project Manager for Secure World Foundation, where she oversaw and supported projects dealing with international security and legislative issue areas. Tiffany previously worked for the Center for American Politics and Public Policy at UCLA, where she assisted the Director and Administrative Director with research projects and program logistics. She interned with the Monterey Institute for International Studies' Center for Nonproliferation Studies (CNS) in Washington, DC where she provided research support on a wide array of topics including export control issues.

Tiffany also served as the U.S. National Point of Contact for the Space Generation Advisory Council (SGAC) and was involved with SGAC's Space Safety and Sustainability Working Group. She is an alumna of the International Space University's (ISU) 2012 Southern Hemisphere Summer Space Program held in Adelaide, Australia. She holds a Master's in International Relations from Johns Hopkins University's School of Advanced International Studies (SAIS) and a Bachelor's in Political Science and European Studies from the University of California, Los Angeles (UCLA).

# References

Bilimoria, K. D., & Krieger, R. A. (2011, September 11). Slot architecture for separating satellites in sun-synchronous orbits. Paper presented at the American Institute of Aeronautics and Astronautics Space 2011 Conference and Exposition, Long Beach, California.

Bradley, A.M., & Wein, L.M. (2011). "Space debris: Assessing risk and responsibility." *Advances in Space Research*, 43(2009), 1372-1390.

Hardin, G. (1968) "The Tragedy of the Commons," *Science*, 162(3849), 1243-1248. Available from: http://www.sciencemag.org/content/162/3859/1243.full

Iridium Communications, Inc. (2014). 2014 Annual report. Retrieved from: http://files.shareholder.com/downloads/ABEA-3ERWFI/241799381x0x818513/963FB60D-1E2B-4A1A-AD81-D556803DE47D/IRDM_10-K_Wrap_2014_.PDF

Jakhu, R.S. (2007) "Legal issues of satellite telecommunications, the geostationary orbit and space debris." *Astropolitics*, 5, 173-208.

Jakhu, R.S. (2010). "Iridium-Cosmos collision and its implications for space operations." In Schrogl, K.-U., Rathgeber, W., Baranes, B., & Venet, C. (Eds.), *Yearbook on Space Policy 2008/2009* (pp. 254-275). Vienna, Austria: Springer.

Macauley, M. (1994) "Close encounters of the trash kind." *Journal of Policy Analysis and Management*, 13(3), 560-564.

Macauley, M. (2003) "In pursuit of a sustainable space environment: economic issues in regulation space debris." In J.A. Simpson (Ed.), *Preservation of the near-earth space for future generations* (pp. 147-158). Cambridge University Press.

Ostrom, E. (2000). "Collective action and the evolution of social norms." *The Journal of Economic Perspectives*, 14(3), 137-158. Retrieved from: http://tuvalu.santafe.edu/events/workshops/images/b/b3/Ostrom2000.pdf

Ostrom, E. (1998) *Governing the Commons*. Cambridge: Cambridge University Press

The Space Foundation. (2014). The Space Report 2014: Overview. Retrieved from: http://www.spacefoundation.org/sites/default/files/downloads/The_Space_Report_2014_Overview_TOC_Exhibits.pdf

Union of Concerned Scientists. (2014) UCS satellite database. Updated 1 August2014. Retrieved from: http://www.ucsusa.org/nuclear_weapons_and_global_security/space_weapons/technical_issues/ucs-satellite-database.html

Weeden, B. (2012a). "2007 Chinese antisatellite test fact sheet." Secure World Foundation. Retrieved from: http://swfound.org/media/9550/chinese_asat_fact_sheet_updated_2012.pdf

Weeden, B. (2012b). 2009 "Iridium-Cosmos collision fact sheet. Secure World Foundation." Retrieved from:

http://swfound.org/media/6575/swf_iridium_cosmos_collision_fact_sheet_up
dated_2012.pdf

Weeden, B. & Chow, T. (2012). "Taking a common-pool resources approach to space sustainability: A framework and potential policies." *Space Policy*, 28(3), 166-172.

Weeden, B., & Chow, T. (2013). "Engaging all stakeholders in space sustainability governance initiatives," IAC-13-E.3.4.2, presented at 64th International Astronautical Congress, Beijing, China, September 23-27, 2013. Retrieved from: http://swfound.org/media/119718/IAC-13,E3,4,2,x18131_TC.pdf

Weeden, B. & Shortt, K. (2008, May). "Development of a sun-synchronous zoning architecture to minimize conjunctions." Paper presented at AIAA SpaceOps 2008, Heidelberg, Germany. Retrieved from http://swfound.org/media/1725/sunsync-bw-spaceops-2008.pdf

Wolf, J. (2009, February 12). "Iridium says in dark before orbital crash." *Reuters*. Retrieved from: http://ca.reuters.com/article/topNews/idCAN1244243120090212

CHAPTER 8

# ASSURING A SAFE, SECURE AND SUSTAINABLE SPACE ENVIRONMENT FOR SPACE ACTIVITIES:
## CONSIDERATIONS ON A CONSENSUAL ORBITAL DEBRIS REMEDIATION SCHEME

PROF. DR. LESLEY JANE SMITH
LEUPHANA UNIVERSITY LÜNEBURG
WEBER-STEINHAUS & SMITH, BREMEN

## A HOLISTIC VIEW ON SECURITY AND COLLABORATION IN SAFE SPACE

This chapter looks at the interaction between the classic rules of Outer Space Law as they apply to space debris. It transcends the traditional legal debate, which focuses primarily on impact-related damage, particularly on Earth. Today, with longer term concerns for a safe, secure and sustainable space environment at stake, there appears to be a strong case for aligning the debris codes, rules about preventive and precautionary measures for outer space activities, with the traditional liability system. The ultimate goal is to impact positively on debris control in Outer Space, enable its reduction, and agree on a modus vivendi that ranges from debris mitigation to include debris remediation.

The chapter also considers how the current space law regime could enable practical and recognisable means of ensuring compliance with the debris rules and debris mitigation, possibly through incentive based systems. With the advent of space surveillance tracking systems, the Space Security Index, and various other thematic programmes, a new matrix combining existing rules to shape sustainability is at the top of the regulatory agenda. The precarious state of the outer space environment is known to all, yet final agreement on the action required is still under discussion. This chapter highlights some essential aspects of what could be done to initiate these remedial projects, and in doing so it draws attention to salient aspects of law and policy.

## I. INTRODUCTION

### I.1  LEGAL CHALLENGES OF DEBRIS MITIGATION AND REMEDIATION

The consequences of outer space debris on the sustainability of outer space activities have been known for more than a decade. There is concern that 'without changes to the way that missions are performed, regions of near-earth space will become so cluttered by debris that routine operations will no longer be possible'.[1] Yet causing space debris is not and never has been a breach of any international legal obligation, or legally reprehensible.

---

[1]    See H. Klinkrad, Policy position paper on orbital debris 1993, cited in H. Klinkrad, IAA Study Group 5.10, Policy, Legal and Economic issues of orbital debris removal, chap. 1.d, in: IAA Cosmic Study on the subject of Space Debris Environment Remediation, 2015, *forthcoming*.

Hence, debris mitigation and particularly remediation require a paradigm shift and established mechanisms to make sustainability possible.

The monitoring efforts of the space surveillance network confirm that the current situation requires mission designs to move on from dealing with space debris as part of mission disposal plans, to include proactive measures to remove debris. The risks of debris to aero-space traffic are evident, and have now led to proposals for space traffic management.[2] The low-earth orbit (LEO) is currently the most debris-impacted of all orbits,[3] and mission designs must therefore now ensure that launch systems and payloads do not deteriorate the space debris environment further. Considerations are underway, for example, for On-obit servicing models for remediation in GEO.[4]

Reliable reports indicate that the removal of one piece of debris per year could significantly improve the current dangerous state of the outer space environment.[5] Nevertheless, the international community requires advance agreement on a way forward to move active debris remediation (ADR) and on-orbit servicing (OOS) projects towards realisation.

A major uncertainty within this discussion is that the international debris mitigation guidelines are not binding legal norms.[6] The legal status of remediation under international space law remains open to debate,[7] and while compliance with debris mitigation standards has become part of national technical mission disposal requirements, these are not formal international legal instruments. This compounds negatively with the lack of punity for debris creation. Whether and how a transition should take

---

[2]   See Corinne Contant-Jorgenson, Petr Lála, Kai-Uwe Schrogl (eds), IAA, Study on Cosmic Space Traffic Management, 2006; the EU is involved in efforts relation to space traffic management, see Decision 541/2014/EU of 16 April 2014 establishing a Framework for Space Surveillance and Tracking Support, OJ L 158/227 of 27.5.2014

[3]   See T. Flohrer, H. Krag, H. Klinkrad, ESA's process for the identification and assessment of high-risk conjunction events, in: Advances in Space Research, Volume 44, Issue 3, 2009, 355–363

[4]   A. Krolikowski, E. David, Commercial On-Obit servicing, National and International Policy Considerations Raised by Industry Proposals, New Space. 1(1) 2013, 29-41

[5]   IAA, Study on Cosmic Space Traffic Management, n. 2 above, 2006, ibid.

[6]   The sources of international are contained in the Statute of the International Court of Justice, Art 38 (1), which lists four categories of binding international conventions; international custom; the general principles of law recognized by civilized nations; judicial decisions and the teachings of the most highly qualified publicists of the various nations, as subsidiary means for the determination of rules of law.

[7]   UNCOPUOS, Report of Legal Subcommittee, 53rd session, 15th April 2014, A/AC.105/1067, at 98; for a list of the various international guidelines, see further below, p. 4, n. 22

place for the various sets of international debris mitigation guidelines to become binding rules of black letter law is part of this chapter's focus, and some of the direct and indirect methods of remediation are looked at in terms of the law.

## I.2. METHODS AND PERSPECTIVES OF REMEDIATION

There are two main mechanisms to counteract the increase in space debris: first, active debris remediation (ADR), and second, on-orbit servicing (OOS). The former involves actual rendez-vous and docking to another space object to enable its removal and disposal through burn up on re-entry; the latter involves repair, re-fuelling and upgrading while in or near operational orbit.[8] The techniques and scenarios currently under discussion range from the use of tethers, directed energy beams (space based/airborne based) to other forms of sweeping that 'catch' the debris, and include de-orbiting sails.[9] Some of these techniques involve direct contact with the debris, others indirect approaches through capture and removal, with the debris burning on re-entry. While outer space manouevers are known in practice – they are undertaken, for example, to ensure the safety of the International Space Station faced by the threat of debris collision[10] - the remediation activities described may nevertheless create potential interference for third states. There are policy issues requiring governmental support and validation by the parties concerned, but the international non-proliferation commitments of state parties, as well as the export control rules applicable to dual use, make remediation a sensitive issue.

Debris remediation projects – as opposed to OOS - are still experimental, and no remediation mission has yet started. Studies show that it is not yet economically feasible to undertake these operations as purely commercial efforts, an option for debris remediation that had been considered at an earlier stage. Space agencies such as ESA are pursuing the subject in the light of their own immediate concerns for their space

---

[8]   A. Krolikowski. E. David, "Commercial on-Orbit satellite Servicing," n 4. above, at 29

[9]   Full details of the exact techniques required to undertake such missions are beyond the scope of this paper but can be seen in e.g. Brian Weeden, How Do I Ask Permission to Engage With A Piece of Space Debris?; further, Alexander Soucek, Active space debris removal and export control, ESA/ CNES, 3rd European Workshop of Space *Debris Remediation,* June 16-18, *2014*

[10]  For details of a debris-related manouver of the ISS, raising its altitude for approximately half a mile to avoid a spent Ariane upper stage, in early April 2014, http://www.nasa-usa.de/content/station-conducts-debris-avoidance-maneuver/#.U_8Ai2Pj_1U

objects[11] which should help the community towards developing conventions and good practices for the future.

## II. FROM DAMAGE AND LIABILITY TO REMEDIATION AND PRECAUTIONARY MEASURES

### II.1. ABSENCE OF LEGAL INTEREST OF OUTER SPACE ENVIRONMENT

The provisions governing the freedom of use of outer space (Art I OST) and the freedom of exploration and use of outer space (Art III OST) are prescriptive norms, formulated to enable states equality of access and peaceful use of outer space.[12] The accompanying provisions of Arts VI, VII OST, as well as Art IX OST, dictate that such use should have regard to the interests of all other State parties to the Treaty. Only Art IX (cooperation and mutual assistance) specifically addresses the issue of avoiding harmful contamination,[13] with no reference to the legal consequences of such activities. Remedies for breach of international obligations, or breach of international space law can take various forms, based on the principle of *restitutio in integrum*, whether reparation, restitution or compensation.[14]

The environment of outer space has not been accorded a legal interest or right to protection under international space law, and even if the creation of debris were deemed to be a breach of international law, which it is currently not, the protection of the environment of outer space is not included within the provisions on responsibility and liability of the main outer space treaties, OST and LIAB. The definition of "damage" under the space treaties relates solely to damage to persons or spacecraft caused by a space object; deterioration of the outer space environment does not fall

---

[11]  See ESA presentation by H. Klinkrad, February 2011, "Space Debris Mitigation Activities at ESA," http://www.oosa.unvienna.org/pdf/pres/stsc2011/tech-40.pdf (accessed 1st July 2015)

[12]  See R. L. Spencer, International Space Law; a basis for international regulation, in: Ram Jakhu, National Regulation of Space Activities, Berlin, Springer, 2010, 1-22.

[13]  See Cologne Commentary on Space Law, CoCoSL, Eds. Hobe, Schmidt-Tedd, Schrogl, vol. I, Outer Space Treaty (OST), Arts I (Hobe), Art VI (Gerhard), Art VII (Smith/Kerrest), Art IX (Marchisio), Cologne, Heymann, 2009

[14]  See International Law Commission, Articles Responsibility of States for Internationally Wrongful Acts 2001, Arts 34-36, Text adopted by the Commission at its fifty-third session, in 2001, available at http://legal.un.org/ilc/texts/instruments/english/draft%20articles/9_6_2001.pdf (accessed 1st July 2015). Art 37 provides for alternative satisfaction, should the remedies under Arts 34-36 not be available.

within the definition of damage.[15] This lack of legal interest has fed the
call to protect outer space as part of the global commons, in an effort to
secure its long-term sustainability.[16] [Editor's note: Please see Chapter 7
for a detailed discussion of this topic.] However, this status alone is
insufficient pro-active legal protection to remediate today's cluttered outer
space environment or create new sanctions for contributing to its further
deterioration.[17]

## II.2. LIABILITY

Liability cannot exist without a legal duty. It is an ex-post norm regulating
the duty to pay compensation in the form of damages for space-object
induced damage; under outer space law, the launching state(s) are liable
under Art V LIAB, jointly and severally, to pay compensation for any
impact-related damage. The general provisions of Art VII OST, and the
special provisions of Art II (absolute liability for damage to earth), and Art.
III LIAB (fault-related liability for damage in outer space) are well known
provisions. Art. II and III LIAB are, respectively, limited to damage on
Earth, or damage to the persons or property aboard a space object in outer
space, where the damage is caused by another space object. It is irrelevant
for the purposes of LIAB whether the object is functional or not.[18] This
should be recalled when considering ADR and OOS.

The general principles of peaceful use of outer space and precautionary
measures do not provide actionable rights for outer space environmental
damage caused by space-related debris. However, one of the pressing
questions for today is whether uncontrolled debris left in outer space might

---

[15] Smith/Kerrest, Art VII OST, in CoCoSL vol I, n.13 above; Smith, Kerrest, UN
Convention on Liability for damage caused by space objects (LIAB), Art II, in
Cologne Commentary on Space Law, CoCoSL Eds. Hobe, Schmidt-Tedd, Schrogl,
vol II, Cologne, Heymann, 2013.

[16] Cf. Manfred Lachs, The law of outer space, an experience in contemporary law
making: 'Though the notion has also been adopted by international law, one can
hardly argue that outer space and celestial bodies, through physically the latter may
be reminiscent of some parts of our globe, can be encompassed by this term. None
of them being a *res*, they cannot in fact become *res extra commercium* or *communis*.
Outer space and celestial bodies are to be viewed as spheres of states activities, as an
environment subjected to a special legal regime and enjoying the particular
protection of the law. 46, Nijhoff 2010. For a fuller discussion of what is meant by
the Global Commons, see Introduction, United Nations Environmental Programme,
UNEP, Division of environmental law, available
http://www.unep.org/delc/GlobalCommons/tabid/54404/Default.aspx (accessed 1st
July 2015).

[17] Smith/Kerrest, Art III LIAB, in CoCoSL, vol II, n. 15 above, nos. 144-146

[18] Smith/Kerrest, Art VII OST, in CoCoSL, vol I, n. 13 above, nos. 29-31

in future be seen to constitute the fault or negligence that is a pre-requisite to triggering international state to state liability under Art II, Art III LIAB. This question can be raised in relation to a) whether it causes damage to other functional space objects and/or b) whether it causes further debris, and thereby endangering the environment for ongoing space activities.[19] Since Art XI.2. LIAB does not exclude the jurisdiction of domestic courts, this question may easily be put before a national court to decide. The increase in commercial space activities means that a decision on fault at that level of court cannot be excluded.[20]

## III. DEBRIS: PRELIMINARY ISSUES OF DEFINITIONS, OWNERSHIP AND JURISDICTION

### III.1. DEFINITIONS

The mitigation and remediation discussion is compounded by the fact that much of the outer space debris is no longer functional, and an even greater part is, by nature, fragmented. Liability for outer space activities can only ever arise for damage caused by a space object. Space debris, whether components, parts, or remnants resulting from the activities of a manmade spacecraft that are attributable to the space object, fall *prima facie* within the sparse treaty definition of a space object under Art VII OST and Art I OST.

Neither the Outer Space Treaty nor the Liability Convention provides any detail beyond this as to what constitutes a space object (Art. VII OST; Art. 1.d LIAB/Art. 1.b REG). There are no limits to its size, shape, or volume. A separate but workable definition for space debris is that of the Inter-Agency Space Debris Coordination Committee (IADC).[21] This defines space debris as "all manmade objects including fragments and

---

[19] Fault liability (*dolus*) is required for damage in outer space. Lyall, Larsen, Space Law, a Treatise, 2009, indicate correctly that much debris is caused by military satellites, and notably ANTISAT activities, at 525-526; the limits of military action in outer space are given by the UN Charter and the requirement of peaceful purposes, but at the same time, the right to (proportionate) self-defence, id., 526.

[20] Such a decision could fall to be determined by arbitration at the Permanent Court of International Arbitration under the optional rules applicable to disputes arising out of outer space activities.

[21] The Inter-Agency Debris Coordination Committee first established its terms of reference in 1993 which have been updated over time. Its Debris Mitigation Guidelines were prepared in 2007, see further, http://www.iadc-online.org/index.cgi?item=docs_pub

elements thereof, in earth orbits or re-entering the atmosphere that are non-functional."[22]

Debris-remediation missions, irrespective of the debris size, should take into account that the state on whose registry the space object appears retains jurisdiction and control over the object while it is in outer space (Art. VIII OST). Ownership of these space objects – even if in debris form – is not affected while in space.[23] There is no legal presumption of abandonment in outer space law, even though much of space debris is abandoned property, largely in the form of fragments.[24] This is a further challenge to remediation projects which require an advanced, consensual system that condones justified interference in third party property, even if dysfunctional. Introducing a presumption of abandonment, if only for debris fragments, would require clear tools of interpretation in the light of Art. VIII OST. One basis for a consensual protocolled proposition could be to rely on the concept of title, allowing consensual relinquishment of ownership for dysfunctional and fragmented debris. A de facto transfer of property in the debris through the remediation effort could be accompanied by the practice of notifying the register that dysfunctional debris has been removed, as part of the mechanisms to be agreed on within a validation process among stakeholders.

### III.2. DISJUNCTION BETWEEN RESPONSIBLE AND LIABLE STATES

Generally speaking, states have little economic or practical interest in dealing *ex-post* with their dysfunctional debris: as indicated, much of it is fragmented in size. This is compounded by the fact that the ownership of space objects and the liability of the launching state are not aligned in the law of in outer space. The rules of liability focus on the launching state(s) as the addressee of claims during what is the most dangerous part of outer

---

[22]  UNCOPUOS Space Debris Mitigation Guidelines, 2007; Inter-Agency Debris Code (IADC) 2007; the EU Code of Conduct (2004, amended 2008); the ITU Recommendations (ITU-R S.1003.2); ISO Standard 24113, also incorporated in the system of European Cooperation in Space Standards (ECSS), and ESA Policy March 2014; for details, see Legal Subcommittee, Compendium of space debris mitigation standards adopted by States and international organizations, A/AC.105/C.2/2014/CRP.15 LSC, Vienna, 24 March-4 April 2014.

[23]  Some governments are looking to distinguish between functional and dysfunctional space object, by limiting their duty of effective control under Art VI OST to objects that can still be controlled, and placing the continuing duties on the party exercising effective control, normally the operator. This may further bi-furcate the distinction between state responsibility and liability. See text of the Amendment to Belgian Law on outer Space Activities of 2005, January 2014, Moniteur Belge, 15th January 2014.

[24]  F. Lyall, P. Larson, Space Law, A Treatise, n. 19 above, 311

space activities, the launch (Arts II, III LIAB). The fault liability system for damage in outer space stems from the logic that each party bears its own system-related risks, incurring liability only where these risks are no longer carefully controlled, and there is damage to third parties.[25] It is the only practical approach to hazardous activities. This is why collision or impact in outer space, notably with or through man-made debris, requires fault, and thereafter, causation to be proved, before any liability can arise (Art III LIAB). The legal question for today is whether fault liability can occur when these risks are no longer controllable, as is the case with debris, where states and operators relinquish their continuing interests in their property; as indicated above, the LIAB does not distinguish between functional and dysfunctional objects.

There is no legal advantage in introducing further definitions of debris to cater for dysfunctional craft or fragmented debris pieces – these already fall under Arts VI, VII OST. Without binding rules relating to debris control and mitigation, new definitions cannot alter the existing matrix of responsibility and liability. Transfer of control and ownership are issues that may require perfectioning between the state of registry and the new state of jurisdiction and control under Art VIII. UNGA Resolution 62/101 indicates that the state of registry, as well as the appropriate state under Art VI, should cooperate in notifying the date of change in supervision, the identity of the new owner and any change or orbital position or function of the object.[26]

Meanwhile, the only effective link in international space law is to ensure that the responsible – and appropriate - state of nationality under Art VI OST carries out its duties of monitoring and continued control. This is the background to the ongoing call by UN COPUOS for the introduction of national space legislation.[27] It is a means of ensuring that states actively undertake their duties of supervision and control, of both the governmental and non-governmental sector, in accordance with international obligations.

---

[25] Damage to earth or air flight incurs absolute liability of the launching state (Art II LIAB), see Smith/Kerrest, Art II LIAB in CoCoSL, vol II, no. 15 above, nos. 83-88.

[26] UNGA Resolution 62/101, Recommendations on enhancing the practice of states and international intergovernmental organisations in registering space objects, 17 December 2007.

[27] UNCOPOUS has recently commented on the continuing importance of national legislation, see UNGA Resolution 68/74, and A/AC.105/1067 of 16 December 2013; see further, Tronchetti, Fundamentals of Space law and Policy, ISU/ Springer, 2013, p. 26-32

National law cannot reduce or water down the content of obligations existing by virtue of international treaty law.[28]

## III.3 SHIFTING THE FAULT PARADIGM

The causation of debris and its existence over the past fifty years has not been seen to constitute fault under outer space law.[29] Rebutting this presumption may be a challenge, but it is also an important goal. Fault is a measure of sub-standard conduct that leads to damage. Defining fault requires recognition and adherence to technical standards,[30] and a link to legal rules that support the interpretation of what constitutes fault.

Until such time as the debris guidelines and the failure to control debris are recognised as binding and recognised practice, the existence of debris will not be held to constitute sub-standard or negligent practice in the future. In addition, while the launching state liability cannot be transferred to the operator, the change of ownership interests can be registered.[31] Even though existing debris can claim state of the art defence, in that it is not illegal, more recent activities such as ANTISAT (anti-satellite) tests remain highly sensitive.[32] It seems appropriate to ensure compliance with debris mitigation, as is being done at national level, and agreement to move towards remediation systems. Agreement on an interpretation of fault for the future would not require an amendment to long-standing systems settled by outer space treaty law, but it would require agreement on the status of the debris guidelines. The subject remains on the agenda of both the Technical and Scientific Committee, as well as the Legal Subcommittee

---

[28]  F. Tronchetti, id.

[29]  Liability can only arise if there is a failure to maintain an objective obligation under international law; in its absence, it is difficult to prove fault.

[30]  The landmark case of Donoghue v Stevenson [1932] UKHL 100, especially the dicta of Lord McMillan, page 8, are often cited as an indication that 'the categories of negligence are never closed.'

[31]  Two issues arise here; one, clarification of the relationship to the original launching state(s), notably because of the maxim, 'once a launching state, always a launching state;' further, transfer of information contained on the registry is possible, in so far as the appropriate state under Art VI OST and the (existing) state of registry undertake the notification and new registration.

[32]  The various ANTISAT tests, both by the Chinese in 2007, and the US in 2008, constitute one of the greatest sources of space debris, on debris and the UN Resolution on Transparency and confidence building measures in outer space, UN GA RES/69/38 of 11th December 2014. See Lyall, Larsen, Space Law, on non-aggressive military use of outer space and the inherent right of self-defence, n. 19 above, 525-7

of UNCOPUOS, which is looking to achieve "mechanisms that will contribute effectively to the sustainability of outer space activities."[33]

## IV. REMEDIATION SCENARIOS

### IV.1. INTERFERENCE IN (FOREIGN) PROPERTY, IMMUNITY, NECESSITY AND CONSENT

The conditions enabling remediation to take place are still to be mapped out at international level.[34] Removal of a state's own property in space is not an issue, in so far as the operations comply with the rules of international law of outer space, are authorised, and cause no third party damage.[35]

Interference with foreign national property over which jurisdiction is (still) asserted – and where consent is lacking and ownership not abandoned, are likely to constitute a breach of international law.[36] Because states are immune from suit in any international forum unless they agree to jurisdiction or arbitration, the LIAB and subsequent international arbitration rules for space activities provide for mechanisms for settlement of international compensation claims.[37] Whether the difficult "doctrine of necessity" – which entitles interference by third states where there is imminent danger – applies to functional space objects, or whether unauthorised but necessary interference would be seen to constitute fault liability, requires detailed consideration and consensus.[38] Intermediate categories of authorised, justified or appropriate interference in the

---

[33] Id. UN GA Resolution 69/38 on Transparency and confidence building measures in outer space

[34] A. Krolokowski and E. David, n.4 above, cite examples of government refusal to support the activities.

[35] This point is discussed below in relation to the subject of non-proliferation

[36] See Art 34, International Law Commission, Draft Articles on Responsibility of States for Internationally Wrongful Acts 2001, n.14 above.

[37] The Liability Convention provides for settlement of the dispute before a Claims Commission, Art XIV; the Liability Convention does not exclude the jurisdiction of domestic courts and the right to raise proceedings there, see Art XI.2.; see further, the Permanent Court of Arbitration Optional Rules for the Arbitration of Disputes Relating to Outer Space Activities, available http://pca-cpa.org/shownews.asp?ac=view&pag_id=1261&nws_id=323

[38] See J. Crawford, The International Law Commission's Articles on State Responsibility – Introduction, Text and Commentaries (Cambridge University Press, 2002), 178; J. J. A. Salmon, 'Faut-il codifier l'état de nécessité en droit international', in. J. Makarczyk, (ed.), Essays in Honour of Judge Manfred Lachs (Martinus Nijhoff, 1984) 235-70, 244

property in outer space, despite its size, should be considered in this question.[39] While analogies have been drawn with the international law of the sea, where vessels not flying their flags may be mounted in specific situations,[40] and include considerations for possible salvage mechanisms on which to base possible remediation service models, the principle of consent should apply in advance – *ex ante* – of any remediation mission. Otherwise such missions are not in compliance with international law of outer space.

The question is therefore how to ensure consultation, and/or validation of method, particularly in the cases of fragmented debris where ownership is by definition unclear.[41] The prevailing attitudes among governments, and their policies towards remediation, will largely determine the viability of space activities in the future.[42]

## IV.2. NATIONAL AUTHORISATION AND EXPORT CONTROL

Debris remediation operations require national authorisation.[43] The existence of national space legislation is a response to the duties under Art VI OST to ensure monitoring and control of outer space activities through such authorisation systems. Many states require adherence to the debris mitigation guidelines as part of the national technical requirements.[44] As indicated above, few, with the exception of France, have turned these codes into black letter law at national level, imposing sanctions for non-compliance.[45]

---

[39]  See the pertinent judgment of Judge Learning Hand in United States v. Carroll Towing Co. 159 F.2d 169 (2d. Cir. 1947), where he relies on calculus of probability to determine liability as follows; where B is burden of taking precautions, P the probability of harm and L the injury liability depends upon whether B is less than L multiplied by P: i.e., whether B <PL.

[40]  On the law of the High Seas, see Brownlie, Principles of Public International law, 8ᵗʰ ed., Oxford, 2012, 223ff at 238-9

[41]  The relevant space register should be consulted whether the object is registered as dysfunctional or still catalogued.

[42]  A. Krolikowski, E. David, refer to examples where governments have declined support for remediation, see n. 4. above

[43]  Remediation operations on third state property would call for both national authorization, as well as the consent of the state(s) of registry and nationality.

[44]  See Compendium of space debris mitigation standards adopted by States and international organizations, A/AC.105/C.2/2014/CRP.15 LSC, Vienna, 24 March-4 April 2014, n. 22, above

[45]  France has implemented the provisions of the Debris Mitigation Guidelines into its 2008 national space law, see Philippe Clerc, Consequences of the French space law on space operations (FSOA) on CNES's mission as a contracting space agency, in: Smith/Baumann, Contracting for Space, Contract Practice in the European Space Sector, Farnham, Surrey, Ashgate, 2011, 117-131.

A more recent amendment at the level of national space law, and possibly one of greater concern, brings potential national disparities in the interpretation of the term space activities to light. This seeks to exclude state responsibility and liability for dysfunctional, as opposed to functional objects. Whether this is indicative of a new trend is unclear.[46] It would however be against the spirit and the letter of outer space law and international law for a state to adopt this approach, to limit its international obligations in this way.

## V. FOREIGN POLICY, EXPORT CONTROL OVER SENSITIVE TECHNOLOGY AND REMEDIATION

### V.1 REMEDIATION SCENARIOS

Remediation can be carried out unilaterally, in relation to a state's own spacecraft, or on the basis of agreement and with the consent of the state of ownership. To the extent that contact with foreign property is involved – this can happen through approach and capture operations – issues of technology classification arise in terms of national or international arms control and export regulations. Space technology and equipment may, but need not, always be combined in equipment for use in or launched out of a foreign jurisdiction different to where produced.[47]

Once in outer space, there may be conjunctions or approaches with foreign property, raising issues of export as well as arms control, which are fed by national security and foreign policy. Debris remediation may, as is explained below, involve the creation of an export scenario, even although almost fictive, as it relates to property in outer space. There are no territorial or claims of sovereignty in outer space, yet since ownership and jurisdiction over spacecraft remain unaffected, concepts of adjacent foreign property arise. The exact interpretation of export in terms of armaments and export control requires to be considered carefully in the context of developing an acceptable remediation plan or protocol.[48]

---

[46] See details relating to Belgian national space law reform, n. 23, above
[47] EU Export Regulation 428/2009, OJ L 134 of 29.5.2009, 1, as amended by EU Regulation 599/2014 OJ L 173/79 of 12.6. 2014; ITAR §123.9 (a) requires that 'the Country designated as the country of ultimate destination on an application for a license, .... Or ... where an exemption is claimed under this subchapter, must be the country of ultimate end-use.'
[48] Term coined by Brian Weeden, in n. 9, above

## V.2 EXPORT OF DUAL USE TECHNOLOGY

There are various unilateral, bilateral or multilateral agreements governing non-proliferation, the most important of which is notably the UN Treaty on Disarmament relating to the use of nuclear weapons.[49] While military presence in outer space is not prohibited, the use of weapons of mass destruction is forbidden under Art IV OST; the use of outer space is subject to the provisions of Art 2(3) UN Charter that requires peaceful settlement of international disputes.[50]

The US ITAR regulations are the best known among the main arms control regulations. These implement the Arms Export Control Act, which governs the export of defence articles and services, and includes the United States Munitions List, USML.    Parallel to this are the Export Administration Regulations (EAR) that implement the  These regulate the export of commercial and dual use items, which are compiled in the Commercial Control List (CCL).

The most significant provisions of ITAR are its extra-territorial effect.[51] Items subject to ITAR continue to remain subject to these rules, where there is re-sale or distribution to a country other than the one for which the export license is granted.[52] ITAR also restricts re-exports by US or foreign persons, by virtue of the above provisions.  This extends to satellites covered by the US Munitions List (USML), whether or not in the US.  As a result, it can be safely assumed that the ITAR provisions continue to apply in the context of remedial operations, even if activities in outer space are to be seen to be non-territorial or non-national.

---

[49]   The international agreements relating to limitation of arms and technology are contained in the following international agreements: the Missile Technology Control Regime (MCTR 1987), a 34 state organization, details available at http://www.mtcr.info/english/partners.html,; the Wassenaar Agreement 1995; for its member states and details of export control of arms, see http://www.wassenaar.org/ and the Hague Code of Conduct against Ballistic Missile Proliferation (HCOC), 2002, see http://www.hcoc.at/

[50]   Art 2(3). All Members shall settle their international disputes by peaceful means in such a manner that international peace and security, and justice, are not endangered. The provisions of the Charter are available at http://www.un.org/en/documents/charter/

[51]   ITAR § 123.9 (a) – (e)

[52]   ITAR § 123.9 (a), requires that "The written approval of the Directorate of Defense Trade Controls must be obtained before reselling, transferring, reexporting, retransferring, transshipping, or disposing of a defense article to any end-user, end-use, or destination other than as stated on the export license, or in the Electronic Export Information filing in cases where an exemption is claimed under this subchapter, except in accordance with the provisions of an exemption under this subchapter that explicitly authorizes the resell, transfer, reexport, retransfer, transshipment, or disposition of a defense article without such approval."

Remediation may involve approaches and proximity to foreign technology. Any person dealing with US origin items, components and technology, irrespective of location, falls under the Export Administration Regulations, EAR. In short, because the US rules apply to exports and re-exports of US goods on the particular military and commercial control lists, the impact of ITAR and EAR cannot be discounted in the context of debris remediation. The extent to which such operations can be exempted from its provisions is therefore important.[53]

EAR, unlike ITAR, contains a *de minimis* rule, whereby foreign manufacturers may benefit from some relief for military commodities that contain less than 25% US components.[54]

Since debris remediation involves transferring control of the object, disclosing sensitive technical data, and further, the exact applicability of the above agreements needs to be examined in advance in the context of the registered state, the components in operation and the mission involved. The type of technology, as well as a decision as to whether ADR docking constitutes an 'export' for which exemption from control may be available, remains important from the arms and export control perspective. The interaction between the above agreements, the ongoing reform of the US ITAR and EAR regimes, together with the United States Military Lists contained in USML XV, are all questions that directly impact on security and foreign policy of the interested states, and with this, on the immediate practicalities of space remediation. Clearly the combined impact of these rules means that national service operators will be preferable to international commitments.[55]

## V.3. THE VALUE OF CODES, STATE PRACTICE AND STATE RESPONSIBILITY

Various sets of debris mitigation guidelines have been internationally agreed by several recognised authorities in the international space community. These include the international UN authorities responsible for managing the radio frequency spectrum, the ITU, as well as the UN Committee body responsible for the continuing application and supervision of space law, UN COPUOS. This speaks for securing their approval and validation at international level.[56]

---

[53]  ITAR§123.16 makes provision for exemptions of general applicability.
[54]  50 USCA §2404(m) (National Security Controls)
[55]  A. Krolikowski, E. David, n. 4, above
[56]  For a list of the various codes, see n. 22 above.

These provisions are frequently referred to with the post-modernist title of "soft law."[57] The guidelines have not been delivered in treaty form, nor are they yet seen to constitute customary norms of international law. Their status as soft law is clear, and the rules are welcomed as an additional source of authority to secure the future sustainability of space operations. However, if remediation is to become viable, granting these codes greater value can achieve greater compliance for the future.

State practice can lead to the development of recognised norms of customary international law.[58] It has been stated by noted scholars that customary international law of outer space may develop over a relatively short time span.[59] In the case of outer space, this can be influenced by the pace of technological development. Technology may therefore be the special factor that enables custom to be recognised in relation to outer space, provided there is evidence of adherence to its terms.[60] The debris mitigation guidelines already directly influence the practice of space operations at state and international level, as the UN Compendium now clearly shows.

UN COPUOS has issued regular statements of open support for these rules. This was repeated in the context of its Resolution 68/74 on the importance of national space law.[61] Often framed as general principles, General Assembly resolutions are not binding on member states, but are seen as a basis "for the progressive development of the law and the speedy consolidation of customary rules."[62] In short, the individual resolution relating to the principle in question can only be assessed from a legal perspective on the basis of all surrounding information. In relation to the role of debris mitigation, there are strong indications that state practice already confirms support of these rules.

Custom can also constitute a source of international law under Art 38 (1) (b) of the Statute of the International Court of Justice. Art. 38 (1) (c) continues with the sources thereafter, referring to 'the general principles of law recognised by civilised nations.' Finally, Art 38 (1) (d) refers to sources of law under the category of 'subsidiary means' of ascertaining the

---

[57] I. Marboe, *Soft Law* in Outer Space: The Function of Non-binding Norms in International Space Law; F. Lyall, P. Larson, Space Law, a Treatise, n. 19 above

[58] Art. 36 (c) Statute ICJ

[59] I. Brownlie, *The Principles of Public International Law*, 8th ed., n.40, above, 'Provided the consistency and generality of a practice are proved, no particular duration is required', at p.7.

[60] For a full discussion of State practice in international law, see the North Sea Continental Shelf Case, ICJ Reports (1969), 3

[61] GA Res. 68/74 A/RES/68/74 of 16 December 2013, 68th session, agenda item 50

[62] Brownlie, n. 40 above p. 15

law through opinio juris.[63] Increased support of the international regulatory community through their adherence and calls for implementation of the guidelines into conventional law could help move an agreed procedure for remediation forward.

## VI. PROGNOSIS/ CONCLUSION

Adopting a consensual approach to an interpretation of the debris guidelines, according them the status of rules of binding international law, as developed through state practice, which are also recognised as such by leading sources and practitioners, has merit to it. The judicial statements made by members of the ICJ on the development of international customary law in the past are directly relevant to the development of these rules.[64] The order of magnitude of collision and danger in outer space is such that active removal of mass from orbit has already been predicted as the only way to prevent collision resulting from cascading debris resulting from the Kessler syndrome.[65] This imminent danger tallies with the requirements placed by the International Court of Justice court that, in the face of imminent danger, appropriate measures taken by states may require some form of legitimate action.

Until such time as debris mitigation becomes a binding rule of international law, there is little likelihood of feasible business models developing to take on the pending task of remediation. The limitations of maritime salvage as a prototype are recognised; unlike abandoned sea vessels on the high seas, for which there is no liability, the provisions of LIAB only impose liability on the launching state for debris that causes physical damage in outer space if fault is shown to exist.[66] As has been

---

[63] Art 38 (1) (d) 'subject to the provisions of Article 59, judicial decisions and the teachings of the most highly qualified publicists of the various nations, as subsidiary means for the determination of rules of law'.

[64] North Sea Continental Shelf, ICJ (1969), 3.

[65] D. J. Kessler, B.G. Cour-Palais; *Collision Frequency of Artificial Satellites: The Creation of a Debris Belt*; NASA Johnson Space Center, Houston, 1978.

[66] The contractual side of removal should be mapped out on the basis of the existing international agreements between agencies, to include the widely practiced system of cross-liability – or interparty waivers, limitation of Third Party Liability. Whether TPL is to be covered by insurance will depend on the volume of involved. See Smith, The Principles of International Space Law and their relevance to Space Industry Contracts, in: Smith, Baumann (eds), Contracting for Space, n. 45, above, 2011, 45-58.

shown, the existence of debris without damage to third parties is not actionable in its own right.

A road map – already described in this context as a protocol[67] – is now needed that links the debris remediation project to the consent of the state(s) on whose register the object is registered, with a procedure for international notification, and an appropriate authority supervising the operations. Given the attention already devoted to this subject by the main international agencies, by the ITU, the UNCOPUOS and the IADC, one of these agencies should be given responsibility for the main coordination of the rules. UNCOPUOS appears the most suited, given its broad remit to maintain and encourage the maintenance of the system of the law of outer space.[68] The system of notification should also require states that are no longer interested in the debris to agree to its removal. State consent could be integrated into and adapted by state registries to include details of "post-life time disposition," at least as far as the removal of related fragmented and dysfunctional debris are concerned.

As with all such debates, the exact formulation will take time. But as has been said before, it might then be a little too late.

•••

[67] B. Weeden, n.9, above
[68] Smith/ Kerrest, CoCoSL, vol I, Art VII Outer Space Treaty n. 13 above, nos. 7-10; UNCOPUOS, Report of Legal Subcommittee, 54<sup>th</sup> session, 13-24th April 2015, A/AC.105/1090

## PROF. DR. LESLEY JANE SMITH

Lesley Jane Smith was appointed Professor of International and European law at the Leuphana University Luneburg, Germany in 1996, and is a Visiting Professor of space law at the University of Strathclyde, Glasgow, as well as a member of the London Institute of Space Law and Policy. She is a solicitor and partner in the law firm Weber-Steinhaus & Smith in Bremen.

Lesley Jane has an extensive command of the field of national and international space law, including developments within Europe. She is an expert on commercial space activities and earth observation, and a contributor to the leading commentaries on these subjects. She is co-editor of 'Contracting for Space', which outlines contract practice in the European space sector (2011).

Lesley Jane is on the editorial board of the Brill series *Studies in Space Law* and a member of the editorial board of the *Journal of Air and Space Law*. She is a corresponding member of the International Academy of Astronautics (IAA), a board member of the International Institute of Space Law (IISL) and a former board member of Women in Aerospace Europe (WIA-E). In 2014, she was appointed General Counsel to the International Astronautical Federation in Paris (IAF).

### BIBLIOGRAPHY

A. Krolikowski, E. David, Commercial On-Orbit Satellite Servicing: National and International Policy Considerations Raised by Industry Proposals, New Space, Volume 1, March 2013.

A. Soucek, Active space debris removal and export control, ESA/ CNES, 3rd European Workshop of Space Debris Remediation, Paris, June 2014.

B. Weeden, How Do I Ask Permission to Engage With A Piece of Space Debris?, 3rd European Workshop on Space Debris Modelling and Remediation, Paris, June 2014.

Cologne Commentary on Space Law, CoCoSL, Eds. Hobe, Schmidt-Tedd, Schrogl, vol. I, Outer Space Treaty, Arts I (Hobe), Art VI (Gerhard), Art

VII (Smith/Kerrest), Art IX (Marchisio), Cologne, Heymann, 2009.

Cologne Commentary on Space Law, CoCoSL, Eds. Hobe, Schmidt-Tedd, Schrogl, vol. II, UN Convention on International Liability for damage caused by space objects, Cologne, Heymann, 2013.

C. Contant-Jorgenson, P. Lála, K.-U. Schrogl (eds), IAA, Study on Cosmic Space Traffic Management, 2006, Paris.

D. J. Kessler, B.G. Cour-Palais; Collision Frequency of Artificial Satellites: The Creation of a Debris Belt; NASA Johnson Space Center, Houston, 1978.

F. Lyall, P. Larson, Space Law, A Treatise, Ashgate, 2009.

F. Tronchetti, Fundamentals of Space law and Policy, ISU/ Springer, 2013.

H. Klinkrad, IAA Study Group 5.10, Policy, Legal and Economic issues of orbital debris removal, chap. 1.d, in: IAA Cosmic Study on the subject of Space Debris Environment Remediation, 2015.

I. Brownlie, Principles of Public International law, 8th ed., Oxford University Press, 2012.

I. Marboe, Soft Law in Outer Space: The Function of Non-binding Norms in International Space Law, Böhlau Wien, 2012.

J. Crawford, The International Law Commission's Articles on State Responsibility – Introduction, Text and Commentaries, Cambridge University Press, 2002.

J. J. A. Salmon, 'Faut-il codifier l'état de nécessité en droit international', in. J. Makarczyk, (ed.), Essays in Honour of Judge Manfred Lachs, Martinus Nijhoff, 1984.

K. C. Little, S. D. Reifman, A. J. Dietrick, U.S. Export Controls Apply Extraterritorially, Circumstances In Which Foreign Persons Are Subject To U.S. Export Laws And Regulations, Vinson & Elkins LLP, Washington, 2007.

M. Lachs, The law of outer space, an experience in contemporary law making, Nijhoff, 2010.

P. Clerc, Consequences of the French space law on space operations (FSOA) on CNES's mission as a contracting space agency, in:

L.J. Smith/ I. Baumann, Contracting for Space, Contract Practice in the European Space Sector, Farnham, Surrey, Ashgate, 2011.

Smith, The Principles of International Space Law and their relevance to Space Industry Contracts, in: Smith, Baumann (eds), Contracting for Space, Contract Practice in the European Space Sector, id.

R. L. Spencer, International Space Law; a basis for international regulation, in: Ram Jakhu, National Regulation of Space Activities, Berlin, Springer, 2010.

T. Flohrer, H. Krag, H. Klinkrad, ESA's process for the identification and assessment of high-risk conjunction events, in: Advances in Space Research, Volume 44, Issue 3, 2009.

UNCOPUOS, Report of Legal Subcommittee, 53rd session, 15th April 2014, A/AC.105/1067.

UNCOPUOS, Report of Legal Subcommittee, 54th session, 13-24th April 2015, A/AC.105/1090.

CHAPTER 9

# LEGAL ISSUES RELATING TO UNAUTHORIZED SPACE DEBRIS REMEDIATION

JOYEETA CHATTERJEE
MCGILL UNIVERSITY, INSTITUTE OF AIR AND SPACE LAW

Following loss of contact with the earth observation satellite Envisat on April 2012, the European Space Agency declared the end of its mission a month later in May 2012 after failed attempts to restore control. This minibus-sized satellite weighing 8 metric tons is currently drifting uncontrolled in the Low Earth Orbit (LEO). The increasing proliferation in the population of uncontrollable man-made objects in the Earth orbit poses severe navigational threats to functional satellites and other space assets. Studies conducted on achieving long-term security and sustainability of outer space activities reflect the consensus of the scientific community that space debris remediation in the form of active removal of debris is essential to prevent cascading collisions between the space debris in orbit.

This chapter explores the body of space law and its implications on space debris remediation. Relying on the example of Envisat, it will be demonstrated that the existing framework of international space law does not authorise interception with space objects without the prior consent of the State of Registry. In the case of a removal of an object without the authorisation of the State of Registry, it would constitute an internationally wrongful act. This chapter will further draw attention to the need to effectuate unambiguous interpretation of the existing provisions of international space law and the need for close cooperation between members of the international space community for the smooth operation of space debris remediation. Finally, it will conclude that the current provisions of international space law are adequate to address any potential legal controversies arising in this context and there is no need for any amendment or reform in the current legal framework by concluding a new treaty.

## INTRODUCTION

The international space community has been cognisant of the growing threat of orbital congestion since the 1980s. However, concerted international action to address the problem did not begin until the establishment of the Inter-Agency Space Debris Coordination Committee (IADC) by the various national and regional space agencies in 1993 to foster dialogue across nations.[1] The IADC adopted a set of guidelines for space debris mitigation measures in 2002.[2] With a view to expediting the international adoption of voluntary debris mitigation measures, a Working Group of the United Nations Committee on the Peaceful Uses of Outer Space (UN COPUOS) collaborated with the IADC to update and revise the IADC guidelines on debris mitigation. Finally, the agreed upon guidelines

---

[1]  Terms of Reference for the Inter-Agency Space Debris Coordination Committee (IADC), online: http://www.iadc-online.org/index.cgi?item=torp_pdf. "The primary purpose of the IADC is to exchange information on space debris research activities between member space agencies, to facilitate opportunities for cooperation in space debris research, to review the progress of ongoing cooperative activities and to identify debris mitigation options."

[2]  IADC Space Debris Mitigation Guidelines (2002) (hereinafter IADC Guidelines), online: http://www.iadconline.org/docs_pub/IADC-101502.Mit.Guidelines.pdf.

were adopted[3] and subsequently endorsed by COPUOS in 2007,[4] as the Space Debris Mitigation Guidelines of the Committee on the Peaceful Uses of Outer Space.[5]

Since the launch of Sputnik I in 1957,[6] space debris in the form of uncontrollable man-made objects in the Earth orbit continue to pose increasing navigational threats to functional satellites and other space assets, including human space flight and robotic missions.[7] The International Space Station has had to perform more than a dozen collision avoidance manoeuvres in the last decade.[8]

It is clear that the preventive measures taken during the last decade in the form of voluntary non-binding debris mitigation guidelines have not been able to effectively address the impending catastrophic situation. Based on scientific analysis and the projections made by various technical models, the only way to ensure secure and sustained access to and long-term utilization of space is through space debris remediation in the form of active removal of debris and on-orbit satellite servicing.[9]

Unlike space debris mitigation which aims to arrest the generation of further debris, space debris remediation refers to actively remedying the congested nature of outer space. Remediation activities can include

[3]  *Report of the Scientific and Technical Subcommittee on the Work of its Forty-fourth Session*, UNCOPUOS, 50th Sess, UN Doc A/AC.105/890 (2007) at para 99.
[4]  *Report of the Committee on the Peaceful Uses of Outer Space*, UNGAOR, 62nd Sess, Supp No 20, UN Doc A/62/20 (2007) at para 118-119.
[5]  Space Debris Mitigation Guidelines of the Committee on the Peaceful Uses of Outer Space, as annexed to UN doc. A/62/20, Report of the COPUOS (2007) at 1.
[6]  Michael Stoiko, *Soviet Rocketry: Past, Present, and Future* (Holt, Rinehart & Winston, 1970) at 79.
[7]  Interagency Report on Orbital Debris (Office of Science and Technology Policy, U.S. National Science and Technology Council, Washington, DC, 1995); *Technical Report on Space Debris*, text of the Report adopted by the Scientific and Technical Subcommittee of the United Nations Committee on Peaceful Uses of Outer Space, UN Doc A/AC.105/720 (New York: United Nations, 1999) (hereinafter *Technical Report on Space Debris*).
[8]  "International Space Station Again Dodges Debris" (2011) 15 Orbital Debris Quarterly News 1, online: http://www.orbitaldebris.jsc.nasa.gov/newsletter/pdfs/ODQNv15i3.pdf
[9]  J.-C. Liou, N.L. Johnson, "Instability of the present LEO satellite populations" (2008) 41 Adv. Space Res. 1046; J.C. Liou & Nicholas L. Johnson, "Risks in Space from Orbiting Debris" (2006) 311 Science 340-341, online: http://www.sciencemag.org/content/311/5759/340.full Generally, see J.C. Liou, "A Note on Active Debris Removal" (2011) 15 Orbital Debris Quarterly News 7, online:http://www.orbitaldebris.jsc.nasa.gov/newsletter/pdfs/ODQNv15i3.pdf at 7-8.

retrieval of a space object from the outer space environment or from a particular orbit, repairing/servicing a space object, refuelling missions to extend the life of the space object or salvaging a space object for recycling or other purposes. On-orbit servicing and salvaging operations remediate space debris by repairing and restoring maneuverability in an object or removing it to avoid collision with a functional satellite. The following sections will study the implications of the existing framework of international space law and public international law on space debris remediation.

## 2. DEFINITION OF SPACE DEBRIS FOR ACTIVE REMEDIATION

The objective of this section is to study the question: is 'space debris' equivalent to a 'space object' *ad infinitum*?[10] It is important to draw a distinction between a 'space object' and a piece of 'space debris' because the absence of a clear legal definition introduces severe ambiguity in the enforcement of the rights and obligations assigned to States in relation to the objects they have launched in space or the debris created by their activities in outer space.

To understand the legal milieu in which space debris are sought to be regulated, it is necessary to study the definition of 'space debris.' First, this section will chronologically discuss the international legislative attempts to define a 'space object.' It will then address the current definition of 'space debris' with its origin in 'soft law' and its implications in the operation of space activities. Finally, it will comment on the legal uncertainties surrounding the status of objects in space vacillating between that of a 'space object' and/or 'space debris' by relying on the example of the decommissioning of the Envisat satellite by ESA.

The current regime of international space law, consisting of the five United Nations treaties and five Declarations, does not contain any definition of 'space debris.' The operative terminology used in those instruments is a 'space object,'[11] which has been rather obliquely defined.

---

10  For distinction between 'space object' and 'space debris,' see Luboš Perek, "Ex Factor Sequitur Lex: Facts which Merit Reflection in Space Law in Particular with Regard to Registration and Space Debris Mitigation" in Marietta Benkö & Kai-Uwe Schrogl, *Space Law: Current Problems and Perspectives for Future Regulation* (Utrecht: Eleven International, 2005) at 40-43.

11  Armel Kerrest, "Liability for Damage Caused by Space Activities" in Marietta Benkö & Kai-Uwe Schrogl, *Space Law: Current Problems and Perspectives for*

Article VII of the Outer Space Treaty[12] lays down that the launching State will be held internationally liable for damage caused by an object launched into outer space or its component parts. This principle is echoed in Article II of the Liability Convention[13] which states that: "A launching State shall be absolutely liable to pay compensation for *damage caused by its space object* on the surface of the Earth or to aircraft in flight." (emphasis added) Further, Article III of this Convention emphasizes this criterion again to determine liability for damage caused elsewhere than on the surface of the Earth. Hence, the concern over the absence of a proper definition of 'space object' is aggravated by the fact that "the basis of liability is that the damages or injury is caused by a space object."[14]

## 2.1 DEFINING A 'SPACE OBJECT'

Even prior to the promulgation of any of the space law treaties, the Convention for the Establishment of a European Organization for the Development and Construction of Space Vehicle Launchers (ELDO) defined a 'space vehicle' as "a vehicle designed to be placed in orbit as a satellite of the Earth or of another heavenly body, or to be caused to traverse some other path in space..."[15]

In the 1963 Declaration of Legal Principles[16] which serves as the precursor to the 1967 Outer Space Treaty, a space object has not been defined but has been referred to as "object launched into outer space and ... their component parts." Adopting this language, the 1967 Outer Space Treaty has alluded to a 'space object' in Articles VII and VIII as "an object launched into outer space," including "objects landed or constructed on a celestial body."

---

*Future Regulation* (Utrecht: Eleven International, 2005) at 97-98; S. Gorove, "Legal and Policy Issues of the Aerospace Plane" (1988) 16 J. Space L. 147 at 154; Julian G. Verplaetse, "On the Definition and Legal Status of Spacecraft" (1963) 29 J. Air L. & Com. 131.

[12] Treaty on Principles Governing the Activities of States in the Exploration and Use of Outer Space, including the Moon and Other Celestial Bodies, 27 January 1967, 610 UNTS 205 (hereinafter Outer Space Treaty).

[13] Convention on International Liability for Damage Caused by Space objects, 29 March 1972, 961 UNTS 187 (hereinafter Liability Convention).

[14] S.B. Rosenfield, "Where Air Space Ends and Outer Space Begins" (1979) 7 J. Space L. 137 at 145.

[15] Annex to art. 19, UNTS 507 at 205. Also, see J.A.C. Gutteridge, "The United Nations Committee on the Peaceful Uses of Outer Space" in *Current Problems in Space Law: A Symposium* (British Institute of International and Comparative Law, Holland, 1986) at 36.

[16] Declaration of Legal Principles Governing the Activities of States in the Exploration and Use of Outer Space, UN GA Res. 1962 (XVIII) 13 December 1963.

The Liability Convention was the first international agreement which attempted to define a 'space object' as "component parts of a space object as well as its launch vehicle and parts thereof."[17] The Registration Convention adopted this definition in its Article I(b).[18] This description fails to define the term exhaustively while merely providing a vague inclusive boundary for the term. Strikingly enough, it does not include functionality as a decisive criterion.[19]

The term 'space object' has not yet been defined in international space law. More importantly, it is also silent as to when, if at all, a space object or its component or fragmented parts, ceases to be a 'space object.' Assuming that there is no change in the status of such fragmented space objects and are still continued to be regarded as 'space objects' under international space law, then *de jure* jurisdiction and control will be retained by the launching State on whose registry the space object is carried.[20]

The definition for a 'space object' prescribed by Baker in his excellent treatise on the legal status of space debris is of particular importance. He postulates that a 'space object' –

Means
any object
intended for launch, whether or not into orbit or beyond;
launched, whether or not into orbit or beyond; or
any instrumentality used as a means of delivery of any object as defined in 1(a); and
Includes
any part thereof or
any object on board which becomes detached, ejected, emitted, launched or thrown, either intentionally or unintentionally, from the moment of ignition of the first-stage boosters.[21]

In the spirit of the Liability Convention as an example of victim-oriented law, it is suggested that the interpretation of space object ought to

---

17  Liability Convention, art I(d). See Bess C.M. Reijnen, *The United Nations Space Treaties Analysed* (Editions Frontieres, 1992) at 182-83.
18  Convention on Registration of Objects Launched into Outer Space, 29 November 1971, UN GA Res. 3235 (XXIX) (hereinafter Registration Covention).
19  Mathias Forteau, "Space Law" in James Crawford, et al (eds.), *The Law of International Responsibility*, (Oxford University Press, 2010) at 906.
20  Outer Space Treaty, art VIII.
21  H.A. Baker, "Liability for Damage Caused in Outer Space by Space Refuse" (1988) 12 Ann. Air & Sp. L. 183 at 225.

be "liberal...in favour of an innocent victim."[22] Hence, 'space objects' should be given a broad interpretation to include objects constructed or assembled in outer space under the regime of the Liability Convention to ensure that States do not ignore the law by constructing or assembling their space objects in outer space.[23] This is important to address issues arising from the status of satellites whose components have been derived from functional parts of 'space debris' salvaged or serviced in outer space. It is not a technologically distant dream because the goal of the Phoenix program under the aegis of the Unites States Defence Advanced Research Projects Agency is focused on recycling space assets by 2015.[24]

With the above understanding of the legal definition of a 'space object,' the following sub-section will focus on the definition and attributes of space debris, for the purposes of performing active debris remediation.

## 2.2  DEFINING 'SPACE DEBRIS'

Unanimously adopted at its 66th conference in 1994, the International Law Association's International Instrument on Space Debris[25] was the first international attempt to provide a legal definition of 'space debris.' In the first article on definitions, space debris has been defined in paragraph (c) as, "man-made objects in outer space, other than active or otherwise useful satellites, when no change can reasonably be expected in these conditions in the foreseeable future."[26]

---

[22]   T.E. Wolcott, "Some Aspects of Third Party Liability in Space Shuttle Operations" (1980) 13 Akron L.R. 613 at 617.

[23]   Bruce A. Hurwitz, *State Liability for Outer Space Activities in Accordance with the 1972 Convention on International Liability for Damage Caused by Space Activities*, (Martinus Nijhoff, 1992) at 23-24. This conclusion is supported by the 1980 NASA Authorization Act which defines "space vehicle" as "an object intended for launch, *launched or assembled in outer space*, including the Space Shuttle and other components of a space transportation system [the official designation of the Shuttle], together with related equipment, devices, components and parts." National Aeronautics and Space Administration Authorization Act, 1980, Pub. L. No. 96-48, 93 State. 348 (1979), Section 308 – Insurance and Indemnification at Sec. 308(f), quoted by G.J. Mossinghoff, "Managing Tort Liability Risks in the Era of the Space Shuttle" (1979) 7 J. Space L. 121 at 127-128. Emphasis added.

[24]   David Barnhart, Program Manager, Tactical Technology Office, "DARPA's Phoenix Project" presented at the NASA Second International Workshop on On-Orbit Satellite Servicing (May 2012), online: http://ssco.gsfc.nasa.gov/workshop_2012/McGuirk_final_presentation_2012_workshop.pdf.

[25]   The ILA Finalizes its International Instrument on Space Debris in Buenos Aires, August 1994, (1995) 23 J. Space L. 47.

[26]   For the text of the instrument, see Karl-Heinz Böckstiegel, "ILA Draft Convention on Space Debris" (1995) 44 ZLW 29.

The Technical Report on Space Debris was published in 1999 as a product of the multi-year work plan 1996-1998 of the Scientific and Technical (S&T) Subcommittee of the UN COPUOS. It was one of the earliest United Nations documents on space debris, which served as a basis for further deliberations on the topic of congestion in the space environment. It reports the following definition proposed at the 32nd session of the S&T Subcommittee for the sake of a common understanding of the term 'space debris.'

"Space debris are all manmade objects, including their fragments and parts, whether their owners can be identified or not, in Earth orbit or re-entering the dense layers of the atmosphere that are non-functional with no reasonable expectation of their being able to assume or resume their intended functions or any other functions for which they are or can be authorized."[27]

In 2002, pursuant to its charter, the IADC developed the 'IADC Space Debris Mitigation Guidelines' based on the fundamental principles present in the national policies of the member agencies and were agreed to by consensus.[28] The definition of space debris contained therein was an abbreviated form of the above-mentioned definition, which was later borrowed verbatim in the United Nations Space Debris Mitigation Guidelines. The publication of the IADC Guidelines prompted the S&T Subcommittee of the UN COPUOS to create a Space Debris Working Group,[29] which produced a draft set of "high-level qualitative guidelines" based on the work of the IADC.[30] This draft was adopted by the UN COPUOS in 2007 and endorsed by the General Assembly later that year through Resolution 62/217.[31] The definition of space debris provided in the UN COPUOS Guidelines is as follows:

"All man-made objects including fragments and elements thereof, in Earth orbit or re-entering the atmosphere, that are non-functional."[32]

27  *Technical Report on Space Debris*, note 7 at 2, para. 6.
28  IADC Mitigation Guidelines, note 2. These have been elaborated upon by *Support to the IADC Space Debris Mitigation Guidelines* (2004), online: http://www.iadconline.org/docs_pub/IADC.SD.AI20.3.10.2004.pdf.
29  *Report of the Scientific and Technical Subcommittee on the Work of its Forty-First Session*, UN COPUOS, UN Doc. A/AC.105/823, 2004 at 20.
30  *Progress Report of the Working Group on Space Debris, Submitted by the Chairman of the Working Group*, UN COPUOS, UN Doc. A/AC.105/C.1/L.284, 2006, at 2.
31  GA Res 62/217, 21 December 2007, 'International cooperation in the peaceful uses of outer space,' para 26. In GA Res 63/90, 5 December 2008, the General Assembly invited States to 'implement' these Guidelines (para 26).
32  Space Debris Mitigation Guidelines, note 5.

It is interesting to note that the definition of 'space debris' is not contained in any of the actual Guidelines but it is included in the introductory section entitled 'Background' of the document. Further, it is important to bear in mind that this definition is explicitly limited to the purpose of this document by a preceding proviso.[33]

Although the General Assembly has declared that the UN Guidelines "reflect the existing practices as developed by a number of national and international organizations," the legal status of the Guidelines are amply clear insofar as it states, in no uncertain terms, that "They are not legally binding under international law."[34] It further states that "Member States and international organizations should *voluntarily* take measures...to ensure that these Guidelines are implemented."[35] (emphasis added) It is evident that these Guidelines reflect technical best practices. The technical nature of the Guidelines is underscored over its legal implications by the fact that they were adopted solely by the S&T Subcommittee without any involvement or contribution from the Legal Subcommittee.

Thus, the definition of space debris enshrined in the UN Guidelines can be classified as 'soft law.'[36] Although soft law is said to lack "the requisite normative content to create enforceable rights and obligations,"[37] it is, nonetheless, capable of producing certain legal effects.[38] It is not only considered as an "expression of emerging notions of an international public order,"[39] but it also constitutes "an important element in the progressive institutionalization of international cooperation."[40] Hence, the definition of 'space debris' contained in these Guidelines reflect a relatively less obligatory approach, which helps to balance the conflicting priorities of the space players[41] and to establish a minimal standard of care for States in the realm of debris mitigation and remediation measures.

---

[33] Ibid.

[34] Ibid., section 3, para. 2.

[35] Ibid., section 2.

[36] Joseph Gold, "Strengthening the Soft International Law of Exchange Agreements" (1983) 77 AJIL 443; Christine Chinkin, "A Hard Look at Soft Law" (1988) Proceedings ASIL 371, at 389; C. Schreuer, "Recommendations and the Traditional Sources of International Law" (1977) 20 German Yearbook of Int'l Law 103.

[37] Francesco Francioni, "International 'Soft Law': A Contemporary Assessment" in Vaughan Lowe and Malgosia Fitzmaurice (eds.), *Fifty Years of the International Court of Justice, Essays in Honour of Sir Robert Jennings* (Cambridge University Press, 1996) at 168.

[38] "A Hard Look at Soft Law" (1988) 82 Am. Soc'y Int'l L. Proc. 371

[39] Ibid.

[40] Francioni, "International 'Soft Law', note 37, at 178.

[41] "Some writers have enthusiastically endorsed this normative category, highlighting the need for flexibility and responsiveness to the contemporary need for

## 2.3 DECOMMISSIONING OF ENVISAT

On 8 April 2012, ESA lost contact with Envisat, the largest non-military earth observations satellite in orbit.[42] After several failed attempts to regain control of the satellite, ESA declared the end of its mission on 9 May 2012.[43]

It is currently drifting uncontrolled in a sun-synchronous polar orbit and is being tracked by the U.S. Joint Space Operations Centre. Its enormous size – ten metres in length and five metres in width, with an even larger solar array and weighing 8 tons – aggravates the concern of its collision with other functional space objects.[44] It has been estimated that given its orbit and area-to-mass ratio, it will take 150 years for natural decay through atmospheric drag.[45] ESA has calculated a 30 percent collisional probability with other orbital debris in this duration.[46] Therefore, it is potentially an ideal candidate for removal from orbit.[47]

In this case, the question arises whether Envisat can be qualified as 'space debris.' Although it is drifting uncontrolled and is no longer manoeuvrable due to loss of communications, it is otherwise an intact satellite. Further, if technological development allows re-establishing communications with it, as in the case of the Intelsat Galaxy-15 satellite, then Envisat can be re-commissioned back to service as a 'space object.'

## 2.4 ANALYSIS

It has been rightly pointed out by the 2006 IAA Cosmic Study on Space Traffic Management that "no legal distinction is made between valuable

---

accommodation between competing interests in a diversified and conflictual world community." Ibid, at 168.

[42] Tariq Malik, Huge Satellite Loses Contact with Earth (16 April 2012), online: Space.Com, http://www.space.com/15290-huge-satellite-envisat-contact-lost.html

[43] ESA Declares End of Mission for Envisat (9 May 2012), online: ESA, http://www.esa.int/Our_Activities/Observing_the_Earth/Envisat/ESA_declares_end_of_mission_for_Envisat.

[44] Mike Wall, Huge Dead Satellite May Be Space Junk for 150 Years (11 May 2012), online: Space.Com, http://www.space.com/15640-envisat-satellite-space-junk-150years.html.

[45] Envisat To Pose Big Orbital Debris Threat for 150 Years, Experts Say (23 July 2010), online: Space News, www.spacenews.com/civil/100723-envisat-orbital-debris-threat.html/.

[46] Space Risks: A New Generation of Challenges, An Insurer's Perspective from Allianz Global Corporate & Specialty, online: http://www.agcs.allianz.com/assets/PDFs/white%20papers/1844%20Allianz%20Space%20White%20Paper%20201o.pdf at 5.

[47] For an excellent factual summary of the operation and the life-span of Envisat, see Martha Mejía-Kaiser, "ESA's Choice of Futures: Envisat Removal or First Liability Case" (2012) 55 Proc. Colloq. on the Law of Outer Sp.

active space-craft and valueless space debris."[48] It further recommended the UN COPUOS to "start discussing whether or not space debris are space objects in the sense used in space law. If it is decided that space debris are space objects, an additional protocol should be elaborated stating what provisions of the treaties apply to valuable spacecraft and which provisions apply to space debris. If it is decided that space debris are not space objects, the protocol should determine under what conditions space debris may be removed or re-orbited in order to prevent collisions or close encounters with valuable spacecraft."[49]

The formulation of a "transparent and reasonable selection matrix on the basis of which objects are targeted"[50] is a prudent method to ascertain which space objects can be designated as targets for removal. In the wide gamut of views put forth by experts,[51] the consensual opinion seems to be based on the common denominator of "the ability of the man-made instrumentality to traverse in outer space."[52] Hence, the maneuverability or functionality of the space object is key to determining its status as space debris so that it can be classified as a target for remediation.

While a fresh legislative endeavour in the form of an additional protocol or a separate treaty to address this situation is the easiest and ideal solution,[53] our current geo-political environment is not conducive for such an approach due to the competing interests and priorities of different States. Hence, it is essential to investigate a pragmatic alternate resolution to this problem through optimal utilization of the already available resources, that is, to effectuate a broader interpretation of the existing legal principles in order to accommodate the rapidly changing commercial and environmental realities of activities conducted in outer space.

---

[48]  Corinne Contant-Jorgenson, Petr Lála, Kai-Uwe Schrogl (eds.), Cosmic Study on Space Traffic Management (Paris: International Academy of Astronautics, 2006) online: http://iaaweb.org/iaa/Studies/spacetraffic.pdf, at 40.

[49]  Corinne Contant-Jorgenson, Petr Lála & Kai-Uwe Schrogl, "Report: The IAA Cosmic Study on space traffic management" (2006) 22 Space Policy 283 at 287.

[50]  Jan Helge Mey, "Space Debris Remediation: Some Aspects of International Law Relating to the Removal of Space Junk from Earth Orbit" (2012) 61 ZLW 251 at 271.

[51]  S.M. Beresford, "Requirements for an International Convention on Spacecraft Liability" (1963) 6 Proc. Colloq. L. Outer Sp. 1 at 11; G.D. Schrader, "Space Activities and resulting Tort Liability" (1963) 6 Proc. Colloq. L. Outer Sp. 1 at 2.

[52]  Hurwitz, State Liability for Outer Space Activities, note 23, at 23.

[53]  Thierry Senechal, "Orbital Debris: Drafting, Negotiating, Implementing a Convention," Master's thesis, Massachusetts Institute of Technology (2007)

## 3. STATE RESPONSIBILITY FOR SPACE DEBRIS REMEDIATION

Due to the absence of a legal status granted to space debris, orbital remedial activities give rise to a plethora of regulatory complexities and unanswered legal questions. Imagine the following hypothetical scenario: Conjunction analysis has identified an uncontrolled satellite, X belonging to State A as a high-probability threat to a functional satellite, Y belonging to State B, which attempts to deorbit X without authorization from State A. Due to technical anomalies, it erroneously incapacitates another satellite belonging to State A. In the meanwhile, State A manages to successfully revive satellite X and manoeuvre it back to its allotted orbit.

Is State A under an international legal obligation to avoid causing damage to another State's space assets? Is State B justified in exercising jurisdiction and control over satellite X to avoid collision with its own space asset? What are the legal implications of unauthorized active debris removal?

State responsibility has been viewed as "a legal construct that allocates risk for the consequences of acts deemed wrongful by international law to the artificial entity of the State."[54] The distinction between State responsibility and liability lies in the fact that the prerequisite to the former is an act breaching international law and to the latter, the harmful effects of an activity, which is not per se a violation of international law.[55] In international space law, while responsibility applies to a "State's obligation to regulate and control space activity both in the present, and in the future, to assure compliance with not only the letter but the spirit of the Outer Space Treaty principles," liability on the other hand refers to an "obligation of a State to compensate for damages."[56]

As has been observed by Cheng, international state responsibility in the outer space field arises the moment a breach of an international obligation is produced and not when the State is seen to have failed in its duty to prevent or repress such breach, for the State is immediately accountable for

---

[54]  Christine Chinkin, "A Critique of the Public/Private Dimension" (1999) 10 EJIL 387 at 477

[55]  Rebecca M. M. Wallace, *International Law*, (Sweet & Maxwell, 2003) at 203.

[56]  W. B. Wirin, "Practical Implications of Launching State – Appropriate State Definitions", (1994) 37 Proc. of Colloq. on the Law of Outer Sp. at 109.

the breach on the international plane as if it itself had breached the international obligation. [57]

## 3.1 International Responsibility: Article VI, Outer Space Treaty

The vital question of responsibility over space objects is addressed in *lex spatialis*, first in the 1963 Declaration of Legal Principles and then in the 1967 Outer Space Treaty. At the time of its adoption, the Outer Space Treaty represented "the lowest common denominator of issues on which consensus existed in COPUOS."[58] This sentiment was reflected in the views of the then U.S. Secretary of State, who had described the legislative efforts behind the conclusion of the Outer Space Treaty as an "outstanding example of how law and political arrangements can keep pace with science and technology."[59] As of 1 January 2014, the Outer Space Treaty has been ratified by 103 States and signed by 25 signatories.[60] It is noteworthy that all spacefaring States so far have ratified the Treaty which indicates that some of its provisions have likely crystallized into customary international law.[61]

The possible involvement of private enterprises in outer space and the attribution of responsibility for such private activities to the States had been one of the controversial issues between the U.S.A. and the erstwhile Soviet Union during the development of a legal regime governing outer space

[57] Bin Cheng, "Article VI of the 1967 Space Treaty Revisited: 'International Responsibility', 'National Activities' and 'The Appropriate State'" (1998) 26 J. Sp. L. 7 at 15.

[58] Nicolas Mateesco Matte, "Outer Space Treaty" in R. Bernhardt (ed.), *Encyclopedia of Public International Law*, Vol. 1 (Elsevier, 1992) at 838. "Containing general principles for the peaceful exploration and use of outer space, including the moon and other celestial bodies, it was not to deal with all contingencies that might arise from their exploration and use. It is not a perfect instrument. Some of its principles are obscurely stated and its terms lack precision and definition. Nevertheless, it represents the most important source of space treaty law."

[59] Dean Rusk, "Letter of Submittal from Secretary Rusk to President Johnson" (27 January 1967) in *Hearings on Treaty on Outer Space Before the Senate Committee on Foreign Relations* (1967), 90th Cong., 1st Sess., at 112.

[60] Status of International Agreements Relating to Activities in Outer Space as at 1 January 2014, A/AC.105/2014/CRP.7, online: United Nations Office of Outer Space Affairs http://www.oosa.unvienna.org/pdf/limited/c2/AC105_C2_2014_CRP07E.pdf.

[61] Bin Cheng, "The 1967 Outer Space Treaty: Thirtieth Anniversary" (1998) 23 Air & Sp. L. 156; Bin Cheng, "United Nations Resolutions on Outer Space: 'Instant' International Customary Law?" (1965) 5 Indian J. Int'l L. 23; Vladlen S. Vereshchetin & Gennady M. Danilenko, "Custom as a Source of International Law of Outer Space" (1985) 13 J. Sp. L. 22.

activities.[62] Principle 5 of the United Nations General Assembly Resolution 1962 (XVIII) reflected the compromise reached between the two parties by allowing private participation in space activity subject to the control of the "appropriate State" and imposing consequent international responsibility on the State for such activities.[63] It was later incorporated in Article VI of the 1967 Outer Space Treaty. On deconstructing this article, it is clear that the following obligations are imposed on States:[64]

- To bear responsibility for national activities in outer space regardless of whether such activities are carried out by public or private entities;
- To assure that national activities are conducted in conformity with the Outer Space Treaty and, through Article III, with international law;
- To authorize and continually supervise, where appropriate, the activities of nongovernmental entities in outer space; and
- To share international responsibility for the activities of international organizations of which the State is a participant.

The scope of this chapter is to examine the space behaviour of States as subjects of public international law and *a fortiori*, international space law. The regulatory concerns about the activities of private actors will not be addressed because ultimately, States shall "bear international responsibility" for such activities, which "require authorisation and continuing supervision" by the appropriate State under the dictate of Article VI of the Outer Space Treaty. Hence, this chapter explores the duties and responsibilities of States as members of the international space community and their legal rights and obligations for space debris remediation conducted under their national jurisdiction and control.

The extent of obligation as far as damage to third parties is concerned is the international responsibility of the obligation to control; in particular to make sure that the obligations set by Article III (activities must be

---

[62] While the U.S.A. urged for private participation in space ventures by arguing that outer space should be used as freely as the high seas and not limited to use by sovereign State actors, the Soviets asserted that only States should participate in space activity and that "to give private companies a free hand in outer space could lead to chaos and anarchy." U.N. Doc. A/AC.105/C.2/SR.28 (9 July 1963) at 13.

[63] Carl Q. Christol, *The Modern International Law of Outer Space*, (New York: Pergamon Press, 1982) at 65.

[64] Ricky Lee, *Law and Regulation of Commercial Mining of Minerals in Outer Space*, (Springer, 2012) at 128.

carried on according to international law, including the Charter of the United Nations as *lex generalis*) and Article VI (activities must be carried on according to the Outer Space Treaty as *lex specialis*) of the Outer Space Treaty are implemented.[65]

In the event of a space debris remediation activity, it can be inferred from Article VII of the Outer Space Treaty that although the remediation might be conducted by a third party, the launching State of the space object in question would continue to incur international responsibility for any damage caused by it. While international law does not explicitly impose an obligation to avoid causing damage to another State's space assets, there is an underlying duty to observe a standard of care or due diligence in performance of its activities. With a view towards balancing the conflicting State interests in its 1978 report,[66] the Working Group to the International Law Commission noted that "the essential obligation owed by a State in such a context has tended to be conceived as one of moderation, or of care or due diligence, in relation to its own activities or of private activities within its jurisdiction or control."[67] It was emphasized in the Special Rapporteur's report that "treaty regimes of a universal character, dealing with acts not prohibited by international law, had been established in relation to," among other issues, the regulation of "space objects."[68]

It is stated in Special Rapporteur Baxter's first report on international liability for injurious consequences arising out of acts not prohibited by international law in 1980:

"Depending upon the circumstances, the standard of reasonable care or due diligence may well require a standard more exacting than its own as

---

[65] Armel Kerrest, "Liability for Damage Caused by Space Activities" in Marietta Benkö & Kai-Uwe Schrogl, *Space Law: Current Problems and Perspectives for Future Regulation* (Utrecht: Eleven International, 2005) at 107.

[66] "On the one hand there is the benefit to be obtained by the State conducting the activity, but on the other hand there is the injury inflicted on the foreign State as a result of the conducting of that same activity." Hurwitz, *State Liability for Outer Space Activities*, note 23, at 147.

[67] ILC Yearbook 1978, vol. II, part two, at 151 (Para 19)

[68] Preliminary report on international liability for injurious consequences arising out of acts not prohibited by international law. Doc A/AC.4/344 and Add. 1 and 2. Reprinted in ILC Yearbook 1980, Vol. II, part one (Para 4). The law of outer space was included within the category of "recent materials that are, or may be, relevant to the development of a new topic." ILC Yearbook 1978, vol. II, part two, at 150 (Para. 12) Also, see Setsuko Aoki, "The Standard of Due Diligence in Operating a Spacecraft" (2012) 55 Proc. of Colloq. on L. of Outer Sp.

part of a special regime of protection that includes guarantees of redress for the potential victims of any hazard that cannot be wholly eliminated."[69]

He goes on to clarify the controversy regarding the absence of a standard of care in space law with the following remarks:

"[T]he regime of absolute liability provided in the [Liability Convention] may be regarded not only as an applicable conventional rule, but also as evidence of the standard of care which the authors of the Convention believed to be reasonable in relation to that particular activity."[70]

## 3.2   NEED FOR CONSENT

The existing framework of international space law does not authorize interception with space objects without the prior consent of the launching State. In the case of a removal of an object without the authorization, it would constitute an internationally wrongful act.

However, prior consent obtained from the launching State, or the State of registry in the case of multiple launching States, would constitute a circumstance precluding the wrongfulness of conduct that would otherwise not be in conformity with the international obligations of the State performing the remedial activity. It has been opined by the ICJ that the existence of such a circumstance does not annul or terminate the obligation; rather it provides a justification or excuse for non-performance while the circumstance in question subsists.[71]

Article 20 of the International Law Commission's Articles on State Responsibility reflects the basic international law principle of consent:

"Valid consent by a State to the commission of a given act by another State precludes the wrongfulness of that act in relation to the former State to the extent that the act remains within the limits of that consent."

In accordance with this principle, consent by a State to particular conduct by another State precludes the wrongfulness of that act in relation to the consenting State, provided the consent is valid and to the extent that the conduct remains within the limits of the consent given. Validity of the consent must be assessed to ensure that it is freely given and clearly established. It must be actually expressed by the State rather than merely

---

[69]   Preliminary report on international liability for injurious consequences arising out of acts not prohibited by international law, by Mr. Robert Q. Quentin-Baxter, Special Rapporteur, A/CN.4/334 and Add.1 & Corr.1 and Add.2, reproduced in *ILC Yearbook* (1980) Vol. II (1) at 252.

[70]   Ibid.

[71]   *Gabcikovo-Nagymaros Project (Hungary/Slovakia)*, I.C.J. Reports 1997, at 39, para. 48.

presumed on the basis that the State would have consented if it had been asked. It must also not be vitiated by the influence of error, fraud, corruption or coercion.[72]

## 4. STATE JURISDICTION AND CONTROL OVER SPACE OBJECTS

The term 'jurisdiction' has been described as "the lawful power of a State to define and enforce the rights and duties, and control the conduct, of natural and juridical persons."[73] It is "the power of the state under international law to regulate or otherwise impact upon people, property and circumstances and reflects the basic principles of state sovereignty, equality of states and non-interference in domestic affairs."[74]

Eminent jurist Judge Manfred Lachs has defined jurisdiction as "a basic attribute of a State, whereby it exercises fundamental powers as a subject of international law."[75] He has qualified the limits upon the exercise of such jurisdiction as "determined by the rights of other States and the requirements of cooperation in international relations."[76]

This section contains a survey of the identical and uniform treatment bestowed on the twin concepts of 'jurisdiction and control' over space objects in international space law followed by some additional comments on related concepts such as ownership and registry of space objects.

### 4.1 JURISDICTION AND CONTROL

Article VIII of the Outer Space Treaty relates to jurisdiction and control over a space object by a State through launching of the space object. It provides that:

---

[72] ILC Articles on State Responsibility, Commentary to Art. 20, at 175

[73] Bernard H. Oxman, "Jurisdiction of States" in R. Bernhardt (ed.), *Encyclopedia of Public International Law*, Vol. 1 (Elsevier, 1992) at 55.

[74] Malcolm N. Shaw, *International Law*, 6th ed., (Cambridge University Press, 2008) at 645. Generally, see M. Akehurst, "Jurisdiction in International Law" (1972) 46 BYIL 145; F. A. Mann, "The Doctrine of Jurisdiction in International Law" (1964) 111 HR 1; F. A. Mann, "The Doctrine of Jurisdiction in International Law Revisited After Twenty Years" (1984) 186 HR 9; D. W. Bowett, "Jurisdiction: Changing Problems of Authority over Activities and Resources" (1982) 53 BYIL 1; I. Brownlie, Principles of Public International Law, 6th edn, Oxford, 2003, chapters 14 and 15; O. Schachter, *International Law in Theory and Practice*, Dordrecht, 1991, chapter 12; R. Higgins, *Problems and Process*, Oxford, 1994, chapter 4.

[75] Manfred Lachs, *The Law of Outer Space* (Leiden: Sijthoff Publishers, 1972) at 69.

[76] Ibid. Also, see Manfred Lachs, "The International Law of Outer Space" (1964) 113 RdC at 58.

"A State Party to the Treaty on whose registry an object launched into outer space is carried *shall retain jurisdiction and control* over such object, and over any personnel thereof, while in outer space or on a celestial body."[77] (emphasis added)

Some commentators have suggested a conceptual distinction between 'jurisdiction' and 'control' insofar as describing 'control' in terms of a separate technical function – "a separate concept, to mean not only observation (passive) but, in the first place, an obligation for the State of Registry, to active guidance of the space object; and a prohibition of interference with the space object by a third (non-Registry) State."[78] The Soviet authors have further expanded the concept to include "activities of special services of the State of Registry aimed at monitoring the technical condition of the space object during the launching and putting into orbit, as well as its functioning in outer space and during the landing."[79] It is unnecessary to dissect the twin concepts of 'jurisdiction and control' that have received identical and uniform treatment throughout international space law instruments. Hence, it has been rightly pointed out that "jurisdiction should induce control and control should be based on the jurisdiction."[80]

In the context of this discussion, it is important to simultaneously take into account the provisions of the Registration Convention because it is viewed as an attempt towards further elaboration of Article VIII of the Outer Space Treaty.[81] Article II(2) of the Registration Convention provides that:

"Where there are two or more launching States in respect of any such space object, they shall jointly determine which one of them shall register the object..., bearing in mind the provisions of article VIII of the [Outer Space Treaty], and without prejudice to appropriate agreements concluded or to be concluded among the launching States on jurisdiction and control over the space object and over any personnel thereof."

In order to exercise legitimate jurisdiction, it is essential for the State to identify a "sufficient nexus between itself and the object of its assertion of

---

77　Outer Space Treaty, Article VIII.
78　Bess C.M. Reijnen, *The United Nations Space Treaties Analysed* (Editions Frontieres, 1992) at 119.
79　Ibid.
80　Gabriel Lafferranderie, "Jurisdiction and Control of Space Objects and the Case of an International Intergovernmental Organisation (ESA)" (2005) 54 ZLW 228 at 231-232.
81　Registration Convention, preamble.

jurisdiction."[82] There is wide scholarly consensus that registration of space objects establishes such a link between the State and the space object.[83] In case a space object is not registered, it has been observed that ownership serves as the determining factor to ascertain which State could exercise jurisdiction and control.[84]

However, some authors do not consider registration as a "legal confirmation of ownership" or a "binding legal commitment of liability" on the ground that the State of registry may not be the launching State.[85] The State of registry has been defined in the Registration Convention as "a launching State on whose registry a space object is carried...."[86] It follows that the State of registry, therefore, has to be one of the launching States, that is, a State which launches or procures the launching of a space object or a State from whose territory or facility a space object is launched.[87]

In the wake of increasing international collaborative space ventures and private participation, the election of a State of registry among multiple launching States for the purpose of retention of jurisdiction and control is likely more complicated than it may appear. The State whose national is the owner of the payload/satellite will be more interested in acquiring legitimate jurisdiction and control rather than the State from whose territory/facility the launch had taken place. Although State practice with respect to the registration of space objects is sometimes sketchy and

---

[82] Bernard H. Oxman, "Jurisdiction of States" in R. Bernhardt (ed.), *Encyclopedia of Public International Law*, Vol. 1 (Elsevier, 1992), at 56. "The requisite contacts with a State necessary to support the exercise of jurisdiction differ depending on the nature of the jurisdiction being exercised."

[83] "Registration of space objects seem *ipso facto* to be sufficient to provide the link between these objects of international law and the subjects of international law." Stephan Hobe, "Spacecraft, Satellites and Space Objects" Max Planck Encyclopedia of Public International Law; "This link has a double intention. On the one hand, it assures to the spacecraft the protection by the State; on the other hand, the interests of third persons are protected by the fact that the State will be responsible for the spacecraft belonging to this State." I.H.Ph Diederiks-Verschoor, "Registration of Spacecraft" in E. McWhinney & M.A. Bradley (eds.), *New Frontiers in Space Law* (Leiden, 1969) at 125.

[84] "Failing registration, the act of launching and the ownership of such space objects seem to provide a sufficient link." Stephan Hobe, "Spacecraft, Satellites and Space Objects," ibid.

[85] Henry R. Hertzfeld & Ben Baseley-Walker, "A Legal Note on Space Accidents" (2010) 59 ZLW 230 at 233

[86] Registration Convention, Art. I(c)

[87] Liability Convention, Art. I(c); Registration Convention, Art I(a); Outer Space Treaty, Art. VII.

seemingly inconsistent, clarifying declarations by spacefaring States help to eliminate the ambiguities.[88]

From the above discussion, it is apparent that public international space law is silent about the legality of remediation when it relates to assuming or transferring legal jurisdiction and control of a particular space object. In the event of a remediation carried out by a State or a State licensed actor, it will be considered legitimate if the State retains *de jure* jurisdiction and control of that space object or obtains explicit authorization from the State of registry. Thus, no legal complications are anticipated when a State seeks to remediate its own space objects. However, when a State or State licensed actor seeks to remediate a space object that it did not carry on its registry, the question will arise whether there can be an exception to this general rule of jurisdiction and control on grounds of the public policy goal of facilitating space debris remediation to avoid orbital congestion and ensure long-term sustainability of outer space.

## 4.2   TRANSFER OF REGISTRATION

Neither the Outer Space Treaty nor the Registration Convention contains any provisions for the transfer of the registration of a space object. Consequently, this has generated extensive academic debate about the validity of such transfer agreements. The process of privatization of the International Maritime Satellite Organization (INMARSAT) had highlighted this issue.[89]

Before proceeding to examine this issue in greater detail, it is important to take note of the language in Article II of the Registration Convention,

---

[88]   Kenneth Hodgkins, U.S. Adviser to the 57th Sess, of the UN General Assembly, *International Cooperation in the Peaceful Uses of Outer Space*, Remarks on Agenda Item 75 in the Fourth Committee of the United Nations General Assembly, New York, 9 October 2002, online: http://2001-2009.state.gov/g/oes/rls/rm/2002/14362.htm. "We intend to include on the U.S. registry all space objects that are owned or operated by U.S. private or governmental entities whether launched from inside or outside U.S. territory. In general, the United States will not include on its registry non-U.S. payloads that are launched from U.S. territory or facilities. It is our view that such non-U.S. payloads should be included on the registry of the State of the payload's owner/operator because that State is best positioned to exercise jurisdiction and control. In addition, we will continue our practice of including certain non-functional objects on the U.S. Registry."

[89]   David W. Sagar, "The Privatization of Inmarsat" (1998) 41 Proc. of the Colloq. on the Law of Outer Sp.; David W. Sagar, "The Privatization of Inmarsat – Special Problems" (1999) Proceedings of the Third ESA/ECSL Colloquium on International Organizations and Space Law – Their Role and Contributions, Perugia, Italy.

which lays down that space objects can be registered by launching states only.

Several commentators have argued in favor of an amendment to the Registration Convention to resolve the challenges arising from transfer of registration of a space object. However, existing State practice demonstrates otherwise where non-launching States have successfully registered space objects over which they retain jurisdiction and control pursuant to Article VIII of the Outer Space Treaty. This was evident in the transfer of satellites registered in the United Kingdom to China as a consequence of the handover of Hong Kong in 1998.[90] This is consistent with Article II because it does not prohibit subsequent transfers of jurisdiction and control rights among launching States.[91]

However, the Registration Convention does not explicitly regulate subsequent transfers of jurisdiction and control rights to non-launching States. The *note verbale* submitted by the Netherlands to the UN COPUOS to register the transfer of ownership of satellites from New Skies Satellites is particularly interesting because it expressly renounces the status of the launching State or the State of Registry and consequently rejected its obligation to furnish information under Article IV of the Registration Convention. However, by virtue of the in-orbit transfer of ownership, it assumed international responsibility under Article VI of the Outer Space Treaty and also claimed the retention of jurisdiction and control under Article VIII of the Outer Space Treaty.[92]

It is also noteworthy that the principle of 'treaty stipulations in favor of third States' is well-recognized in customary international law. It allows States to enter into agreements conferring actual rights of their own to a third State, which can then exercise such a right upon compliance with the conditions of its exercise. It has been codified in Article 36 of the Vienna

---

[90] Information Furnished in Conformity with the Convention on Registration of Objects Launched into Outer Space, *Note verbale* dated 27 March 1998 from the Permanent Mission of the United Kingdom of Great Britain and Northern Ireland to the United Nations (Vienna) addressed to the Secretary-General, UN Doc. ST/SG/SER.E/333 – Notification of the removal of AsiaSat-1 (1990-030A), APSTAR-I (1994-043A), Asiasat-2 (1995-064A) and APSTAR IA (1996-039A) from national register effective 1 July 1997. Also see UN Doc. ST/SG/SER.E/334 for notification of addition of above named satellites to the register of the Hong Kong Special Administrative Region of the People's Republic of China effective 1 July 1997.

[91] Ricky J. Lee, "Effects of Satellite Ownership Transfers on the Liability of the Launching State" (2000) 43 Proc. Of Colloq. On Law of Outer Sp. 148.

[92] UN Doc. A/AC.105/806 (22 August 2003).

Convention on the Law of Treaties[93] and has been substantiated by international jurisprudence espoused by the world courts[94] and juristic opinion in favor of it.[95] Therefore, launching States may enter into specific agreements with non-launching States to lawfully transfer the right to jurisdiction and control over a space object.

The language in Article II of the Registration Convention unambiguously imposes a positive obligation on launching States to register the space object. However, in the event of transfer of ownership to a non-launching State, such a right to register the space object can be found in Article VIII of the Outer Space Treaty for domestic registrations and General Assembly Resolution 1721B (XVI)[96] for registration with the United Nations. Hence, this eliminates any need for an amendment of the Registration Convention and the transfer of 'jurisdiction and control' can be carried out under the existing framework of space law.

## 4.3  OWNERSHIP

Under the current legal regime, ownership of space objects is not co-extensive with the jurisdiction and control over such objects. Article VIII of the Outer Space Treaty states that:

---

[93]  Vienna Convention on the Law of Treaties, 1155 UNTS 331; 8 ILM 679 (1969), Article 36.
Treaties providing for rights for third States
1. A right arises for a third State from a provision of a treaty if the parties to the treaty intend the provision to accord that right either to the third State, or to a group of States to which it belongs, or to all States, and the third State assents thereto. Its assent shall be presumed so long as the contrary is not indicated, unless the treaty otherwise provides.
2. A State exercising a right in accordance with paragraph 1 shall comply with the conditions for its exercise provided for in the treaty or established in conformity with the treaty.

[94]  *Free Zones of Upper Savoy and the Dictrict of Gex (France v. Switzerland)* [1932] P.C.I.J. (ser. A/B), No. 46 at 147.

[95]  E Jimenez de Arechaga, "Treaty Stipulations in Favour of Third States" (1956) 50 Am. J. Int'l L. 338; M Fitzmaurice, "Third Parties and the Law of Treaties" (2002) 6 Max Planck YUNL 37; G Napoletano, "Some Remarks on Treaties and Third States under the Vienna Convention on the Law of Treaties" (1977) 75 Italian Ybk. Int'l L. 75.

[96]  International co-operation in the peaceful uses of outer space, UN GA Res. 1721 (XVI), online: United Nations Office of Outer Space Affairs, http://www.oosa.unvienna.org/oosa/SpaceLaw/gares/html/gares_16_1721.html. "Calls upon States launching objects into orbit or beyond to furnish information promptly to the Committee on the Peaceful Uses of Outer Space, through the Secretary-General, for the registration of launchings"

"Ownership of objects launched into outer space, including objects landed or constructed on a celestial body, and of their component parts, *is not affected* by their presence in outer space or on a celestial body or by their return to the Earth." (emphasis added)

While 'jurisdiction and control' is clearly geo-spatial in nature as it can be retained "while in outer space or on a celestial body," 'ownership' is in perpetuity as it "is not affected by their presence in outer space or on a celestial body or by their return to the Earth."[97] The law is silent about the temporal factor of 'jurisdiction and control' as to when can a State relinquish *de jure* jurisdiction and control. This is particularly important in cases when a State of registry has lost *de facto* control over a space object due to a technical anomaly which has rendered the space object non-functional and consequently, a potential target for remediation.

It is important to bear in mind that Article VIII of the Outer Space Treaty enjoins the State of Registry to retain its jurisdiction and control over the space object. More so, it cannot be abandoned after the expiry of its functional phase because Article VIII grants ownership in perpetuity, which ties the State of Registry to bear international responsibility and liability for any damage caused by its space object, pursuant to Article VII of the Outer Space Treaty, even though it is no longer operational or controllable.

While this provision has been alleged as an impediment towards space debris remediation activities,[98] it is, in fact, not an inhibiting factor as States can enter into separate agreements for the transfer of ownership of space objects as discussed in the preceding section. Thus, although international space law does not contain explicit provisions for the transfer of registry, public international law jurisprudence coupled with contemporary State practice have circumvented that lacuna through conclusion of bilateral or multi-lateral agreements. Therefore, it would be misleading to make an unequivocal assertion that space debris remediation activities are being thwarted by the 'ownership' clause in the Outer Space Treaty.

---

[97] Outer Space Treaty, art VIII.

[98] Matthew Schaefer, "Analogues between Space Law and the Law of the Sea/International Maritime Law: Can Space Law Usefully Borrow or Adapt Rules from These Other Areas of Public International Law?" (2012) 55 Proc. Of Colloq. On Law of Outer Sp.

## 5. CONCLUDING REMARKS

From the above discussion, it has been observed that public international law jurisprudence developed over the years can effectively resolve the unanswered questions arising from space debris remediation and principles from public international law can be relied upon to address the lacunae in the legal fabric of international space law.

The next step is for the international community, particularly the established space actors, to engage in discourse for developing State practice and legal and policy guidelines on space debris remediation. Given the lack of political will on the international level towards encouraging remedial activity, it might be prudent for the major space players to undertake unilateral action and also proactively encourage responsible space behaviour amongst their licensed private entities to expedite organizational and operational aspects of space debris remediation.

•••

## JOYEETA CHATTERJEE

 Joyeeta Chatterjee is a graduate of the Institute of Air and Space Law, McGill University, where she was awarded the Nicolas Mateesco Matte Prize for Outstanding Performance in Space Law and where she held the Erin J. C. Arsenault Fellowship in Space Governance. She is a winner of the prestigious Diederiks-Vershoor Award of the International Institute of Space Law. She is also an alumnus of the International Space University. She has gained professional experience during internships at Arianespace and the Indian Space Research Organization. She is a member of the Space Business practice team of the law firm Dentons at their New York office.

CHAPTER 10

# DIRECTED GENETIC MANIPULATION AND ENGINEERING:
## REGULATORY ISSUES FACING AN INCIPIENT INDUSTRY SHAPING TRANSHUMANS AND POST HUMANS IN SPACE

DR. GEORGE S. ROBINSON

A seemingly *de minimus* aspect of the United States and, indeed, international and future global space programs, is the evolving genetic nature of human*kind* anticipated for long duration and particularly permanent habitation in numerous Earth-alien and synthetic life support environments. This, in turn, requires a focus on the current and evolving genetic manipulation and genetic engineering laws faced by a variety of commercial businesses including universities and other non-profit entities, as well as private commercial operations.

The focus of the ensuing discussion is on the developing need for legal regulation addressing the biotechnological and the subsequently genetically engineered "transhumans" and "post humans" occupying off-Earth space. The developing capacity to manipulate the human genetic structure and, indeed, the entire human genome, has precipitated a rush to set guidelines and rules both for the type of research involved and its potentially directed efforts both for humans and other biotic species remaining on Earth, and for the projected increase in human*kind* population of near and deep space in synthetic life support systems, or those genetically engineered for certain "natural" environments in interstitial space and on other celestial bodies.

For the most part, genetic manipulation of the human Deoxyribonucleic Acid and Ribonucleic (DNA and RNA components of cellular structure) is in its early stages; primarily the subject of basic research in university and pharmaceutical laboratories. Nevertheless, if uncontrolled and without proper and sensitive standards of usage normally embraced in policy and regulatory measures, the inclination of various types of entrepreneurs can lead to abuses with irretrievably negative consequences for societies, civilizations, and the human species itself.

Further, the next evolving steps embrace biotechnological integration of *Homo sapiens sapiens* functioning in space, referred to by certain individuals as "transhumans," and then ultimately as "post humans" reflecting artificial intelligence *in extremis* and the ability to self-replicate and metabolize. This may well constitute a "species" or technological entity of a unique independence not subject to the jurisprudence and implementing positive laws of its predecessor creators; not subject to the principles of Andrew G. Haley's "metalaw" precept of "do unto others as they would have you do unto them" when confronting or interacting with alien life forms, regardless of their relative state of evolutionary development.

## A.  TRANSHUMANS AND POST HUMANS

Numerous definitions of "transhuman" and "transhumanism," and variations in the interpretations of those definitions, still embrace certain commonalities of central themes, values, and interests that paint the concept with a relatively distinct identity. Generally, "transhumanism"

asserts that *Homo sapiens sapiens*, or modern humans, both can and should
... perhaps *must* ... evolve and develop to higher levels of cognizance, of
sentience or abstract reasoning capability; perhaps as an evolutionarily
survival dictate. Transhumanism has been characterized as a philosophic
construct forcing human evolution.

As Nick Bostrum, a Swedish philosopher and professor at St. Cross
College, Oxford University and Director of the Future of Humanity
Institute, asserted fifteen years ago that, "'transhumans' may be considered
as the earliest manifestation of new evolutionary beings, on their way to
becoming post humans." Further, a "post human would no longer be a
human being, having been so significantly altered as to no longer represent
the human species."[1] The basis of this view is a belief that *Homo sapiens
sapiens,* i.e., modern humans, in their current biochemical form does not
represent the end of the species' evolutionary development. Rather, it's just
the beginning of the next step in biological and biotechnological evolution.
While we stand on the shoulders of our single cell ancestors, based upon
the patterns and unfolding lessons of biological and biotechnological
evolution, post humans will be a totally independent and separate species
from its evolutionary predecessor, i.e., *Homo sapiens sapiens.*

## B. THROWING AWAY ALL THE RULES

Humankind evolution through advanced artificial intelligence (and more
likely through directed human genetic engineering) may well result in
humans able to download an entire consciousness into a biotechnological
entity, thereby paving the way for a "post human." But the evolution both
of transhumans and post humans occupying near and deep space may well
take a totally different approach, i.e., a genetically manipulated, designed,
and engineered entity totally separate and distinct biologically from its
designer and manufacturer. And this is where the current concern injects
itself relating to design engineering of the human DNA/RNA and
ultimately genome, that puts a new species into the increasingly fuzzy fold
of taxonomy. This is where the growing awareness about advances in
genetic manipulation and directed design engineering of *Homo sapiens
sapiens* creates policy concerns and perceived regulatory needs relating to
potential commercialization of this selective methodology. Its potential

---

[1]  Bostrum, Nick. *The Transhumanist FAQ: A General Introduction.* 2003. Accessed
at: http://www.nickbostrom.com/views/transhumanist.pdf

and provocative applications to evolving humankind in a space ambience is rather clear.

In short, once again, the law generally, and commercially oriented space law in particular, has not caught up with technology. The rather unique implications for evolving humankind in space can either be helpful, or frighteningly destructive.


## C. GENE MANIPULATION AND DIRECTED GENETIC ENGINEERING:
## THE SEGUE FOR A NEW COMPONENT OF SPACE JURISPRUDENCE

Laws relating to transhumans and post humans as evolutionary occupants and settlers of space, i.e., space law, likely will find their genesis both in established and evolving domestic and international law relevant to the evolving genetic manipulation research and incipient medical applications. It is a growing commercial industry.

In focusing briefly on the genetic research activities, the results of which ultimately are likely to be applicable to long-duration and permanent humankind space inhabitants, basic definitions are important to understand the current and anticipated domestic and international laws that impact the various commercial entities, such as those researching and selling pharmaceutical remedies and behavioral modifications. For example, genetics and genomics are two terms frequently and incorrectly used interchangeably. Genetics relates to the study of single genes and their roles in the manner in which specific traits are passed from one generation to the next. On the other hand, genomics is the term used to describe the investigation of all parts of an organism's genes as a system of the complete set of DNA within a cell.

Genetic manipulation is also referred to as genetic modification, or genetic engineering. If conducted toward a specifically defined objective, the manipulation/modification is referred to as directed genetic engineering, i.e., the process of transforming genes of an organism toward a specific objective using modern biotechnological techniques. Genetic manipulation is used to control the results of gene activity. Genetic modification is the process of artificial isolation, manipulation, and reintroduction in order to manipulate coding and sequencing, etc., usually conducted outside the organism's natural reproductive process. Genetic engineering, also called genetic modification, is the direct human manipulation of an organism's genetic material in a way that does not occur under natural use of recombinant DNA techniques.

## D. REGULATING GENETIC MANIPULATION AND THE DIRECTED GENETIC ENGINEERING INDUSTRY

Genetic enhancement raises a host of ethical, legal and social questions. What is meant by normal? When is a genetic intervention "enhancing" or "therapeutic?" How should the benefit from a genetic enhancement be calculated in comparing its risks and benefits? As noted above, there has been speculation that genetic enhancement might affect human evolution. Philosophical and religious objections also have been raised, based on the belief that to intervene in such fundamental biological processes is "playing God" or attempting to place us above God. People from various perspectives believe that any interference with the random offerings of nature is inherently wrong, and question our right to toy with the product of years of natural selection.

Genetic engineering is, at best, a debatable branch of science. For some, it's the wave of the future: a method for perfecting the human genome, discarding flaws from infants before they're even born and ensuring they live longer, healthier lives. For others, it's an abomination: a method of circumventing what nature gave you, and wholly unnatural. Which one is right?

Genetic engineering can also be used to move humans above and beyond their normal restraints...into a wholly independent post human, such as might be anticipated for enhancing survivability of certain components of the human genome that embrace and perpetuate the "essence" of what it means to be human or human*kind*.

There are serious evolutionary policy considerations that must be addressed in law, particularly as they relate to genetically engineering a potentially new subspecies...or series of species, either self-perpetuating or newly created...for long duration and permanent off-Earth habitation in Earth-alien life support environments. Although they might have a positive effect, new organisms created by genetic engineering could present an ecological problem, even in an outer space environment.

The changes that a genetically engineered species would make on the environment of a region are unpredictable. Just like an exotic species, the release of a new genetically engineered species would also have the possibility of causing an imbalance in the ecology of a region, for example, on another planet where extraterrestrial contamination rules and agreements apply, and where the "Metalaw" principle (the Interstellar Golden Rule of "Do unto Others as They Would Have You Do Unto Them") of Andrew G. Haley might apply.

Genetic engineering will ultimately raise many additional concerns, ranging from ethical issues to a lack of knowledge about the effects genetic engineering may have. One major concern is that once an altered gene is placed in an organism, the process cannot be reversed. "Playing God" has become a strong argument against genetic engineering in a very immediate context. But despite all of these current concerns, the potential for genetic engineering of a post human species for long-duration and permanent space habitation is tremendous. Technologically competent societies may be ready to carry out directed genetic engineering, regardless of whether for transhumans and post humans intended for permanent space habitation, but they and the global community may not be ready for the jurisprudential responsibility it brings.

The right to personal genetic identity also is being addressed in the international legal arena; all of which is reflected in the issue of when does a genetically engineered post human in space assume a totally different persona and related jurisprudence distinct from that of its "manufacturers"?

## E. THE INTERNATIONAL LAW OF HUMAN GENETIC MANIPULATION: REGULATION OF THE HUMAN GENOME

The underlying objective behind the international law of human genetic manipulation and the regulation of the human genome is the protection of the genetic identity of both the human individual and the human species. Moving to the present and observing the current regulation of the human genome, two international instruments assume particular relevance: the Universal Declaration on the Human Genome and Human Rights (UDHGHR) and the Oviedo Convention on Human Rights and Biomedicine (Oviedo Convention).

The Oviedo Convention emphasizes the protection of the dignity and identity of all human beings, allowing only interventions that seek to modify the human genome for preventive, diagnostic, or therapeutic purposes. The UDHGHR, while stating in its first article that the human genome "underlies the fundamental unity of all members of the human family, as well as the recognition of their inherent dignity and diversity," contributes to the international legal framework by declaring the human genome as the Common Heritage of Mankind ... a singularly operative phrase in all current space treaties and conventions. These international instruments are committed to ensuring the preservation of the human species by defending it from scientific and technological practices that may violate its integrity and common identity. By preventing possible modifications to the human genome, these instruments seem to have the objective of ensuring that humans remain humans in light of scientific advancements.

The regulatory legal framework of the human genome is, nonetheless, deeply flawed and inconsistent, and it hides a number of problems and erroneous assumptions. The underlying general problem cutting across the flaws concerns the unresolved tension between individual and collective interests, namely the current imbalance between an alleged human species identity and every other individual identity. These flaws and problematic issues demonstrate how the right to personal identity has been undermined by a collective right to the preservation of an alleged human identity.

The situation is different in today's legal framework. The Oviedo Convention, although it explicitly refers to the protection of the integrity of human beings, does not enshrine a general right to genetic integrity. Along the same lines, the final version of the UDHGHR, despite some hesitation that occurs during its drafting phase, does not make any reference to the integrity of genetic heritage and limits itself to forbidding a restricted number of practices which are contrary to human dignity. This document, rather than focusing upon the idea of integrity, insists upon the link between human genome and dignity.

Nevertheless, international human rights law is not entirely coherent or uniform in this matter. While the Oviedo Convention explicitly precludes genetic interventions aiming at introducing any modifications in the genome of any descendants (art 13), the UDHGHR is slightly more flexible, as it does not explicitly prohibit interventions upon germ-line cells. The latter international instrument, in fact, "only" states that "practices which are contrary to human dignity, such as reproductive cloning of human beings, shall not be permitted."

Another important factor, present in the regulation of the human genome, that may undermine the right to personal identity is the declaration of the human genome as Common Heritage of Humanity. The HGHRD declared the status of the human genome as a "common" heritage to which we all have a claim. In this light, art 1 confirms that the "human genome underlies the fundamental unity of all members of the human family, as well as the recognition of their inherent dignity and diversity. In a symbolic sense, it is the heritage of humanity." Hence the query:

Will genetically independent post humans remain the genetic heritage of *Homo sapiens sapiens* at large?

## F. CONCLUSION

The idea of "species integrity," which is behind the value of non-intervention and preservation of the human genome, along with the conceptualization of the human genome as common heritage of humankind, the designation of humanity as a subject of rights and, ultimately, the humanity's collective right to identity, are all misleading legal constructions. In this regard, one should realize that species are not static collections of organisms, and that their genetic complexions shift evolutionarily across time and space. Scientifically speaking, the idea of integrity of the human species is, as a matter of fact, a slippery and relativistic concept ... certainly a constantly moving target for space law and space lawyers. Its immediate impact on commercial research and development leading to transhuman and post human spacekind should be obvious.

•••

## DR. GEORGE S. ROBINSON

Dr. Robinson has been in the private and public practice of law since 1963, and has taught or lectured in space law at numerous universities around the world. After serving at NASA and the Smithsonian Institution in Washington, DC, for thirty years, he retired into private law practice and concentrates primarily on space matters. Dr. Robinson has served on hospital boards of directors, as well as various science and space-related boards of trustees and governmental and private advisory committees. He earned his AB degree from Bowdoin College, an LL.B. degree from the University of Virginia School of Law, an LL.M. degree from the McGill University Graduate Law Faculty, Institute of Air and Space Law, and the first Doctor of Civil Laws degree in space law awarded by that Institute. He presently is in private practice with his two sons and daughter-in-law. He is also author of Chapter 4 in this volume.

# PART III

# NATIONAL PERSPECTIVES ON SPACE SUSTAINABILITY

## INTRODUCTION TO PART III

How do individual nations approach the issues and challenges of space sustainability?   Here we examine a wide variety of perspectives and policies from many nations, as well as the important theme of environmental responsibility and the ecological risks that space launches present to those on the ground.

CHAPTER 11

# NATIONAL SPACE POLICIES AND THEIR IMPORTANCE IN ENSURING THE LONG TERM SUSTAINABLE USE OF SPACE

VICTORIA SAMSON
SECURE WORLD FOUNDATION

Given the increase of new actors to and users of space, it is important that the space domain remain usable and reliable over the long-term, and that there is confidence in continued access to space assets and capabilities. Making publicly accessible a national space policy or strategy – some sort of written documentation from the national government that spells out national goals and priorities for space – is one way to undertake a transparency and confidence-building measure (TCBM). This demonstrates intentions and priorities for a national space program, and is included in several international initiatives currently underway to work toward the long-term sustainable use of space, such as the Group of Governmental Experts on space TCBMs, the draft International Code of Conduct, and the draft guidelines on the long-term sustainability of space from a working group by the United Nations Committee on Peaceful Uses of Outer Space (COPUOS). Five countries' space guidance are used here to illustrate the wide spectrum of forms that a national space policy/strategy can take: Australia, Brazil, China, Russia, and the United Kingdom. This TCBM can thus help a country advance its own space capabilities while at the same time working collaboratively with the international community to ensure a safe, secure, and sustainable space environment.


## I.    INTRODUCTION

Given the increasing democratization of access to space, the concurrent swelling of new actors to space (both in terms of nations with new space capabilities and new types of organizations manipulating space technologies), and the ways in which space is being more and more tightly woven into daily lives, economic development, and national security, it is fully apparent that the international community must come together to ensure a safe, secure, and sustainable space environment.

Part of this required stability rests upon reliable and predictable access to space assets, and confidence in the ability to use these space assets as warranted. Because the actions by one space actor can affect how others use space, it is important to demonstrate that one has good intent in regards to its space activities so not to alarm others.

Making publicly accessible a national space policy or strategy – some sort of written documentation from the national government that spells out national goals and priorities for space – is one way to go about doing this. This demonstrates intentions and priorities for a national space program, gives an idea of how much budgeting may go into a nation's space activities, raises transparency, and thus is a confidence building measure. It also forces a government to go through the process of having an intergovernmental discussion about priorities and goals for its space program, information which can then be used to inform national and international discussions.

Using the creation of national space policies as a transparency and confidence building measure (TCBM) can help establish the long-term sustainable use of the space environment, which is already apparent in several international initiatives working toward that goal. It can be seen in recommendations of the recent report of the Group of Governmental Experts (GGE) on space TCBMs, given to the United Nations Secretary-General in the summer of 2013. It is part of the draft International Code of Conduct on Outer Space Activities (CoC). In addition, there are parts of the draft guidelines on the long-term sustainability of space being developed by a working group by the United Nations Committee on Peaceful Uses of Outer Space (COPUOS) which delineate goals that could be met by the publication of national space policies or strategies.

Several active space stakeholders are highlighted to give examples of types of national space policies/strategies, including Australia, Brazil, China, Russia, and the United Kingdom, which together demonstrate the wide spectrum of options available to space stakeholders who may otherwise believe that the only way to demonstrate the intent and goals of a national space program is through an arduously and expensively built bureaucratic and legislative architecture.

## II.    SPACE POLICY AS A TCBM

### DEMONSTRATING GOOD INTENT

A satellite that conducts Earth observations could be used to help a country determine crop yields, and it could also be used to gather intelligence for military and security purposes. Due to this dual-use nature of space technologies, the mere possession of a capability cannot be used to determine what the stakeholder intends to do with that capability, as the technology itself cannot be the sole clue given for what a country's intentions are in space.

There are norms of behaviour that nations which mean no deliberate harm follow, and while deliberately misrepresenting what one intends to do on orbit is not necessarily illegal, it does raise the question about how trustworthy that actor is and if they are going to undertake activities that could negatively affect others' abilities to utilize and access space. In fact, merely neglecting to openly share what one intends to do with a technical capability leaves one open to speculation, which often leads to the worst, most threatening type of explanation.

For example, the United States' refusal to discuss what the X-37B space plane is doing or any specifics about it, have led others to believe, perhaps wrongfully, that it is intended to test out a space weapon capability that would be used against those who are not actively supporting U.S. interests internationally.

Alternatively, a published national space policy or strategy can help answer questions regarding intent for space capabilities, and possibly belay concerns about strategic intentions. Furthermore, by willingly sharing this information and participating in joint efforts to improve relations across various space stakeholders, a nation demonstrates that it recognizes that there are some responsibilities that come with space activities and that it is ready to face them, which in turn is a sign of good faith efforts on orbit. This TCBM can thus help to make the space environment a more predictable, reliable domain which will help lead to its long-term sustainability and peaceful use.

## INTERNATIONAL INITIATIVES AND SPACE POLICY AS TCBMs

While the idea of simply publishing a national space policy or strategy may appear basic, it has been nevertheless been included as a recommendation in several international initiatives attempting to help establish the long-term sustainable use of space.

## GGE ON SPACE TCBMs

The United Nations Group of Government Experts (GGE) on Transparency and Confidence-Building Measures in Outer Space Activities was first called for by Russia in 2010, and was then established formally by UN Secretary-General Ban Ki-Moon in 2011.[1]   The goal of the GGE was to create a consensus-driven report of recommendations on ways in which to improve the overall stability and security of space.   The GGE was comprised of 15 member states: the five permanent members of the UN Security Council, and then ten geographically diverse and representative nations selected by the UN (Brazil, Chile, Italy, Kazakhstan, Nigeria, Romania, South Africa, South Korea, Sri Lanka, and Ukraine).[2]   While the participating experts were chosen by the member states, they were expected to be neutral in their discussions of space TCBMs.

---

[1]   Christopher Johnson, *The UN Group of Governmental Experts on Space TCBMs: A Secure World Foundation Fact Sheet*, Secure World Foundation, updated April 2014, http://swfound.org/media/109311/swf_gge_on_space_tcbms_fact_sheet_april_2014.pdf

[2]   Johnson, updated April 2014.

The group met three times over the summers of 2012 and 2013 and delivered its report in July 2013, which then made its way to the UN General Assembly, which received and endorsed it in December 2013.[3]

The report noted that "the world's growing dependence on space-based systems and technologies and the information they provide requires collaborative efforts to address threats to the sustainability and security of outer space activities."[4] It went on to say, "Transparency and confidence-building measures can reduce, or even eliminate, misunderstandings, mistrust and miscalculations with regard to the activities and intentions of States in outer space."[5] The report urged "the development and implementation of voluntary and pragmatic measures to ensure the security and stability of all aspects of outer space activities."[6] Recommendation 37, called "Exchanges of information on the principles and goals of a State's outer space policy," explained that,

"States should publish information on their national space policies and strategies, including those relating to security. States should also publish information on their major outer space research and space applications programmes in order to build a climate of trust and confidence between States worldwide on military and non-military matters. This should be carried out in line with existing multilateral commitments. States may provide any additional information reflecting their relevant defence policy, military strategies and doctrines."[7]

## DRAFT INTERNATIONAL CODE OF CONDUCT ON OUTER SPACE ACTIVITIES

The draft International Code of Conduct (CoC) on Outer Space Activities is a non-legally binding document which intends to elucidate norms of behavior that space actors can voluntary agree to follow. These best

---

3   *Resolution Adopted by the General Assembly on 5 December 2013, Res. 68/50,* http://www.un.org/ga/search/view_doc.asp?symbol=A/RES/68/50

4   *General and complete disarmament: transparency and confidence-building measures in outer space activities: Group of Governmental Experts on Transparency and Confidence-Building Measures in Outer Space Activities, Res. 68/189, July 29, 2013, p. 2.* http://www.un.org/ga/search/view_doc.asp?symbol=A/68/189

5   *Group of Governmental Experts on Transparency and Confidence-Building Measures in Outer Space Activities,* July 29, 2013, p. 2.

6   *Group of Governmental Experts on Transparency and Confidence-Building Measures in Outer Space Activities,* July 29, 2013, p. 14.

7   *Group of Governmental Experts on Transparency and Confidence-Building Measures in Outer Space Activities,* July 29, 2013, p. 16.

practices have emerged throughout the years as actors continue to gain experience in space activities. The first draft of this document was released publicly for discussion in 2010. Discussion meetings were held in Vienna, Austria, in June 2012; Kiev, Ukraine, in May 2013; Bangkok, Thailand, in November 2013; and Luxembourg in May 2014. Negotiations continue on it to this day. The European Union, which started the process and has been largely responsible for shepherding it along, hopes to open it up for signature by interested parties by 2016.

The draft CoC opens up with the statement, "The purpose of this Code is to enhance the safety, security, and sustainability of all outer space activities pertaining to space objects, as well as the space environment."[8] It also calls for a "comprehensive approach to safety, security, and sustainability in outer space."[9]

Section 6.1 brings up national space policies:

"The Subscribing States resolve to share, on an annual basis, where available and appropriate, information with the other Subscribing States on: their space strategies and policies, including those which are security-related, in all aspects which could affect the safety, security, and sustainability in outer space; their major outer space research and space applications programmes; their space policies and procedures to prevent and minimise the possibility of accidents, collisions or other forms of harmful interference and the creation of space debris; and efforts taken in order to promote universal adoption and adherence to legal and political regulatory instruments concerning outer space activities."[10]

## Draft LTS Guidelines

COPUOS's Scientific and Technical Subcommittee (STSC) began an initiative called the Long Term Sustainability of Outer Space Activities (LTS) Working Group in 2010. Its goal was to create a consensus report of best practices for space actors, coming at it from a bottoms-up approach, in order to build toward the safe and sustainable use of outer space over the long term. Four expert groups have examined different aspects of space

---

[8]   *Draft International Code of Conduct for Outer Space Activities*, VERSION 31 March 2014, http://eeas.europa.eu/non-proliferation-and-disarmament/pdf/space_code_conduct_draft_vers_31-march-2014_en.pdf

[9]   *Draft International Code of Conduct for Outer Space Activities*, VERSION 31 March 2014.

[10]  *Draft International Code of Conduct for Outer Space Activities*, VERSION 31 March 2014.

sustainability [Editor's note: Please see Chapter 4 for a detailed discussion of the four expert groups.]

1. Sustainable space utilization supporting sustainable development on Earth;
2. Space debris, space operations, and tools to support space situational awareness sharing;
3. Space weather; and
4. Regulatory regimes and guidance for new actors.[11]

Initially they had hoped to have agreed-upon guidelines to present to the COPUOS plenary session – which meets every June in Vienna, Austria – by 2014, but now it looks like it will not be completed until 2016.

Various expert groups have put together a series of draft LTS guidelines that chair Peter Martinez of South Africa is in the process of combining and correcting for redundancies as he waits for any more input by interested states. The current draft of LTS guidelines points out that "implementation of national and international frameworks for space activities not only provides assurance to users of the space environment, but also facilitates bilateral and multilateral cooperation in the peaceful uses of outer space and thereby contributes to the safety and stability of outer space."[12]

Along those lines, draft guideline 9 recommends that nations, "Adopt national regulatory frameworks suitable for space activities that provide clear guidance to actors under the jurisdiction and control of each State."[13] Draft guideline 10 suggests that states, "Encourage advisory input from affected national stakeholders in the process of developing, refining and implementing national regulatory frameworks governing space activities."[14] Interestingly, the guideline goes on to explain:

---

[11] Tiffany Chow, *UNCOPUOS Long-Term Sustainability of Space Activities Working Group Fact Sheet*, Secure World Foundation, updated June 2013, https://www.gov.uk/government/publications/civil-space-strategy-2012-to-2016

[12] *Proposal for a draft report and a preliminary set of draft guidelines of the Working Group on the Long-term Sustainability of Space Activities*. United Nations Committee on the Peaceful Uses of Outer Space. Vienna: United Nations Office for Outer Space Affairs. Retrieved Sept. 21, 2014, from http://www.oosa.unvienna.org/pdf/limited/c1/AC105_C1_L339E.pdf, p. 4.

[13] *Proposal for a draft report and a preliminary set of draft guidelines of the Working Group on the Long-term Sustainability of Space Activities*, p. 8.

[14] *Proposal for a draft report and a preliminary set of draft guidelines of the Working Group on the Long-term Sustainability of Space Activities*, p. 9.

"By allowing early advisory input, the State can avoid unintended consequences of regulation that have an adverse impact on key stakeholders... States with developing space capabilities should identify the essential components of a national regulatory framework after advisory input from, or consultation with, relevant stakeholders."[15]

"In instances in which the State has not previously attempted to legally control or regulate space activities, the State may wish to consider other States' space legislation or, by analogy, other national laws, as a guide to drafting."[16]

This recommendation is something for new actors to keep in mind if they opt to follow this TCBM.

Finally, draft guideline 12 builds on the above by recommending, "When adopting or implementing national regulatory frameworks, consider the long-term sustainability of outer space activities."[17]

## III. SELECTED NATIONAL SPACE POLICIES AND STRATEGIES

Some space actors are reluctant to put together a publicly accessible space policy or strategy, as they worry that it would limit their freedom of action or be prohibitively expensive to prepare. Nevertheless, there are a range of responses that national governments have undertaken in an attempt to create written guidance for their national space programs, and this section will examine several different types of national space policy/strategies of space stakeholders.

### AUSTRALIA
Australia has adopted a creative way to guide governmental action regarding space. The Australian government does not control any satellites, and the few Australian satellites which exist are commercial

---

[15] *Proposal for a draft report and a preliminary set of draft guidelines of the Working Group on the Long-term Sustainability of Space Activities,* p. 9.

[16] *Proposal for a draft report and a preliminary set of draft guidelines of the Working Group on the Long-term Sustainability of Space Activities,* p. 9.

[17] *Proposal for a draft report and a preliminary set of draft guidelines of the Working Group on the Long-term Sustainability of Space Activities,* p. 10.

communications satellites. But the country depends heavily on space assets and is using its unique geographic location to increase its involvement in space activities with allies.

After many years of discussion, the Australian government released its Satellite Utilisation Policy in April 2013. It notes that, "The purpose of this policy is to articulate Australia's space interests and objectives, identify existing and emerging opportunities and Australia's competitive advantages, and prepare the nation to meet future challenges effectively."[18] The policy points out that Australia benefits from international capabilities and thus it should contribute where possible, namely in "ground infrastructure and in the application of space information to achieve cost-effective outcomes."[19]

It spells out Australia's national goal in space, which is to, "Achieve on-going, cost-effective access to the space capabilities on which we rely," and explains the five key benefits to Australia in doing so: "improved productivity;" "better environmental management;" "a safe and secure Australia;" "a smarter workforce;" and "equity of access to information and services."

The policy then explains principles for achieving this goal:

1. The first is to "focus on space applications of national significance," namely, Earth observations, satellite communications, and position, navigation, and timing.
2. The second is to "assure access to space capability," including access to radiofrequency spectrum.
3. The third is to "strengthen and increase international cooperation."
4. The fourth is to "contribute to a stable space environment," including supporting norms of behaviour, supporting international regulatory frameworks for space, and become involved in international fora where these issues are being discussed.
5. The fifth is to "improve domestic coordination,"
6. The sixth is to "support innovation, science, and skills development."
7. The seventh and final principle is to "protect and enhance national security and economic well-being."[20]

---

[18] Commonwealth of Australia, Australia's Satellite Utilisation Policy, April 16, 2013, http://www.space.gov.au/Documents/Australia%27s%20satellite%20utilisation%20policy%20-%20version%201.1p%20-%2016%20April%202013.pdf, p. 1

[19] Australia's Satellite Utilisation Policy, April 16, 2013, p. 1.

[20] Australia's Satellite Utilisation Policy, April 16, 2013, pp. 6 – 16.

## BRAZIL

Brazil's national space efforts are guided by its National Program for Space Activities (PNAE) for the Brazilian Space Agency. The PNAE is currently in its fourth iteration, and covers the years 2012–2021.

The PNAE indicates its underlying interest in using space for national development with the following statement: "A country's sovereignty and autonomy are proportionally related to its capacity for technical development. Space technology is undoubtedly the most far reaching in this scenario."[21]

It highlights Brazil's wish to be more involved in space industry in order to "drive industrial progress" and asks what it calls the "developed world" if Brazil could "cooperate with joint technological development, mutual interests and shared benefits."[22]

Eight strategic guidelines are given to try to meet the PNAE's priority of using space to enhance Brazil's industrial sector:

1. Consolidate the Brazilian space industry by increasing its competitiveness and innovation capacity, and also through the use of the State's purchasing power and the partnerships with other countries.
2. Develop an intensive program of critical technologies in order to foster the capacity building in the space sector, with greater participation of academia, S&T governmental institutions and the industry.
3. Expand partnerships with other countries, by prioritizing joint development of technological and industrial programs of mutual interest.
4. Encourage funding of programs based on public and/or private partnerships.
5. Promote greater integration of the space activities governance system in the country, by increasing the synergy and effectiveness of actions among its main players and the creation of the National Space Policy Council, conducted directly by the President of the Republic.
6. Improve the legislation to strengthen space activities, by encouraging and facilitating government purchases, allocating

---

[21] Brazilian Space Agency, *National Program of Space Activities: PNAE: 2012-2021.* Brasilia: Ministry of Science, Technology, and Innovation, 2012, p. 3.

[22] *National Program of Space Activities: PNAE: 2012-2021*, 2012, pp. 7, 5.

more funds for the Space Sectorial Fund, and decreasing taxes to the industry.

7. Encourage the human resources development by training of experts needed in the Brazilian space activities, both domestically and abroad.

8. Promote public awareness on the relevance of the study, use and development of the space activities in Brazil.[23]

The PNAE goes on to identify priority actions for the country to meet its goals, and suggests strategic actions that could help support the Brazilian space industry.

It includes a section specifically on international cooperation on space issues, noting that,

"For us, space cooperation in the fully globalized world of the twenty-first century is much more than a business transaction, it is about promoting joint scientific, technological and industrial development, with trusted partners, based on mutual interest, common effort and sharing benefits."[24]

The PNAE also gives an estimate for how much it will cost to carry out the programs and initiatives suggested by it over the timeframe of 2012-2021: "R$ 9,1 billion, with 47% allocated to satellite mission projects, 17% to space access projects, 26% to space infrastructure and 10% to other special and complementary projects."[25]    [Editor's note: the current exchange rate is approximately 4 Brazilian Reals to one US Dollar, giving an approximate budget of US$2.25 billion.]

## CHINA

In 2000, China released a white paper detailing its five-year plan for its space program. This was updated in 2006 and again most recently in 2011, to guide China's space activities through 2016.

The 2011 white paper begins by noting the new opportunities to China's space program afforded to it by industrial development, but carefully states, "China will work together with the international community to maintain a peaceful and clean outer space and endeavor to

---

[23] *National Program of Space Activities: PNAE: 2012-2021*, 2012, p. 8.
[24] *National Program of Space Activities: PNAE: 2012-2021*, 2012, p. 14.
[25] *National Program of Space Activities: PNAE: 2012-2021*, 2012, p. 16.

make new contributions to the lofty cause of promoting world peace and development."[26]

Purposes given for China's space industry are:

1. Space exploration;
2. Use space for "peaceful purposes" in order to "benefit the whole of mankind;"
3. "Meet the demands of economic development, scientific and technical development, national security and social progress;" and
4. To "improve the scientific and cultural knowledge of the Chinese people, protect China's national rights and interests, and build up its national comprehensive strength."[27]

The white paper sets goals for space applications, primarily Earth observation, communication/broadcasting satellites, and position, navigation, and timing capabilities; space science, Sun-Earth space exploration, lunar scientific research; and space debris mitigation. Major tasks for the next five years include developing "a comprehensive plan for construction of space infrastructure, promoting its satellites and satellite applications industry," undertaking more scientific space research, and continuing the "comprehensive, coordinated and sustainable development of China's space industry."

It also lists some efforts the Chinese government is undertaking to promote Chinese space industrial development, including:

"Strengthening legislative work. To actively carry out research on a national space law, gradually formulate and improve related laws, regulations and space industrial policies guiding and regulating space activities, and create a legislative environment favorable to the development of space activities."[28]

The white paper also has a long section on international exchanges and cooperation. It starts off with the statement that,

---

[26] People's Republic of China, China's Space Activities in 2011, Information Office of the State Council, December 2011, http://news.xinhuanet.com/english/china/2011-12/29/c_131333479.htm
[27] China's Space Activities in 2011, December 2011.
[28] China's Space Activities in 2011, December 2011.

"The Chinese government holds that each and every country in the world enjoys equal rights to freely explore, develop and utilize outer space and its celestial bodies, and that all countries' outer space activities should be beneficial to economic development, the social progress of nations, and to the security, survival and development of mankind."[29]

It goes on to assert that international cooperation should "promote inclusive space development on the basis of equality and mutual benefit, peaceful utilization and common development."[30]

Fundamental policies to guide international cooperation include, "Supporting activities regarding the peaceful use of outer space within the framework of the United Nations," focusing on Asia-Pacific space cooperation, and even more broadly, other regional space cooperative efforts, strengthening cooperative efforts with developing countries, and "appropriately using both domestic and foreign markets and both types of resources, and actively participating in practical international space cooperation."[31]

International exchanges and cooperation are intended to focus on scientific research; using Earth observation satellites for environmental, disaster, and climate change monitoring; applications of communication satellites; cooperating on a space lab and space station; and commercial satellite launch services.[32]

## RUSSIA

Russia's national space activities are currently guided by its Federal Space Program 2006-2015,[33] which was most recently was updated in December 2012.[34] This Program reportedly increased the budget for remote sensing and communications satellites by double.

In January 2013, Roscosmos released a draft document called "Space Activities of the Russian Federation in 2013-2020," which reportedly included a space strategy through 2030.[35] According to Yuri Koptev, head

---

[29]  *China's Space Activities in 2011*, December 2011.
[30]  *China's Space Activities in 2011*, December 2011.
[31]  *China's Space Activities in 2011*, December 2011.
[32]  *China's Space Activities in 2011*, December 2011.
[33]  No official English language version of Russia's space documents exists, to the author's knowledge.
[34]  Zak, Anatoly. Russian space program: a decade review (2010-2019), http://www.russianspaceweb.com/russia_2010s.html, last updated Aug. 28, 2014.
[35]  Zak, *Russian space program: a decade review (2010-2019)*.

of the Rostekh state corporation general director's advisor team and ex-head of RosAviaCosmos, "The key priority is the development and use of means supporting socioeconomic development, defense capacities of the country and issues of daily life of individuals and the entire society."[36] Space access and applications were reportedly the top priorities listed, and the program reportedly set the goal of increasing Russia's portion of the global space industry "from 10.7 percent in 2011 to 14 percent in 2015 and 16 percent in 2020."[37] This draft was updated in April 2013.

As of this writing, an update of the Federal Space Program is being developed that would cover the timeframe 2016-2025, and according to Roscosmos head Oleg Ostapenko, "[W]e are completing the stage of coordination and it will be submitted for approval in the nearest future."[38]

## UNITED KINGDOM

The United Kingdom space strategy is comprised of two parts: a Civil Space Strategy, which was published in July 2012; and a National Space Security Strategy, which was published in April 2014.

The civil space strategy set out some of the roles and responsibilities of the relatively new UK Space Agency, which was created in April 2011. It also discussed five ways for the UK space industry to grow:

1. Through new opportunities via new industries and markets, domestically and internationally;
2. From export, with a stated goal of having the UK get to have 10 percent of the global market by 2030;
3. Innovation supporting growth;
4. Science to underpin growth; and
5. Growth "through smarter government."[39]

Interestingly, the strategy points out, "Regulation can be used as a tool to establish a competitive edge in the international arena," since it could "create an environment which attracts inward investment and encourages

[36] "Russian space industry development aims to solve acute state tasks – expert," *Russia & CIS Defense Industry Weekly*, Jan. 18, 2013.

[37] Zak, *Russian space program: a decade review (2010-2019)*.

[38] "Russia to allocate 8.3bn dollars on International Space Station - deputy premier," *BBC Monitoring Former Soviet Union – Political*, Sept. 23, 2014.

[39] United Kingdom Space Agency, Civil Space Strategy 2012-2016, July 10, 2012,. https://www.gov.uk/government/uploads/system/uploads/attachment_data/file/2862 19/uk-space-agency-civil-space-strategy.pdf, pp. 8-18.

industry to develop new systems and services in the UK."[40]  Very often, industry sees regulation – potential or actual – as a way by which to stymie its efforts, increase bureaucracy, and cut into its profits. Overall, the strategy sets the goal of the UK Space Agency heading up the UK civil space efforts in order to "ensure that our central goal of growth becomes a reality and the potential of space to the twenty-first century economy will be both recognised and realised."[41]

The National Space Security Strategy was published in order to protect "the provision of vital services for our economy and national security."[42]  It hopes to do so via four objectives:

1. To make the United Kingdom more resilient to the risk of disruption to space services and capabilities, including from space weather;
2. To enhance the United Kingdom's national security interests through space;
3. To promote a safe and more secure space environment;
4. To enable industry and academia to exploit science and grasp commercial opportunities in support of national space security interests.[43]

The Strategy defines space security as "having safe, assured and sustainable access to space capabilities, with adequate resilience against threats and hazards."[44]  It also calls for the United Kingdom to work with international partners when possible.  The GGE, COPUOS, and CoC are all mentioned as part of its efforts to work toward a safe and secure space environment.[45]

## IV. CONCLUSION

In order to ensure that current space stakeholders and future users can benefit from space over the long-term, it is important to lay the foundations

---

[40] *Civil Space Strategy 2012-2016*, July 10, 2012, p. 18.
[41] *Civil Space Strategy 2012-2016*, July 10, 2012, p. 20.
[42] United Kingdom, National Space Security Strategy, April 30, 2014, https://www.gov.uk/government/uploads/system/uploads/attachment_data/file/3076 48/National_Space_Security_Policy.pdf, p. 2
[43] *National Space Security Strategy*, April 30, 2014, p. 4.
[44] *National Space Security Strategy*, April 30, 2014, p. 7.
[45] *National Space Security Strategy*, April 30, 2014, pp. 16-17.

for approaching space in a cooperative manner now. Actively undertaking TCBMs can help ensure that space is used responsibly and make the space domain a predictable one that can be depended upon.

Not all TCBMs have to be arduous or invasive. Establishing a national space policy/strategy and making that guidance available publicly can do much to demonstrate good intent for national space activities. It can also help analysts get a better sense of what a country is intending to do in space, particularly if budget numbers are released as well. This can be done in an unclassified form, and not at the cost of a state's national security. The process of going through the creation of a national space policy/strategy (and updating it when warranted) is helpful to enable others to ascertain national priorities, and sheds some light on domestic discussions of budgetary and programmatic decisions.

There are several concurrent international initiatives that are striving to enhance the long-term sustainable and peaceful use of space. The GGE on space TCBMs, the draft CoC, and the draft LTS guidelines all include a recommendation that suggests putting together a national space policy/strategy.

By examining the guidance that Australia, Brazil, China, Russia, and the United Kingdom have given their space programs we can see that there are a wide variety of policy and communication options available to space stakeholders, and that the creation and publication of a national space policy/strategy can be individualized to best suit a state's needs. This TCBM can thus help a country advance its space capabilities while at the same time work with the international community to ensure a safe, secure, and sustainable space environment.

•••

## VICTORIA SAMSON

Victoria Samson is the Washington Office Director for the Secure World Foundation and has nearly twenty years of experience in military space and security issues. Before joining SWF, she served as a Senior Analyst for the Center for Defense Information, where she focused on missile defense, nuclear reductions, and space security issues. Prior to that, she was the Senior Policy Associate at the Coalition to Reduce Nuclear Dangers, a consortium of arms control groups in the Washington, DC area. She previously served as a researcher at the Riverside Research Institute, where she worked on war-gaming scenarios for the Missile Defense Agency's Directorate of Intelligence. Ms. Samson holds a Master of Arts in international relations from the Johns Hopkins Paul H. Nitze School of Advanced International Studies, and a Bachelor of Arts in political science with a specialization in international relations from the University of California, Los Angeles.

CHAPTER 12

# ESA CLEAN SPACE INITIATIVE

JESSICA DELAVAL
CLEAN SPACE COORDINATOR FOR ESA

## 1. THE BENEFITS OF SPACE ACTIVITIES

Space is a strategic asset of fundamental importance for the independence, security and prosperity of Europe. This is well recognised through decisions on the European Space Agency's (ESA) programmes and activities that are primarily grouped according to their contribution to three axes of priorities already identified in 2012:[1]

   (i) Pushing the frontiers of knowledge, by increasing the understanding of our planet and the Universe;

---

[1] Resolution on the role of ESA in sustaining competitiveness and growth in Europe (ESA/C-M/CCXXXIV/Res. 1 (Final)).

(ii) Supporting an innovative and competitive Europe, thus bringing significant impact on Europe's growth and employment; and

(iii) Enabling services, by providing indispensable technologies and services for the knowledge society.

Furthermore, space is an enabling tool that provides Europe's decision-makers the ability to address global challenges such as the preservation of the environment, studying and mitigating climate change, and enhancing security. Equally important, space provides inspiration for humankind, and draws young people to scientific and technical education and careers.

## ADDRESSING GLOBAL CHALLENGES

A reflection on the objectives, contents and optimal selection of space programmes and activities for Europe beyond 2020, when framed within the overall geopolitical context, and taking into consideration the major global trends, has been conducted throughout 2013 and 2014 by ESA in consultation with Member States and with the support of the Bureau of European Policy Advisors (BEPA) and of the Organisation for Economic Co-operation and Development (OECD).[2] This activity has also gathered the views of many relevant stakeholders in Europe's space sector, including the scientific community, industry and operators.

Already today space has developed into a new global economy that reaches out to nearly all sectors of daily life, and the trend is strong towards a growing integration of technologies enabling "space integrated services for society." All modern countries depend increasingly on technologies and infrastructures including space infrastructures, and integration of space development into the innovation efforts and processes of even firms that are not directly in the space sector is increasing. This is having significant impact in the areas of knowledge developing, innovation, and economic competitiveness.

In the coming decades the world is expected to become more densely populated with an increasing demand for limited resources, and will certainly be paying more attention to climate change and to the preservation of the environment, and will also experience increasing requirements concerning communication technologies, mobility and

---

[2]  The Global Economy in 2030: Trends and Strategies for Europe, Centre for European Policy Studies (CEPS) for the European Union, 2013; The Space Economy at a Glance 2014, OECD, 2014.

security, all of which will inevitably involve further integration of space into economic, scientific, cultural, and political affairs of all nations.[3]

Integrated space systems play a key strategic role in this, notably as the only mechanism that can provide the kind of services needed for the sustainable development of very precious ecosystems, such as the Arctic and oceans, which are under stress from climate change and rapid development and overexploitation. Additional challenges are emerging and are expected to further develop in the domains of security of energy, food and water supplies, as well as of cyber security.

While space-based solutions are instrumental in providing essential capabilities, especially in terms of observation, navigation and communications and also involving robotics, life support and technology, a coordinated approach is required to effectively manage the many interactions that exist among such challenges.

## GUARANTEEING THE FUTURE OF SPACE ACTIVITIES

A noteworthy global trend is the increasing awareness towards the environmental impacts of space activities and, in particular, with respect to the necessity of preserving the Earth's orbital environment as a safe zone free of debris as a prerequisite to guarantee a sustainable use of space.

Since 2012, Clean Space is one of ESA's initiatives that directly addresses, through the Agency's technology programmes, some key technological challenges of the 21st century, in particular, sustainability of space activities on Earth and in orbit.

The Clean Space initiative has enabled ESA to become a global pioneer on the sustainable use of space, in particular by bringing a systematic approach that addresses the entire lifecycle of the various Agency's space activities, including its own as well as those performed by European space industry in the frame of ESA programmes from the early stages of conceptual design to a mission's end-of-life, as well as by preparing active debris removal.

## 2. THREATS TO THE BENEFITS

### IN-ORBIT COLLISION RISKS AND ON-GROUND CASUALTY RISKS DUE TO SPACE DEBRIS

Today's space debris environment poses a safety hazard to operational spacecraft as well as a hazard to safety of persons and property in cases of

---

[3]   According to the findings of both BEPA and OECD.

uncontrolled re-entry events. It is worth underlining that our knowledge on the current space debris environment and its evolution is essential to enable the evaluation of the non-compliances to mitigation efforts and the effectiveness of the different mitigation measures.

The Inter-Agency Space Debris Coordination Committee (IADC)[4] has defined two protected regions about the Earth, which have been adopted by the United Nations.[5] The first is the Low Earth Orbit (LEO) protected region which extends from the lowest maintainable orbital altitude up to a height of 2,000 km above the surface of the Earth. The second is the Geosynchronous Orbit (GEO) protected region, which includes the volume of space bounded in altitude by +/- 200 km of the geosynchronous altitude (35,786 km) and in inclination by +/- 15 degrees.

LEO is the most highly congested region in near-Earth space, containing approximately 75% of all catalogued objects. For objects too small to catalogue, the population levels are even greater. For instance, the number of debris between 1 and 10 cm is assessed to be several hundred thousand, and the number between 1 mm and 1 cm is assessed to be in excess of 100 million.[6]

As reported by the IADC to the 50th Session of the Scientific and Technical Subcommittee of the United Nations Committee on the Peaceful Uses of Outer Space (UN COPUOS STSC) in 2013, since 2005 some IADC members have been assessing the stability of the LEO space object population and the need to use active debris removal (ADR) to stabilize the future LEO environment, under a variety of space debris mitigation scenarios.[7] ASI, ESA, ISRO, JAXA, NASA, and UKSA,[8] employed their own environment evolution models under a common set of initial conditions and assumptions, reaching very similar results. The study confirmed the instability of the current LEO object population and that compliance with existing national and international space debris mitigation measures will not be sufficient to constrain the future LEO object population. To stabilize the LEO environment, more aggressive measures, especially the removal of the more massive non-functional spacecraft and

4    The IADC is an international forum of space agencies for the worldwide coordination of activities related to the issues of man-made and natural debris in space (www.iadc-online.org).
5    Space Debris Mitigation Guidelines of the United Nations Committee on the Peaceful Uses of Outer Space (UN COPUOS, 2010).
6    Space Debris, IADC Assessment Report for 2011 (IADC-12-06, April 2013).
7    Stability of the Future LEO Environment (IADC-12-08, Rev. 1, January 2013).
8    The six IADC members that have been principal participants in the study on Stability of the Future LEO Environment.

launch vehicle stages, should be considered and implemented in a cost-effective manner.[9]

The objects in the GEO region are much less numerous than in LEO and reside in about seven times the spatial volume of LEO, but this unique region is the home to more than 400 operational communications and other spacecraft which serve vital purposes for all countries of the world. The total number of catalogued objects in or near the GEO region is in excess of 1,100. The current number of estimated objects as small as 10 cm in or near the GEO region is on the order of 3,000. Only rough estimates are available for debris smaller than 10 cm at these high altitudes.[10]

## REGULATIONS

The increased attention to safeguarding Earth's orbital environment is well reflected by a number of relevant regulations that are being set forward. ESA actively commits to limiting causes of harmful interference in space activities, in particular, through the adoption of ESA's "Space Debris Mitigation Policy for Agency Projects" in March 2014, which includes:

- Preventing fragmentation by carrying out the power and propulsion system passivation of all space assets in-orbit after the end-of-mission.
- In the densely populated LEO environment, limiting the residence time of space objects in altitudes below 2000 km to 25 years after the end-of-mission. Furthermore, for those space objects whose on-ground casualty expectancy in the event of an uncontrolled re-entry exceeds one in 10,000 a controlled re-entry over an un-populated area is mandatory.
- For GEO, a post mission disposal by re-orbiting to a graveyard orbit beyond the geostationary ring.

This Agency-level policy complements national and international regulations on the matter, such as France's Space Operations Act and the above mentioned Space Debris Mitigation Guidelines of the UN COPUOS.

## ENVIRONMENTAL IMPACT OF SPACE ACTIVITIES

The growth in public awareness of environmental impact of human activities has stimulated a generalised call for environmental friendly

---

[9] IADC Presentation to 50th UNCOPUOS STSC (IADC-13-01, February 2013).
[10] Space Debris, IADC Assessment Report for 2011 (IADC-12-06, April 2013).

policies. As a result, implementation or adherence to such policies is beginning to provide a commercial advantage in some industries.

The space sector is not an exception to that global trend. Since 2007, ESA has been performing a systematic process of reflection on a sustainable development strategy, with a view to becoming a 'leading sustainable development organization.' It is worth mentioning the Agency's "Framework Policy on Sustainable Development," presented in March 2010, as an opportunity to align ESA's existing practices on best standards for internal operations but also for projects in core-business activities with the most innovative or qualitative solutions in the sustainability field.

As environmental impact is becoming more and more a criterion for product selection, comprehensive measurement and communication of the environmental impacts is needed for the organisation to consistently quantify, assess and improve its environmental behaviour. The ECOSAT pilot experience in 2009 examined this newly added design dimension: ESA initiated the study of the applicability of environmental design philosophies and assessment methodologies to space systems design, notably by addressing the design of a study case mission, throughout the different processes of its life cycle, under the light of eco-design methodologies. The Life Cycle Assessment (LCA) methodology[11] was then applied to better understanding the environmental impacts of launchers and their sources.

## THE REGULATORY PERSPECTIVE

Environmental laws, rules and regulations are rapidly evolving and becoming more stringent, particularly within the European Union (EU). As prime examples, the EU directives and regulations – RoHS[12] and REACh[13] – have considerable implications for European space activities.

In this evolving regulatory context, ESA, with the Clean Space initiative, aims at supporting European space industry in the timely development of compliant technologies with respect to both space debris mitigation (SDM) and environmental requirements, including minimising

---

[11]   Further detail in Chapter 3.
[12]   Directive 2008/35/EC of the European Parliament and of the Council of 11 March 2008 amending Directive 2002/95/EC on the use of certain hazardous substances in electrical and electronic equipment (RoHS 1) and Directive 2011/65/EU of the European Parliament and of the Council of 8 June 2011 on the restriction of the use of certain hazardous substances in electrical and electronic equipment (RoHS 2).
[13]   EU Regulation on Registration, Evaluation, Authorisation and Restriction of Chemicals.

possible disruption of qualified materials and processes in the European supply chains.

Furthermore, early adoption of eco-friendly approaches such as eco-design and technologies such as green propellants and green manufacturing will contribute to fostering European industry innovation and competitiveness. The use of resource-efficient clean technologies is also expected to save costs by decreasing material inputs, energy consumption and waste. [Editor's note: Please see also the following chapter for a detailed discussion of green propellants.]

## 3. Reaching for Sustainability: Technical Solutions

"Clean technologies for space" are defined by ESA as those that contribute to the reduction of the environmental impact of space programmes, taking into consideration the overall life-cycle and the management of residual waste and pollution resulting from space activities, both in the Earth eco-sphere and in space.

The Clean Space Initiative organizes the technical solutions around four distinct theme-based "branches" for which technology development roadmaps have been prepared and discussed with European stakeholders, notably through dedicated workshops and presentations including the ADR Workshop (2012) in The Netherlands; 6th European Conference on Space Debris (2013) in Germany; Clean Space workshop (2013) in the United Kingdom. European space industry has demonstrated keen interest all through the consultation process, which includes coordination with the European Commission on relevant issues within EU's FP7 and Horizon 2020 Programmes.

The focus during the first three years of activities, until 2015, has been placed on technology developments, with a view to its future systematic use in ESA programmes, as well as to the preparation of a mission for active debris removal. Progress has been made through the selected stepwise implementation approach on all four branches, although with a non-uniform pace across them – a result partially influenced by external factors.

As identified in 2012, the four theme-based Clean Space branches are:
1. Eco-design: The development of tools to monitor and evaluate programmes' environmental impact and compliance with legislation.

2. Green technologies: The development and qualification of new technologies and processes to mitigate the environmental impacts of space activities.

3. Space debris mitigation: The study and development of affordable technologies required for managing the end-of-life of space assets.

4. Technologies for space debris remediation: The study and development of the key technologies for active debris removal.

BRANCH 1: ECO-DESIGN

The objective is to develop the tools to support space projects in assessing their environmental impact and in monitoring compliance with the environmental regulations such as REACh, as well as mitigating the risk of industrial supply chains disruptions. As a longer term goal, tools will be developed to aid future projects with eco-design considering the environmental aspects during the conception of a project, in order to make decisions to reduce the overall environmental impacts of a project.

The global awareness of the need to monitor environmental impacts has increased rapidly, and European industry has pioneered the implementation of environmental regulation and eco-design approaches in various sectors. In the European space sector, ESA has been acting as the focal point to coordinate efforts on Life Cycle Assessment (LCA), in particular carrying out LCA studies on the European launcher family and several complete space missions with industry. Following from this, LCA methodologies and tools specifically adapted to space activities are being developed,[14] with the objective of making them available to all European stakeholders. A part of the ISO 14000 environmental management standards, LCA is instrumental in the assessment of the environmental impact of space activities including benchmarking with terrestrial applications, and fundamental to meeting the increasing requests from institutional and commercial actors, including governmental regulators, for manufacturers and operators to report on the environmental impact of their systems. Furthermore, LCA provides key inputs for eco-design that

---

[14] ESA's Concurrent Design Facility (CDF) ECOSAT study, in 2009, ECOSAT was a pilot experience on eco-design applied to space systems and assessed both Cradle-to-Cradle and Cradle-to-Grave methodologies. This study was the first attempt to use LCA techniques to analyse end-to-end mission life cycle. Later, a complete LCA of the current European family of launchers was carried out, with the involvement of most of the main industrial firms in the European launchers production chain. This endeavour has been further expanded to encompass the whole life-cycle of a space mission.

supports the identification of environmental hot-spots and stimulates innovation in the definition of new solutions.

Already in progress is the gathering of know-how on the monitoring of the impact on space activities of the European environmental legislation, together with the setting up of a specific LCA database for space activities to be made available to all European stakeholders. The process aims at identifying the use of hazardous materials in particular, according to the REACh regulation, and any relevant actions to decrease obsolescence risks or supply disruptions. Several initial studies about the replacement of endangered technologies including chromates free coatings are under way.

Further progress will be made in the following years on the areas of LCA standardisation and a REACh database for space applications, R&D on less hazardous materials, and additional studies such as combustion impacts on the atmosphere, possibly including measurements of Ariane 5 and Vega plumes, to assess ozone depleting substances and their impact through all layers of the atmosphere, and to validate the relevant specific models.

## BRANCH 2: GREEN TECHNOLOGIES

The objective is to develop green technologies allowing:

1. A reduction of the energy consumption across the life-cycle of a space mission;
2. A more sustainable use of resources, in particular by minimising the use of harmful substances for human health and bio-diversity (as identified in Branch 1 activities);
3. A pro-active commitment to environmental legislation (e.g. REACh and RoHS), by anticipating the development of compliant technologies and mitigating the risk of industrial supply chain disruption; and
4. The sustainable management of the waste and polluting substances associated to space activities.

It is worth underlining that the development of green technologies constitutes a significant support to innovation and competitiveness by promoting the streamlining of production processes, with a reduction of overall cost as a result from the more sustainable use of resources and the reduction of energy consumption and waste.

During the first three years of activities (i.e. until 2015), the focus has been on the areas of:

1. Green propulsion (e.g. mono-propellant and bi-propellant alternatives to hydrazine and hydrazine derivatives, which are included in the REACh list of substances of very high concern), with the significant outcome of a green propulsion roadmap being prepared within the Neosat programme[15] in the Telecommunications domain;

2. Advanced manufacturing (friction stir welding, and additive manufacturing (e.g. 3D printing); and,

3. New environmentally friendly materials and processes, which are addressed along three main axis, developing alternatives to materials and substances (e.g. chromates free coatings), improving safety in the manufacturing processes and ground operations (e.g. replacement of nitric acid with citric acid for steels passivation), and fostering innovative developments (e.g. bio-composite structures).

Substantial progress can be noted in the domain of additive manufacturing, which has attracted a keen interest in the European space industry. Taking stock from the comprehensive overview of European activities and some specific technology developments that are in progress, it will be possible to assess the best approach for the application of these technologies in the domain of satellite manufacturing.

As expected, innovation is created when different disciplines meet. Out of the green technologies branch of Clean Space, and from exploratory activities in the metallurgy and material domains, ESA has decided to start a new cross-cutting technology in the area of innovative manufacturing, which is revolutionising the traditional approach to parts design, development and verification.

### BRANCH 3: SPACE DEBRIS MITIGATION

The objective is to develop technologies for the systematic compliance of ESA missions, both spacecrafts and launchers, with debris mitigation requirements, including re-entry or parking in graveyard orbits and passivation. Branch 2 also aims at a continuous improvement of debris modelling and understanding.

Regarding re-entry requirements, a state-of-the-art LEO spacecraft with a mass above 0.5 ton is known to have parts that may survive re-entry.

---

[15] Neosat, element 14 of ESA's Advanced Research in Telecommunications Systems programme (ARTES), aims at developing, qualifying and validating in orbit next-generation satellite platforms for the core satcom market.

Consequently, a direct and controlled re-entry of the spacecraft at its end of life (i.e. de-orbit) is required, and to that end specific systems need to be developed and qualified.

While controlled de-orbit is likely necessary for spacecraft with a mass above 1.5 to 2 ton, the high associated cost both in terms of complexity and of additional spacecraft's mass may drive spacecraft with otherwise a mass below 1.5 ton out of the light-lift launchers such as Vega class launch capability, thus implying a significantly increased mission launch cost. For spacecraft with a mass below 1.5 ton, an alternative to de-orbiting is to design the system so that it contains no parts large enough to pose a risk on ground. This alternative approach is called Design for Demise (D4D).

A series of concurrent engineering sessions on D4D, open to external industrial experts, were organised by ESA in September 2013. These sessions, highly valued by industrial participants, contributed to streamlining the technology activities in this domain. As a result, more than fifteen technology activities on D4D are currently being implemented within the Agency's technology programmes.

When considering the impact that the systematic compliance with space debris mitigation requirements has on a number of sub-systems, it appears that an evolution of the LEO platforms is necessary.

As LEO is becoming more and more a commercial arena, manufacturers that have recurrent platforms and supply chains can attain an important competitive advantage. The development of technologies compliant with SDM requirements, through a coordinated European approach and the use of common building blocks provided by common supply chains compatible with different manufacturers' platforms, is expected to lead to economies of scale. This approach can only be implemented by engaging the interest of space agencies, system integrators and subsystem suppliers in a coordinated European effort.

ESA is grouping the activities linked to debris mitigation in a proposal called *CleanSat* that will encompass the evolution of the LEO platforms for compliance with SDM requirements in such a coordinated European effort. The CleanSat initiative aims at providing industry with an efficient framework to bring innovative products and systems into the marketplace, encompassing the necessary LEO platform evolutions and their associated technologies. In such a way, CleanSat embodies Europe's response to worldwide market demand for SDM compliant solutions through a new generation of LEO platform.

For the new platform products to gain market acceptance, the integration and demonstration of the building blocks in a representative mission is considered to be a key factor. In particular, CleanSat is targeting upcoming Earth observation missions. In this respect, it is worth noting that ESA has already established successful partnerships with industry through initiatives that foster technology development while stimulating new markets, notably in the domain of Telecommunications such as Alphasat,[16] Neosat, and Electra.[17]

Several other technology activities including power passivation, propulsion passivation, drag augmentation devices, and solid motor de-orbiting kits are being pursued in strong coordination with users and, in particular, in the domain of Earth observation.

For instance, reliable measurement of on-board propellant is expected to enable successful disposal of a spacecraft, optimising its operational life-time. The development of accurate propellant gauging devices is key to perform the end-of-life operations efficiently, as well as of robust and reliable passivation concepts to guarantee the proper passivation at end-of-life without increasing the risk during the operational lifetime. The passivation procedures applicable to the currently operated spacecraft and upper-stages are also to be dealt with.

## BRANCH 4: TECHNOLOGIES FOR SPACE DEBRIS REMEDIATION

The objective is to develop technologies for space debris rendezvous, capture and re-entry in order to place European industry in the forefront position on anticipated future active debris removal (ADR) markets. Adopting a system approach, technology developments are being focused around a mission that targets the controlled de-orbit of a heavy object.

Since 2012, ESA's active space debris removal mission called *e.Deorbit* has been developed in a stepwise approach, through system level activities complemented by a comprehensive technology maturation programme. Several activities are ongoing, in particular:

---

[16] Alphasat is the first flight opportunity for the Alphabus platform, a high-power platform that gives European industry a unique position in the world telecom market. As the biggest public–private space project ever made in Europe, Alphasat was primarily designed to expand telecommunications provider Inmarsat's existing global mobile network, and also hosts four technology demonstration payloads developed through ESA's Advanced Research in Telecommunications Systems (ARTES) programme.

[17] Electra is the first project under ESA's ARTES 33 PARTNER programme. Electra will support European satellite industry in developing, launching and validating in orbit a full electric-propulsion telecommunications satellite in the 3-tonne launch mass range.

- At system level, the e.Deorbit Phase A was completed in July 2013 through three parallel industrial studies led by Airbus Defence & Space, Thales Alenia Space and Kayser-Threde, and the Preliminary Requirement Review was performed in September 2014.
- Technology activities, streamlined according to the above mentioned system level studies, addressing key technical issues such as:
  - Guidance, navigation and control: Image recognition, rendezvous algorithms, de-orbit control with rigid or flexible link between chaser and target.
  - Capture mechanisms: Design, breadboard and test of mechanisms such as clamping mechanism, robotic arm gripper, net and harpoon.
  - Target characterization: Target attitude motion measurement campaign and model validation.

As CleanSat with respect to SDM requirements, e.deorbit aims at placing European industry at the forefront of the worldwide space debris remediation effort through the development and implementation of cutting-edge technology in areas such as uncooperative rendezvous and formation flight, capture and control of large uncooperative objects or adaptive guidance, navigation and control. Furthermore these developments are expected to pave the way to the creation of new services and markets such as in-orbit servicing.

The outcome of the technical studies and activities in progress will be further consolidated and will form the basis of a proposal for a e.deorbit programme for a decision at the horizon 2016, according to what is already planned in the ESA Long-Term Plan 2014-2023.


## 4. CONCLUSION

Space is a strategic asset of fundamental importance for the independence, security and prosperity of Europe. Already today, space has developed into a new global economy that reaches out to nearly all sectors of daily life and the trend is for a growing integration of technologies enabling "space integrated services for society," with significant impact on the areas of knowledge, innovation and global competitiveness. In particular, space integrated systems are called to play a key role as the only mechanism that can provide the kinds of services needed for the sustainable development of very precious ecosystems, such as Arctic and oceans, which are under stress from climate change and rapid development and overexploitation. Furthermore, additional challenges are emerging and are expected to further develop in the domains of security of energy, food and water supplies, as well as of cyber security.

A noteworthy global trend in recent years is the increasing awareness of the environmental impacts of space activities, and in particular with respect to the necessity of preserving the Earth's orbital environment as a safe zone free of debris as a requisite to guarantee a sustainable use of space. In this context, Clean Space is one of ESA's initiatives that directly addresses, through the Agency's technology programmes, some key technological challenges of the 21st century – in particular, sustainability of space activities on Earth and in orbit.

The Clean Space initiative has enabled ESA to become a global pioneer on the sustainable use of space, in particular by bringing a systematic approach that addresses the entire lifecycle of the various Agency's space activities including its own as well as those performed by European space industry in the frame of ESA programmes from the early stages of conceptual design to the mission's end-of-life, as well as by preparing active debris removal.

Clean space roadmaps, which group technology activities according to themes, are implemented through different ESA programmes following a stepwise approach that aims at supporting European space industry to effectively use resources, implement regulations and mitigate risks.

With the Clean Space initiative, ESA aims to turn challenges into an opportunity by preparing European space industry to face the threats related to the impact of space activities on the environment and to debris, thus ensuring a competitive advantage through the timely development of technologies compliant to evolving regulations, as well as resource-efficient clean technologies which contribute to reduce costs by decreasing material inputs, energy consumption and waste.

Clean Space organizes the technical solutions around four distinct theme-based "branches" for which technology development roadmaps have been prepared and discussed with European stakeholders, notably through dedicated workshops and presentations.

European space industry has demonstrated keen interest all through the consultation process, which includes coordination with the European Commission on relevant issues within EU's FP7 and Horizon 2020 Programmes.

During the first three years of activities from 2013 to 2015 the focus has been on technology development with a view to its future systematic use in ESA programmes, as well as to the preparation of a mission for active debris removal. Progress has been made through the selected stepwise implementation approach on all four branches, although with a non-uniform pace across them – a result partially influenced by external factors.

At present, ESA is grouping the activities linked to debris mitigation in a proposal called CleanSat that will encompass the evolution of the LEO platforms for compliance with SDM requirements in a coordinated European effort. The CleanSat initiative aims at providing industry with an efficient framework to bring innovative products and systems into the marketplace, encompassing the necessary LEO platform evolutions and its associated technologies. In this way, CleanSat embodies Europe's response to worldwide market demand for SDM compliant solutions through a new generation of LEO platform.

In addition, e.Deorbit is intended to place European industry at the forefront of the worldwide space debris remediation effort through the development and implementation of cutting-edge technology in areas as uncooperative rendezvous and formation flight, capture and control of large uncooperative objects or adaptive guidance, navigation and control. These developments are expected to pave the way to the creation of new services and markets such as in-orbit servicing.

•••

ESA Clean Space website: www.esa.int/cleanspace
Follow Clean Space on Twitter : @ESAcleanspace

## JESSICA DELAVAL

Jessica Delaval is Clean Space Coordinator for ESA, the initiative which aims at guaranteeing the future of space activities by protecting the environment, both on Earth and in space. The goal of Clean Space is to assess the environmental impacts of space activities, develop greener technologies and tackle the space debris issue by developing technologies for mitigating and remediating them.

CHAPTER 13

# CREATION OF A SUSTAINABLE SPACE DEVELOPMENT BOARD, AND OTHER INITIATIVES FOR ENVIRONMENTALLY SUSTAINABLE LAUNCH ACTIVITIES

## OLGA ZHDANOVICH, MSc
SECRETARIAT OF THE EUROPEAN COOPERATION FOR SPACE STANDARDISATION

## INTRODUCTION

The Space Age began in 1957 with the Soviet Union's launch of Sputnik, the first artificial satellite. Since that significant event, many historic achievements have been accomplished: scientific probes landing on other planets in our solar system, human beings walking on the surface of the Moon, space stations orbiting the Earth, and space telescopes scanning the Universe. As we embark on the 21st century, the number of countries with space launch capabilities continues to grow, and emerging countries are committing considerable resources to develop their space programs (*eco*Space executive summary, 2010).

In 1987 the term "sustainable development" was introduced in the UN report "Our Common Future" for the very first time. The report was prepared by the World Commission on Environment and Development with Ms. Gro Harlem Brundtland, former prime minister of Norway, as the Head of the Commission. Since that time this report is known as "Brundtland report." It says, *"Humanity has the ability to make development sustainable to ensure that it meets the needs of the present without compromising the ability of future generations to meet their own needs"* (Bruntland, 1987).

Today, environmentally sustainable practices are being adopted by individuals, states, corporations, and international organizations throughout the globe. Historically, space agencies and the space launch industry have focused on performance and cost as the main criteria for selecting propulsion systems, but unfortunately, in the decision-making process the environmental impact of these systems usually has been neglected.

If the current number of space launches per year remains stable, worldwide launch activities are not expected to have a significant impact on global pollution levels when compared to other industries. However, on a local scale, every rocket launch injects hundreds of tons of chemical by-products into the atmosphere, land, and sea over a period of just minutes.

The toxic nature of some propellants requires the use of elaborate and costly handling procedures to mitigate the risk they pose to human health and the environment. In addition, launch failures and accidents in the manufacturing, handling, and storage of propellants can have a detrimental environmental effect and social impact on local communities. Permanent dropping of rocket stages with remnants of fuel together with launch failures may leave the future generations of local populations affected by rocket launches with significant health deterioration and a polluted and toxic environment.

This chapter presents the findings of the one of the three Team Projects conducted during the Space Studies Program (SSP) of the International Space University held in Strasbourg, France in 2010. Team *eco*Space was comprised of 38 ISU SSP participants from 15 countries, one Teaching Associate and two Chairs.[1] Team members had the common interest in promoting space launches that are sustainable and environmentally friendly, and produced the final report and executive summary "ecoSpace. Initiatives for Environmentally Sustainable Launch Activities."

**Figure 1**
**ecoSpace Executive Summary**
ISU 2010

Here we present the final recommendations of the *eco*Space report as a set of initiatives to promote sustainability of space launches and the use of green propellants. These initiatives are followed by a brief summary of the research conducted by the *eco*Space Team during ISU SSP 2010.

## 1. INITIATIVES DEVELOPED BY THE *ECO*SPACE TEAM

The concept of environmental sustainability has been around for 23 years, but it was only recently that space agencies began adopting sustainable policies to regulate space activities.

For example, NASA developed an Environmental Management System (EMS) based on the ISO 14001 standard to address environmental issues in 2009. These EMS procedures were developed to provide great flexibility to individual NASA Centers while meeting the National

---

[1] A detailed list of the authors is included in the report: "*eco*Space. Initiatives for Environmentally Sustainable Launch Activities," ISU, 2010.

Environmental Policy Act (NEPA) requirements and executive orders. Similar to NASA's EMS, the French CNES recently selected the ISO 14001 standard to certify its launch site in Kourou, French Guiana. ESA initiated sustainable practices by creating a Coordination Office on Sustainable Development and started the Clean Space Program, which evaluate environmentally friendly concepts, life cycle analysis, and green technologies for the launch industry. In 2008, the Russian Space Agency developed the Environmental Safety Standard for Space Technologies. The Russian government established a special environmental system to monitor air, water, and soil quality and to assess human health for regions under the launch corridors and risk zones.

However, while environmental sustainability practices are developing at space agencies, the transition to greener technologies by the launch industry is occurring slowly. Thus, Team *eco*Space proposed the following set of initiatives to accelerate the introduction of green technology in the launch industry (*eco*Space report 2010).

## PROMOTE RESEARCH IN GREEN PROPELLANTS

Team *eco*Space recommends that space agencies consider adapting their procurement policies to promote the research and development of green propellants and environmentally sustainable practices within the launch industry. *Eco*Space Team believes that it is important for space agencies to provide resources to stimulate the research and development of green propellants, and in addition, space agencies should play a key role in promoting the use of green propellants for commercial launches, particularly for suborbital flights, so that the industry will progress along an environmentally sustainable path.

## ENVIRONMENTAL MONITORING AND METRICS FOR THE LAUNCH INDUSTRY

Team *eco*Space recommends the development of a monitoring system with environmental metrics to measure the impact of the launch industry on the environment and human health. The environmental impact of the launch industry can be documented using these metrics. These metrics can be used to identify best practices for a sustainable launch industry.

## ADOPT AND PROMOTE BEST PRACTICES THROUGH OUTREACH AND PUBLIC AWARENESS

*Eco*Space recommends that space agencies develop outreach and public awareness materials to educate space professionals and the general public. The outreach program for space professionals should highlight the efforts in developing and adopting green propellants and environmentally sustainable practices. The collaboration between space agencies and the private sectors would aid in the promotion and adoption of best practices in the launch industry. Human health and safety, as well as environmental concerns, must be taken into consideration at every stage of the rocket launch life cycle. Inclusion of these matters in a risk management analysis is crucial.

## INTERNATIONAL AGREEMENT FOR SPACE LAUNCH SUSTAINABILITY

All space faring nations and nations with space activity and space ambitions should ratify the existing international environmental treaties. It would be beneficial if these nations came together to draft an international agreement for space sustainability, which seeks to protect the Earth environment while nations undertake space activity. In addition, ecoSpace recommends that all space agencies adopt an environmental management system with external third-party audits to ensure compliance of national environmental regulations and implementation of best practices that are environmentally sustainable.

## CREATION OF THE SUSTAINABLE SPACE DEVELOPMENT BOARD (SSDB)

At the international level, all of the initiatives mentioned above can be implemented through the Sustainable Space Development Board (SSDB), an international interagency body coordinating, supporting and promoting sustainable space activities. This interagency body could develop environmental policy for space activity and create standards for environmental and health protection for adoption by the space launch industry

The Sustainable Space Development board should be comprised of senior representatives from industry, academia, and government, and should focus on the environmental sustainability of the launch sector.

The EcoSpace Team estimates that the board can be established over a three year period:

- The first year will focus on gathering support by space agencies, experts, and the launch industry;
- The second year will be dedicated to endorsements by senior management from space agencies and the establishment of an infrastructure to support the board;
- And the third year will be used to implement policies, regulations, or guidelines agreed upon by the board members.

The SSDB should hold annual meetings and forums to encourage collaboration among space agencies, academia, and the launch industry. It should also have an outreach division to promote the sustainable space initiatives.

## EXPAND THE SCOPE OF ICAO TO INCLUDE SPACE ACTIVITIES

Team ecoSpace proposes adopting the recommendations in the IAASS White Book  (IAASS, 2009) to extend the role of International Civil Aviation Organization (ICAO) to encompass commercial space activities, which includes suborbital flights. ICAO has a proven framework for regulating the aviation industry.  Since suborbital flights start from the ground, enter airspace, and transition to outer space, Team ecoSpace believes it is logical for ICAO to regulate commercial launch activities as well.

## 2.  BACKGROUND RESEARCH BY THE ECOSPACE TEAM

To produce its final report, the ecoSpace Team made research during SSP'10 on a wide range of interlinked topics, including the launchers and propellants used, overview of green propellants that are under development, space policy and law of protection of the environment and population affected by launch activities, economics of introduction of green propellants into space industry, as well as public outreach activities to promote sustainable and environmentally friendly launch activities.  In addition the Team received the advice of experts from the space industry, academia, space agencies and ISU Faculty.

Table 1 on the following page presents main terms used in the ecoSpace report.

| Term | Definition |
|------|-----------|
| Sustainability | As outlined in the World Commission on Environment and Development Report, "Humanity has the ability to make development sustainable to ensure that it meets the needs of the present without compromising the ability of future generations to meet their own needs." (Bruntland, 1987) |
| Green Propellants | The term green propellant is a general name for a family of chemical propellants that can be either liquid or solid. Green propellants need to satisfy requirements like low toxicity, low pollution impact, good storability, wide material compatibility, and good performance. |
| Green Technology | Green technology is the application of environmental science to preserve the environment while minimizing the production of pollution and risks to the ecology and human health. The concept of sustainability is central in the development of green technology. |
| Space Launch Industry and Space Activity | "Space launch industry" or "space activity" stand for organizations dealing with the manufacturing, transporting, handling, and storage of rocket propellants, and launch operations for orbital and sub-orbital flights.* |

* Since the ecoSpace Team did not address space debris in the report, the environmental impact of the launch industry is limited to launch operations and dropped stages from multistage rockets. Sounding rockets are not included in the original report because the ecoSpace Team assumed that the pollution from these rockets are negligible.

**Table 1**
**Key terms used in the ecoSpace report**
(ecoSpace, 2010)

## 2.1 LAUNCH VECHICLES AND PROPELLANTS

This section is a brief overview of launch vehicles, propellants used and perspective green propellants that are now under development.

### 2.1.1 LAUNCH VEHICLES

Space launch vehicles can be classified as expendable, fully reusable, and partially reusable. Most launch vehicles are expendable, as their components are discarded after launch. A fully reusable launch vehicle is described as one that can be recovered and re-launched without major refurbishment, but this has not currently been developed. A partially reusable vehicle has a mixture of components of reusable and disposable parts. The Space Shuttle and Pegasus are the closest examples of a fully reusable launch vehicle.

Launch vehicles can be classified by payload lift capacity, propellants used and number of stages employed. The complexity of a launch vehicle increases with the number of stages. Current research trends focus on the development of one or two stage launch vehicles. For a single stage launch vehicle to be as efficient as a multistage launch vehicle, engine performance needs to be improved significantly. Table 2, below, provides a description of the examples of current launch vehicle configurations and typical engine performance (SpaceLaunch Report 2010). Rocket propulsion is characterised by two parameters: 1) Thrust, as amount of force applied to the launch vehicle due to the expulsion of propellant, and 2) Efficiency of propulsion, very often characterised by specific impulse that shows the thrust when burning 1 kg per second of specified propellant (Keys to Space, 1999).

| Rocket | Country | Propellant (B=Booster, S1=Stage1, S2=Stage2, S3=Stage3, U=Upper Stage) | Thrust in Vacuum (Tons) | Specific Impulse in Vacuum (s) |
|---|---|---|---|---|
| Delta II | USA | B: HTPB/AP/AL | 64 | 278 |
| | | S1: Kerosene/LOX | 110 | 301 |
| | | S2: Aerozine (50%UDMH and 50%/$N_2O_4$) | 4 | 319 |
| | | S3: HTPB/AP/AL | 6 | 292 |
| Atlas V | USA | B: HTPB/AP/AL | 127 | 275 |
| | | S1: Kerosene/LOX | 428 | 337 |
| | | S2: $LH_2$/LOX | 10 | 450 |
| Soyuz-FG | Russia | S1: Kerosene/LOX | 100 | 319 |
| | | S2: Kerosene/LOX | 99 | 319 |
| | | S3: Kerosene/LOX | 29 | 325 |
| | | U: UDMH/$N_2O_4$ | 2 | 331 |
| Angara A5 | Russia | S1: Kerosene/LOX | 213 | 337 |

| Rocket | Country | Propellant (B=Booster, S1=Stage1, S2=Stage2, S3=Stage3, U=Upper Stage) | Thrust in Vacuum (Tons) | Specific Impulse in Vacuum (s) |
|---|---|---|---|---|
| (under development) | | S2: Kerosene/LOX | 120 | 367 |
| | | U: UDMH/$N_2O_4$ | 2 | 326 |
| Ariane 5 ECA | Europe | B: HTPB/AP/AL | 650 | 275 |
| | | S1: $LH_2$/LOX | 112 | 432 |
| | | S2: $LH_2$/LOX | 6.5 | 446 |
| Long March 3B | China | S0: UDMH/$N_2O_4$ | 82 | 289 |
| | | S1: UDMH/$N_2O_4$ | 333 | 289 |
| | | S2: UDMH/$N_2O_4$ | 84 | 298 |
| | | S3: $LH_2$/LOX | 16 | 440 |
| Long March 5 (under development) | China | S1: Kerosene/LOX | 273 | 336 |
| | | S2: Kerosene/LOX | 140 | 432 |
| | | S3: $LH_2$/LOX | 16 | 440 |

**Table 2**
**Main Launch Vehicles, Propellants and Engine Performance**
(ecoSpace 2010, Valencia Bel 2014)

Sub-orbital launchers is a class of vehicles that is fundamentally different from orbital launchers. These launch vehicles are currently under development and should be available for commercial flights within the next few years. Compared to orbital launchers, suborbital launchers do not need to achieve the 7.8 km/s velocity necessary to stay in orbit. Virgin Galactic's spacecraft, SpaceShipTwo (Virgin Galactic, 2009), will reach a velocity of about 1 km/s, which is less than 1/50th of the energy required to go into orbit. Suborbital launchers could be useful to evaluate new propellants, as they do not require as much fuel as orbital launchers to achieve their mission (ecoSpace 2010).

### 2.1.2 PROPELLANTS

A propellant is an energetic chemical that converts its chemical energy into thermal energy inside of an engine chamber. The selection of a propellant is based on a tradeoff between propulsion performance characteristics and hazard potentials, cost, and storability.

The propulsion system is required for generating sufficient thrust for a launch vehicle to overcome Earth's gravity and atmospheric drag, and achieve orbital velocity. Most launch systems use a combination of solid and liquid propellant, usually with the main launcher using a liquid propulsion system and the booster, which provides additional thrust, using a solid propulsion system.

According to the study made in 2010 by Ferran Valencia Bel (ESA), around 38,202 tonnes of propellants were consumed during the year in the world (see Figure 2): Russia consumed around 14,838 tonnes of rocket fuel, the US around 12,035 tonnes, Europe (without Russia) 5,009 tonnes and China around 4,362 tonnes.

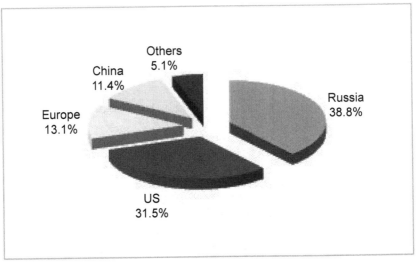

**Figure 2**
**Utilization of Propellants in the World in 2010**
(Valencia Bel, 2011)

In 2010 the following types of propellants were consumed most prominently (Figure 3):

- 36.7% UDMH/N204 (unsymmetrical dimethyl hydrazine/nitrogen peroxide – liquid propellant)
- 37.8% LOX/LH2 (liquid hydrogen/liquid oxygen – liquid propellant)
- 19.2% AP/AL/HTPB (ammonium perchlorate (AP),aluminium (Al), Hydroxy-Terminated Polybutadiene –solid propellants)

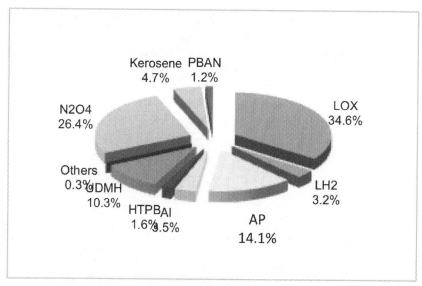

**Figure 3**
**Summary of the Most Used Propellants in 2010 in the World**
(Valencia Bel, 2011)

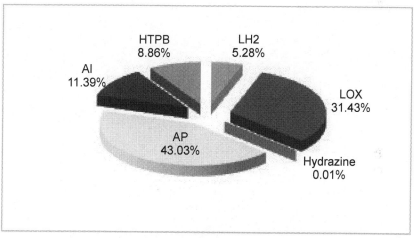

**Figure 4**
**Summary of the Most Used Propellants in Europe, Without Russia in 2010** (Valencia Bel, 2011)

In comparison, Europe (without Russia) consumed more solid propellants 63.28% rather rest of the world (Figure 4).

Upon a review of the performance characteristics, toxicity and combustion products of the aforementioned propellants, data shows that

UDMH/N2O4 is one of the most toxic propellants to human health and the environment. The performance, toxicity and combustion products of the mostly used propellants are specified in Table 3 (*eco*Space 2010).

| Fuel/ Oxidizer | Type | Lethal Concentration (LC) $LC_{50}$ (Inhalation on rat) | | $I_{sp}$, sec (Sea level) (See Note 1) | Density specific impulse kg-sec/l (Sea level) (See Note 1) | Combustion product |
|---|---|---|---|---|---|---|
| | | Fuel | Oxidizer | | | |
| UDMH/ N2O4 | Storable liquid, hypergolic | 252ppm/ 4hr | 88ppm/4 hr | 277 | 316 | CO, CO$_2$, H$_2$O, H$_2$, N$_2$, O$_2$ |
| MMH/ MON3 | | 74ppm/4 hr | 88ppm/4 hr | 280 | 325 | |
| LH$_2$/LOX | Cryogenic liquid, Non-hypergolic | -- | | 381 | 124 | H$_2$O,H$_2$ |
| Kerosene / LOX | Semi-cryogenic Non hypergolic | 5.28mg/L for 4hr | -- | 289 | 294 | CO, CO$_2$, H$_2$O, H$_2$, O$_2$ |
| Al, HTPB/ AP | Solid Propellant | - | Lethal Dose (LD), $LD_{50}$= 4200 mg/kg (oral on rat) | 266 | 469 | HCL, Al$_2$O$_3$, CO, CO$_2$, H$_2$O, O$_2$, N$_2$ |

**Table 3**
**Propellants and Performance Parameters**
(Rocket Propellants, 2008. MSDS, 2010)

- Note 1: Combustion chamber pressure, Pc=68 atm and Nozzle exit pressure Pe=1 atm

- Note 2: MMH is monomethyl hydrazine; MON3 is mixed oxides of nitrogen
- Note 3: $LD_{50}$ (Lethal Dose 50%) $LC_{50}$ (Lethal Concentration 50%) is the median lethal dose required to kill half of the members of a tested population

### 2.1.2.1 Solid propellants

The main chemicals used in solid propellants are Aluminium, Ammonium Perchlorate (AP) and Hydroxy-Terminated Polybutadiene (HTPB). While HTPB is a relatively harmless polymer, AP is considered to be highly hazardous. It contains a powerful oxidizer that must be kept away from heat, sources of ignition, combustibles, organic materials and acids during both manufacturing and transportation (Rocket motorparts, 2010). Thus AP is an expensive material to handle. Aluminium, while requiring specific handling procedures, is less hazardous than AP. A description of the potential effects on the health of humans handling these chemicals is given in section 2.2.1., below.

Different stages of solid propellant manufacturing, as shown in Figure 5 include preparation of raw materials, premixing, final mixing and casting of propellant slurry (mixture of chemicals) into a rocket motor casing. After polymerization, the cast slurry is solidified and is generally referred to as propellant grain. In view of the long processing time, these propellant grains are prepared in advance and are stored in air-conditioned rooms, often with inert gas blanketing.

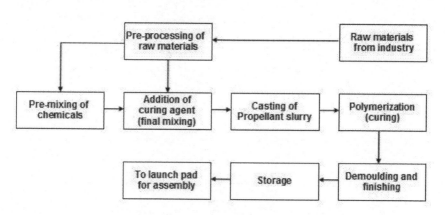

**Figure 5**
**Typical Solid Propellant Processing**
(*eco*Space, 2010)

The US Department of Transport (DOT) classification, handling and storage guidelines for chemicals used in solid propellants are summarized in Table 4. The data indicates that AP (oxidizer) is the most hazardous solid propellant used in the launch industry. Table 8 (section 2.2.1) highlights the potential acute and chronic human health effects of using these solid propellants (*eco*Space 2010).

| Name of the chemical | US DOT Classification | Handling and Storage Guidelines |
|---|---|---|
| Ammonium Perchlorate | Hazard Class 5.1 | A. Keep the materials away from the heat, sources of ignition<br>B. Use respiratory equipment in-case of insufficient ventilation<br>C. Take suitable precautionary measures against electrostatic discharges<br>D. Materials to be grounded properly during storage to avoid electrostatic charging<br>**E.** Keep away from combustible, organic materials and acids |
| Aluminum powder | Hazard Class 4.3 | A. Wash thoroughly after handling<br>B. Prevent water from getting into the container as it reacts violently<br>C. Prevent electrostatic charging<br>D. Store in a cool, dry, well-ventilated area<br>E. Keep away from acidic, alkaline, combustible and oxidizing materials |
| Hydroxyl terminated poly-butadiene | NIL | A. Wash thoroughly after handling<br>B. Avoid contact with eyes, skin and clothing<br>C. Keep the storage container closed<br>D. Prevent water from getting into the container as it reacts |

**Table 4**
**Chemicals Used in Solid Propellant and Handling** (MSDS, 2010)


2.1.2.2 LIQUID PROPELLANTS
Most of the launch vehicles mentioned in Table 2 use liquid propellants for the first and second stages. The following combinations are generally used: LH2/LOX by the Delta IV and Ariane 5, Kerosene/LOX used by the

Falcon 9 and the Soyuz, and UDMH/N2O4 used in the Proton and the Long March family.

LH2/LOX is a cryogenic liquid propellant composed of liquid hydrogen as fuel and liquid oxygen as an oxidizer. The exhaust gases created by the combustion of LH2/LOX are water, oxygen and hydrogen, making the environmental impact of the launch process relatively low. However, there are concerns that carbon footprint of hydrogen during production phase can affect the environment.

Kerosene/LOX is a semi-cryogenic fuel, meaning that while the liquid oxygen has to be maintained at very low temperatures, the kerosene can be maintained at room temperature. The exhaust gases contain both carbon dioxide and carbon monoxide, which can impact the environment, adding to the carbon footprint of the launch process.

UDMH is a storable liquid propellant that is extremely toxic and has proven to be carcinogenic. In the case of a launch failure or traces of UDMH returning to the ground, the environmental impact from this propellant can be a major. The exhaust gases from the combustion of UDMH/N2O4 include carbon dioxide and monoxide, nitrogen and oxygen, which add to the carbon footprint of the launch process.

There are many potential acute and chronic health effects for working with these propellants, which amplifies the need for the development of greener propellants.

The manufacturing process, US Department of Transport classification, and handling and storage guidelines for the fuels described above, are summarized in Table 5. The Table indicates that UDMH and MMH are the two most toxic liquid propellants, as they have a DOT class of 6.1. Table 8 from section 2.2.1 highlights the potential acute and chronic health effects of using toxic substances from liquid propellants (*eco*Space 2010).

| Propellant | Manufactur-ing process | DOT Class | Handling and Storage Guidelines |
|---|---|---|---|
| $LH_2$ | Steam reforming from methane (e.g. for Ariane 5 from methanol) | Hazard Class 2.1 | • Container shall be handled in upright position. Un-insulated piping and storage vessels shall not be touched. Wear cryogenic gloves, safety glasses and full cover with face shield.<br>• Stored in cryogenic containers. Containers and piping shall be equipped with pressure relief and control devices. Store away from oxygen, chlorine and other oxidizers. |
| UDMH | Nitrozation of dimethylamine followed by reduction process. Chloramines react with dimethalamine to produce UDMH | Hazard Class 6.1 | • Ground the container when transferring the materials. Wash thoroughly after handling. Store in a tightly closed container in a cool, dry and well-ventilated area.<br>• Can be stored at room temperature. Keep away from heat, sparks and flame.<br>• Protect eyes, face and skin from liquid splashes. |
| MMH | Produced by modified Raschig process. | Hazard Class 6.1 | • It is more stable than hydrazine<br>• Sensitive to catalytic oxidation<br>• Storage and handling conditions are similar to UDMH |
| $N_2O_4$/ MON3 | Catalytic oxidation of Ammonia with air or oxygen. | Hazard Class 2.3 | • Use self-contained breathing apparatus while handling. Ensure adequate ventilation<br>• Protect eyes, face and skin from liquid splashes<br>• Store in an inert environment in a well-ventilated place at temperatures less than 50°C and avoid its contact with air or humidity. To be stored in stainless steel, as it is highly corrosive |

| Propellant | Manufactur-ing process | DOT Class | Handling and Storage Guidelines |
|---|---|---|---|
| LOX | Fractional distillation of liquid air | Hazard Class 2.2 | • Keep away from ignition sources<br>• Should be stored in cryogenic containers in upright conditions, in dry and well-ventilated areas away from sources of heat |
| Kerosene (RP1) | Oil Cracking. Post-processing of performed to minimize its sulfur content by hydrogenation | Hazard Class 3 | • It can act as a non-conductive flammable liquid and form ignitable vapor air mixture in storage tanks<br>• Prevent static initiated fire or explosion during transfer, storage and handling<br>• Store away from fire, sparks and flame |

**Table 5**
**Liquid Propellant Manufacturing Process, Handling and Storage**
(MSDS, 2010)

### 2.1.2.3 GREEN PROPELLANTS

The use of green propellants can potentially decrease the overall life cycle cost of the launch, reduce the environmental impact, and lessen the risks to human health. There are many green propellants available, including combinations such as LOX – LH2 and LOX – pure hydrocarbons. Others, such as LMP-1035 (ADN based propellants) or AF-M-315E (HAN (hydroxyl ammonium nitrate) based propellants can be used either as monopropellants for attitude control to replace hydrazine, or as oxidizers in solid propellants to replace AP.

Hydrogen peroxide and nitrous oxide can be considered green propellants. Hydrogen peroxide decomposes into oxygen and H2O can be used as an oxidizer and monopropellant, while nitrogen dioxide is a non-toxic oxidizer. Other hydrogen peroxide based bipropellants such as hydrogen peroxide/alcohol are under development. These organic compounds and have the potential to be successful environmentally friendly propellants. More detailed information on prospective green propellants can be found in the *eco*Space report.

Tables 6 and Table 7 compare the overall performance of green propellants versus the traditional propellants (Kubota, 2007). The analysis of Team *eco*Space shows that from a technical point of view, the overall performance of green propellants has reached some level of maturity on its way to replace standard fuels, especially for liquid propellants.

| Properties | Standard Propellant | Green Propellants | Comments |
|---|---|---|---|
| Toxicity | High | Low | |
| LCA Assessment | Open | Open | on-going activity |
| Combustion gases | $N_2$, $H_2$, $NH_3$ | $CO_2$, CO, $H_2O$, $H_2$, $N_2$ | |
| Performance | Average | Average/ Good | |
| Maturity | Very high | Low | PRISMA flying with green propellants |
| Main concerns | Toxicity | High Temperature Combustion / Material Compatibility No possibility to cold start | |

**Table 6**
**Standard Monopropellants vs. Green Monopropellants**
(*eco*Space 2010,Valencia Bel 2014)

| Properties | Standard Propellants | Green Propellants | Comments |
|---|---|---|---|
| Toxicity | High | Low | Toxicity |
| LCA Assessment | Open | Open | on-going activity |
| Combustion gases | $O_2$, $H_2O$, $H_2$, $N_2$, $CO_2$, CO | $H_2O$, $CO_2$ | |
| Performance | Very Good | Very Good | |
| Maturity | Very high | Low | |
| Main concerns | Toxicity | | |

**Table 7**
**Green Bipropellants vs. Standard Bipropellants**
(*eco*Space 2010, Valencia Bel 2014)

Nevertheless, there are technical challenges with the development of greener propellants and engines. Decomposition and combustion pathway of ADN-based monopropellant is more complex than traditional hydrazine propellants because the combustion temperature of ADN is higher (1,500°C versus 900°C) (Anflo, 2004). Higher combustion temperatures associated with these green propellants are issues that require additional research and development, such as the development of new materials for engine components that can withstand higher combustion temperatures (*eco*Space 2010).

## 2.2 IMPACT OF PROPELLANTS ON HUMAN HEALTH AND THE ENVIRONMENT

Team *eco*Space analysed the impacts of rocket propellants on human health and the environment with respect to three categories: pre-launch, successful launches (including dropping stages), and failed launches.

### 2.2.1 PRE LAUNCH

The general health of the population surrounding the launch site and of employees involved in pre-launch operations are important factors to take into consideration. With respect to solid and liquid propellants mentioned in section 2.1.2.1 and 2.1.2.2, the potential human health effects of the following chemical compounds are summarized in the Table 8.

| Chemical | Potential Acute Health Effects | Potential Chronic Health Effects |
|---|---|---|
| Hydrazine | Hazardous to skin & eyes (irritant), ingestion and inhalation. Severe over-exposure can be fatal. | Toxic to blood, kidneys, lungs, the nervous system, mucous membranes. |
| UDMH | Irritant to eyes and digestive tract. Harmful if absorbed through skin. Damage to liver and central nervous system. | Potential carcinogen, can cause liver damage |
| MMH | Irritant to eyes, upper respiratory tract causing burns and digestive tract. May be fatal if absorbed through the skin, swallowed or inhaled. | N/A |

| | Can cause breathing difficulty, coma and pulmonary edema. | |
|---|---|---|
| Ammonium Perchlorate | Irritant to skin and eyes if inhaled or ingested | Toxic to blood, kidneys |
| Kerosene | Irritant to skin and eyes if ingested or inhaled. Severe over-exposure can be fatal. | Toxic to nervous system, blood, kidneys, liver |
| Liquid $O_2$ | Breathing pure oxygen under pressure may damage the lungs and affect the Central Nervous System (CNS). Cold gas or liquid may cause severe frostbite. | No harm expected |
| Liquid $H_2$ | Any effects are due to lack of oxygen. Cold gas or liquid may cause severe frostbite. | No harm expected |
| $N_2O_4$ | Corrosive to eyes. May cause burns or frostbite. Toxic by inhalation. | N/A |
| HTPB | Eye, skin and/or respiratory tract irritation. | May cause nausea, drowsiness, headache and weakness. |

**Table 8**
**Toxicity of Chemicals Used in Current Propellants**
(*eco*Space, 2010)

The toxicity of each chemical varies widely. LOX and LH2 are explosive, and hydrazine is extremely hazardous and toxic. By using less toxic chemicals in the pre-launch operations, the overall risks to health can be reduced for not only those individuals who live around the launch site, but also those who have to be in contact with the propellants at work.

To estimate impact on the environment during the pre-launch phase, Team *eco*Space conducted a case study to find a preliminary estimate of carbon dioxide (CO2) equivalent emissions associated with the manufacturing and transport of three different propellants used in the Ariane 5 launch vehicle (AP, Al, and H2). This was a first-order analysis, to be used for comparison purposes only. The combined quantity of carbon emissions for the three propellants, including only chemical manufacturing and transport is about 20,000 tons. For an average of 60 launches per year, assuming the worst case scenario of solid propellant based launch vehicles, the total carbon emission associated with the pre-launch process of the propellant is 1.2 million tons of CO2. As a comparison, the United States had CO2 emissions 5,000 times this amount in the year 2006 alone (*eco*Space, 2010).

## 2.2.2 SUCCESSFUL LAUNCHES

During successful launch, rocket fuel emissions are exhausted into the atmosphere and stages are dropped prior to payload orbit insertion. In a study made by Trinchero and presented in Table 9, five chemical compounds, Aluminum oxide, carbon monoxide (CO) & carbon dioxide (CO2), hydrochloric acid, hydrogen and nitrogen gas, found in rocket plumes are identified to be hazardous to both human health and the environment. Table 9 provides time of presence of Ariane 5 during launch at different altitudes of atmosphere, ejected mass into the atmosphere by Ariane 5 launcher separated into masses of five chemical compounds that affects human health and environment.

| Altitude (m) | Total Time (sec) | Ejected mass (kg) | $Al_2O_3$ (kg) | CO and $CO_2$ (kg) | HCl (kg) | $H_2$ and $H_2O$ (kg) | $N_2$ (kg) |
|---|---|---|---|---|---|---|---|
| [0-500] | 16.7 | 57,185 | 18,163 | 14,590 | 10,494 | 9,077 | 4,309 |
| [1,000-1,500] | 25 | 17,933 | 5,861 | 4,700 | 3,528 | 2,477 | 1,389 |
| [2,000-3,000] | 33.5 | 42,188 | 7,814 | 6,275 | 4,711 | 3,481 | 1,655 |
| [3,000-5,000] | 42.9 | 35,484 | 11,549 | 9,115 | 6,842 | 5,406 | 2,694 |
| [5,000-7,500] | 51.8 | 31,594 | 10,066 | 8,077 | 6,063 | 4,035 | 2,387 |
| [7,500-10,000] | 59.5 | 28,505 | 9,104 | 7,312 | 5,489 | 4,378 | 2,161 |
| Total | 351 | 250,477 | 74,772 | 59,681 | 44,490 | 34,071 | 17,395 |

### Table 9
### Ariane 5 Emissions at Different Altitudes of the Troposphere
(Trinchero, 2006)

During liftoff, rocket plumes contain fine particles of aluminium or aluminium oxide that can cause pulmonary fibrosis and lung damage if inhaled (Trinchero 2006).

High concentrations of carbon monoxide around the launch pad are also a potential risk. CO poisoning occurs when too much carbon monoxide enters the blood stream, and bonds to haemoglobin in the place of oxygen. This process reduces the oxygen-carrying capacity of the red blood cells and decreases the supply of oxygen to tissue and organs, which can affect heart and brain function (Queensland Government, 2010). A study by ICF Kaiser made in 1999 suggests that the levels of CO around a rocket launch site are within the allowable limits, as fixed by the Occupational Safety and Health Administration of the USA. This study determined the concentration of CO to be approximately 10 mg/m3 of air, with the safe limit being 55 mg/m3.

Air pollutants such as CO and CO2 are known to destroy the ozone layer and contribute to greenhouse effect. Within the first ten kilometres of ascent, more than 30,000 kg of CO is produced (ICF Kaiser Consulting Group, 1999). Although this amount of CO may seem insignificant with the current number of launches per year compared to other industries, it may become significant with the projected increase in sub-orbital commercial flights.

Chlorine and hydrochloric acid have been estimated as accounting for 19 percent of ozone destruction, and nitric oxide contributes 32 percent. Of these amounts, rocket launches make up only .032 percent and .0005 percent respectively. The environmental impact due to rocket launches producing these chemicals is not significant at this point in time (*eco*Space 2010).

## 2.2.2.1 DROPPED STAGES

Dropped stages are an essential part of successful launches. Dropped rocket stages with remnants of fuel create danger to the environment and human health. Approximately 9 percent of the propellant from a launch stage remains in the tank once it is dropped (Valencia-Bel, 2010). The excess fuel can pollute the surroundings by poisoning water, soil, flora and fauna. Thus, the impact areas can be subject to major contamination from booster debris. A study conducted by Kisilev et al. (2001) to determine the amount of material that returns to Earth from different launchers has been summarized in Table 10, showing that tons of debris and fuel are brought regularly to the Earth after each rocket launch. The penetration of contaminants depends on the physical and chemical properties of the soil. As an example, when UDMH is supersaturated in soil it can cause contamination of groundwater and surface water. The heptyl component of UDMH is highly toxic and carcinogenic. Furthermore, plants commonly preserve heptyl, which presents a threat to the wildlife that depend on these plants as a food source. In the territory of Karsakpayskogo in Kazakhstan, heptyl was detected in concentrations exceeding 5,000 times the maximum permissible concentration, as a result of 14 Proton carrier rocket launches in 2003, which spilled 10.5 tons of heptyl and 2 tons of oxidizer into the soil (Sergeeva, 2004).

| Booster | Weight (tons) | |
|---------|---------------|--------------------------|
| | Total at Launch | Amount Returning to Earth |
| Thor- Delta | 55 | 5 |
| Thor- Agena | 125 | 7 |
| Vostok | 290 | 21 |
| Saturn IB | 570 | 53 |
| Titan 3C | 645 | 170 |
| Saturn V | 2,930 | 166 |

**Table 10**
**Material that Returns to the Earth from Historic Launchers**
(Kisilev *et al*, 2001)

Hydrocarbon fuels are stable in soil, but within the first year of contamination the concentration decreases rapidly due to evaporation, and transport to the surface and groundwater. The penetration of fuel into the soil during the first few years is no more than 20 cm and after a few decades will effectively disappear (Sergeeva, 2004).

When rocket stages fall back to the Earth's surface at close to 60-80 m/s, the force of the impact can cause an undesired ignition of the stage, in addition to the possibility of harming property or people. The resulting impact craters can also cause contamination to the ground near the impact areas. Metal fragments created during stage re-entry are often buried in the ground, and some regional authorities have complained about the potential environmental impact of leaving this debris in place. As a result, a few launch states including Russia have decided to create special task forces to collect fragments of stages in the regions that stay under the trajectory of rocket launches from cosmodrome Baikonur (Adushkin et al., 2000). The launch corridor of the Soyuz rocket passes over populated areas of Kazakhstan and Russia (*eco*Space 2010).

When stages crash into the ocean they break up into a number of rigid pieces, which either sink or float on the surface of the water. The large floating debris is less of a danger to wildlife than the smaller objects. The sinking debris is not likely to affect many animals, although a small number of marine organisms may be killed by the impact or consumption of the pieces. The more important issue is the dispersion of the unburned fuel that can cover the surface of the water.

In the case of kerosene, 95 percent of the fuel evaporates from the surface within the first few hours, so the damage done to marine animals is minimal in the open oceans. However if fuel should be spilled near the shore, marine animals such as turtles, seals, and dolphins could be harmed (ICF Kaiser, 1999). UDMH is toxic to aquatic species above the level of 1 milligram per liter (MSDS, 2010). Although UDMH will disperse over a large area on the surface, it will not evaporate.

### 2.2.4 FAILED LAUNCHES

In the case of a failed launch, apart from the explosion itself, effects on human health and environment will be amplified mainly because of the larger quantities of unburned fuel dispersed on the surface of the earth and in the oceans. The effect of failed launches is devastating to the local infrastructure, environment and human health and may result in billions of USD claims for clean-up and recovery measures (*eco*Space, 2010).

## 2.3 LEGAL FRAMEWORK FOR LAUNCHES AND PREVENTION OF ITS IMPACT ON ENVIRONMENT AND HUMAN HEALTH

This section provides a brief overview the environmental legal framework and launch regulations in Russia, the United States, France, China and Brazil as well as supports expansion of ICAO activities for space.

### 2.3.1 US LAUNCH REGULATIONS

As outlined by the National Aeronautics and Space Act of 1958, NASA is responsible for the United States civilian space program. The US Federal Aviation Administration (FAA) is the lead agency in the federal government responsible for regulating all commercial space transportation launches, re-entries, and spaceport licensing activities. This responsibility is delegated by the Secretary of Transportation to the FAA Office of Commercial Space Transportation, as authorized by Title 49 of the United States Code, Subtitle IX Chapter 701. The US Commercial Space Transportation regulations specifically address launch safety in the interest of public safety and safety of property. Launch safety regulations are defined in the regulations 14 Code of Federal Regulations Part 417, which covers five main subparts including Launch Safety Responsibility (subpart B), Flight Safety Systems and Ground Safety.

## US ENVIRONMENTAL LAUNCH REGULATIONS

The National Environmental Policy Act (NEPA) is the law that establishes policy for promoting national environmental goals for the "protection, maintenance, and enhancement of the environment" (US Code, 2007). NEPA also calls for the establishment of the Council of Environmental Quality (CEQ), which directs all federal agencies to develop implementation procedures for NEPA (40 CFR Part 1500).

While NASA is required to meet NEPA regulations, they have established requirements and directives that provide a legal framework for regulating space launches while minimizing the environmental impact of launch activities. Similarly, the FAA has guidelines for considering environmental impacts that describe policies and procedures for meeting all NEPA implementation regulations for FAA actions (*eco*Space, 2010).

## 2.3.2 EUROPE

## ESA & EU SUSTAINABLE DEVELOPMENT POLICY AND PRACTICES

The European Union (EU) and the European Space Agency (ESA) are actively working together to create a unified European space policy. ESA addresses sustainability in three domains: environmental, economic, and social & societal. Although each of these domains has interactions with the others, the specific interactions related to the environment are the most important to *eco*Space initiatives. Taking into account different concerns related to sustainability, ESA created a Coordination Office on Sustainable Development and started the ESA Clean Space Program.

In regard to environmental regulations, the European Union has enacted legislation to protect and to preserve the environment. Article 174 of the Lisbon Treaty states that "policy on the environment shall contribute to pursuit of the following objectives:

- Preserving, protecting and improving the quality of the environment
- Protecting human health
- Prudent and rational utilization of natural resources

- Promoting measures at international level to deal with regional or worldwide environmental problems, and in particular combating climate change"

Article 174 takes the precautionary principle into account and promotes preventative action by each State, and rectification and payment by any polluter. Overall, the Lisbon Treaty establishes a legal framework that ensures the protection of the environment and promotes sustainability of a proposed activity (ecoSpace, 2010).

## FRENCH ENVIRONMENTAL LAUNCH REGULATIONS

French jurisdiction is particularly relevant to the study of launch regulations in Europe due to the launch site in Kourou, French Guiana. Laws for space operations deal with the conditions required to launch an object to space under French jurisdiction. The French Space Agency (CNES) analyzes a launch request and provides notification to the French minister in charge of outer space, who will eventually approve the launch.

One French environmental regulation that can be applied directly to launch activities is the REACH (Registration, Evaluation and Authorization of Chemicals) regulation. The REACH regulation "entrusts evaluation responsibility and substance risk management to producing companies and importers, and no longer to authorities" (French Ministry of Ecology, Energy, Sustainable development and Sea, 2010).

## 2.3.3 RUSSIAN SPACE LAW AND LAUNCH REGULATIONS

The Russian Federal Space Agency (Roscosmos) is the federal executive body responsible for carrying out the Russian Space Program as legislated by the Russian Federal space act, Decree No. 5663-1 "About Space Activity." The program is governed by the Constitution of the Russian Federation, international treaties signed by the Russian Federation and general principles and norms of international law. The current Federal Russian Space Program document has a chapter dealing with spaceports and ground facilities operation. The program has plans for the development of a system for environmental monitoring of administrative regions affected by the launch activities from Russian cosmodromes Baikonur which operates via agreement with Kazahstan, Plesetsk and Svobodnyi (under construction) (Roskosmos 2010).

The draft of the law "Federal Law on Safety of Territories along Launch Trajectories" outlines state regulations for administrative regions affected by dropped stages or falling debris from rocket components of failed launches[2]. Seventeen administrative regions in Russia are located under/along launch trajectories (see Fig.4). The proposed law suggests that regions affected by dropped stages should be registered with the Russian State Land database as special zones with regulated economic activity and special regulations for transportation operations. A draft of the law lists necessary measures to reduce the impact of dropped stages on the environment, such as the detoxification of ground contamination and the monitoring of the environment and human health. Administrative regions affected by launch activity will be compensated for damages caused by the Russian space program.

During the last 50 years, the population and the environment of these regions have been impacted by intensive space launch activities. In the Russian Federation, the Ministry of Ecology is responsible for the environmental protection. This ministry has developed a special annual ecological passport for monitoring of air, water, soil, and human health in regions affected by dropped stages from rockets. In addition, there is a state sponsored program for the clean-up of dropped stages and the local environment.

## AGREEMENT WITH KAZAKHSTAN FOR OPERATION OF BAIKONUR COSMODROME

Following the breakup of the Soviet Union, the Russian Federation and Republic of Kazakhstan signed an agreement for the continued use of Baikonur Cosmodrome for the Russian space program. In 2005 a bilateral agreement between Russia and Kazakhstan came into force, whereby Russia would continue to rent the spaceport from Kazakhstan until 2050.

In 1999, Russia and Kazakhstan concluded an agreement on cooperation in the event of an accident during launch activity from Baikonur cosmodrome. The agreement states that in case of an accident during launch, the Russian Party will terminate all launches of the specific type of rocket until assessment of the reasons for failure. Russia is to inform Kazakhstan about the cause of the accident as well as a list of safety measures for subsequent launches of that same booster type before resuming launches. Additionally, an agreement is in place between Russia and Kazakhstan for financial compensation to be given to Kazakhstan in the event of damage or harm being done to Kazakhstan as well as to pay for associated clean-up costs. For example, on July 26, 2006, a Dnepr booster failed 86 seconds after takeoff, crash-landing south of its launching silo and

---

[2] The "Federal Law on Safety of Territories along Launch Trajectories" was enforced during 2010

spilling toxic propellants over Kazakhstan's environment (Russian Space Web, 2006). The damage was estimated at USD 1.1M (Zhdanovich, 2010).

## ENVIRONMENTAL SAFETY STANDARD FOR SPACE TECHNOLOGIES

For many years the Russian space policy appeared to be quite reactive in nature with the government allocating funds to regions and populations after damage was incurred. In 2008, however, Environmental Safety Standard for Space Technologies was introduced, and is still in effect. Technical specifications, given in tenders by Roscosmos for all new designed space systems, contain mandatory requirements for protection of the environment and humans from harmful impacts of space technology (Zhdanovich, 2010).

## 2.3.4 CHINA

### CHINESE LAUNCH REGULATIONS

In 2000 and in 2006, the State Council Information Office of the People's Republic of China (SCIOPRC) released a "White Paper on China's Space Activities," comprehensively expounding on China's previously carried out space activities, principles, guidelines, policies, as well as the main future objectives and tasks. The 2006 White Paper noted that China's space policy is to develop independently, scientifically and openly (SCIOPRC, 2000 & 2006). However, there are still no national space laws in China, and there are only two regulations concerning registration and launching of space objects: "Measures for the Administration of Registration of Objects Launched into Outer Space" (February 8, 2001) and "Interim Measures on the Administration of Licensing the Project of Launching Civil Space Objects" 27-30.(Li J, Li S, Zhao H& Zhao Y, 2010).

### ENVIRONMENT RELATED PROVISIONS

In 1989, China created the "People's Republic of China (PRC) Environmental Protection Law." On the basis of this law, China established laws on marine protection, prevention of water pollution, prevention of air pollution, prevention of environmental noise and solid waste pollution prevention (China.org.ca, 2010; Baidu, 2010a-c). These laws also apply to China's space activities, requiring that all organizations undertake plans for environmental protection and take effective measures to prevent and control environmental pollution. The application for a launch includes proposed conditions regarding environmental protection, and the launch should not result in irreparable harm to the public or property because of significant negligence or willful action. The relevant documents are submitted to prove a project follows China's environmental protection laws and regulations, protection of public safety and property, system reliability, and pollution prevention.

## 2.3.5 BRAZIL

The Brazilian Space Agency, Agência Espacial Brasileira (AEB), is a civilian organization that promotes the development of space activities in Brazil. The main functions of the AEB are to carry out the National Policy for the Development of Space Activities (PNDAE), to shape the National Space Activities Programs (PNAE), to establish national standards, and to issue licenses and authorizations for regulating space activities within Brazil (AEB, 2010).

The General Space Security Regulation contains rules that characterize Brazil as a launch state and provide protection for people, property, and the environment from potentially hazardous activities during the life cycle of a project such as initial design, analysis, and operation. The AEB's safety regulations are composed of a set of technical regulations, definitions, general rules and requirements.

With regard to environmental protection, the National Environmental Policy seeks to minimize any significant environmental impacts arising from new developments and activities conducted in Brazilian territory. This regulation applies to all activities, products, and services relevant to the space sector that could have significant impacts on the environment. Another important environmental policy is the General Technical Regulation for Environmental Safety in Space Activities, which establishes the authority of the AEB to ensure that licensees comply with existing environmental laws. The AEB also requires that a company obtain an environmental license issued by the National System of Environment, to verify that the environmental impact of the proposed space activity has been evaluated (AEB 2010).

## 2.3.6 EXPANDING THE SPAN OF ICAO TO INCLUDE SPACE ACTIVITIES

In 1995, the US FAA proposed the creation of a new organization named International Space Flight Organization (ISFO) to act the same way as ICAO but for space. Three years later, the FAA initiated a study on controlling vehicles from Earth and LEO and its impact on the air traffic control system. In 2000 ICAO Chairman Dr. Assad Kotaite proposed to increase the span of the ICAO into space by adopting on an international level the internal organization structure of the FAA. In all these examples, one of the main goals is to enable space commercialization, including orbital and suborbital flights. Currently this sector is not a significant market, but could develop very rapidly in the coming years. The aforementioned recommendations and a complete version of the proposal to expand the scope of ICAO are available in the "An ICAO for Space?" White Book (IAASS, 2009).

## 2.4 ECONOMICS OF THE INTRODUCTION OF GREEN PROPELLANTS INTO THE LAUNCH INDUSTRY

The introduction of greener technologies in the launch industry and possible marketing strategies are discussed in this section. The US automobile industry is used as an example, to illustrate how the space launch industry might react to future changes in environmental regulations.

### 2.4.1 ANALOGY WITH AUTOMOBILE INDUSTRY

In 1923-1924, lead was introduced as an additive to regular gasoline to increase the performance of internal combustion engines. This paved the way for the development of high-power, high-compression internal combustion engines (Lewis, 1985). Lead became universal, as a simple yet effective additive. In the 1970s several researchers identified motor vehicles as a major source of urban air pollution, and in the USA in 1973 the sale of leaded fuel declined with the introduction of the catalytic converter, which is not compatible with this additive. In 1996, the Clean Air Act banned the sale of leaded fuel for use in on-road vehicles, and most countries followed this trend (US Environmental Protection Agency, 1996). However, aircraft, boats and racing cars can still use this toxic additive (*eco*Space, 2010).

During the same period, the oil crisis and the introduction of catalytic systems forced the automobile industry to change vehicle designs radically, producing smaller automobiles with updated engines. Figure 7 shows the price of a new automobile as a function of time, in constant 2007 USA dollars.

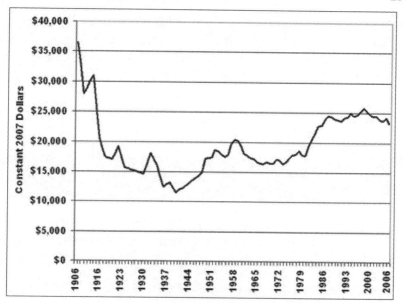

**Figure 7**
**Automobile Value in the 20th Century, in Constant 2007 Dollars**
(USA Department of Energy, 2008)

The increasing cost of vehicles in the 1970s is largely attributed to government regulations including emission reduction and safety. To reduce the air pollution created by lead, three techniques were used (The Regional Environmental Center for the Caucasus, 2008):

- Introducing laws to ban lead fuel
- Increasing high taxes on lead fuel
- Implementing programs to use alternative fuels.

Currently, the space launch industry is not strictly regulated in terms of pollution, but here we consider how the industry could react to the introduction of new regulation regarding environmental issues by presenting different methods the space market might adopt (*eco*Space 2010).

### 2.4.2 LOBBYING FOR ENVIRONMENTAL AWARENESS

Interest in green technologies has been growing in the last few decades as people become more concerned about the environment. To quantify this interest, the Center for Public Integrity has found that there were about 2,340 lobbyists in the USA to influence federal policy in 2008, with a total expenditure of USD 90M. Figure 8 presents the number of lobbyists on climate change as a function of the sector, for 2003 and 2008.

The number of lobbyists has significantly increased in the past few years as concern for the environment has been growing, and stricter laws have been adopted. The rapid rise in public interest regarding environmental concerns may lead to the introduction of new laws and regulations for the entire launch industry (*eco*Space 2010).

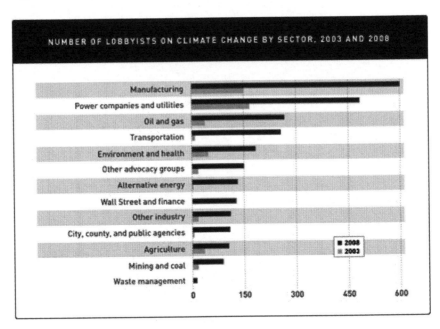

**Figure 8**
**Lobbyists on Climate Change by Sector, 2003 and 2008**
(Lavelle, 2009)

### 2.4.3 CURRENT MARKETING TECHNIQUES TO PROMOTE GREEN TECHNOLOGIES

The space launch industry focuses on three main criteria: performance, safety, and price, and green propellants can be promoted using the fact they may have the same performance with more safety. One of the main reasons for the development of greener propellants is due to the fact that they are safer both in manufacturing and handling. Incorporating a "greener" image in marketing techniques may also be beneficial to a company and to the entire industry.

It is important to distinguish between orbital and sub-orbital flights, as they have different energy requirements for mission success. Orbital flights need a significant amount of energy to reach orbit, whereas sub-orbital flights require less. As a result, less powerful propellants can be used in sub-orbital flight than for orbital flights. Companies such as Virgin Galactic can promote their use of greener propellants such HTPB/N2O, which has a smaller specific impulse than many orbital propellants (*eco*Space 2010).

### 2.4.4 NEW MARKETING TECHNIQUES TO PROMOTE GREEN TECHNOLOGIES

The launch industry should be aware that an increasing number of people are concerned about the environment. The sub-orbital launch industry is affected by this tendency because their customers are people that may care about the environment. This means that companies who advertise a green initiative may be more successful in the sub-orbital business.

Currently, sub-orbital companies do not use the fact that they are using greener technologies as a major marketing technique, but in the last decade an increasing number of people have developed an interest in green products, and according to Wasik (1996), 20 million customers say they are "socially conscious."

To assess whether one technology is greener than another, environmental performance metrics should be used. The most widely known metric is CO2 production, also called "Carbon Footprint" (Carbon Footprint LTD., 2010), widely known due to its use in the automobile industry. In the rocket industry, this metric is not sufficient, as there are several other pollutant by-products from many propellants (HCl, Cl2, Al2O3 and others) that are more toxic and damaging to the environment (Kiselev, 2003).

For space agencies, the situation is different, as they do not sell services to the public, but greener space technologies could still be used to promote space activities among the population as part of an outreach program proposed by Team *eco*Space (see ecoSpace report). In future years, the general public could put a lot of pressure on organizations to develop and implement green initiatives (*eco*Space 2010), and this may eventually have significant impact on the launch industry as well.

### 2.4.5 ESTIMATE OF THE COST TO CHANGE TO GREEN PROPELLANTS

Team *eco*Space made a preliminary estimate of the total costs and timeline required to implement and to test a new rocket with a new propellant, with the assistance of Mr. Ferran Valencia-Bell, an expert in greener propellants at ESA.

The study was based on an engineering methodology using data from Ariane 5 rocket design and validation. The design of the rocket incorporates three engines or stages. The first engine uses solid propellant, while the other two engines use liquid propellant. The development of these engines involves three steps: propellant development, preliminary sizing, and an engineering model design including a test campaign. For each of these steps, an estimate of the time and resources required uses real values from the Ariane 5 rocket (see Table 11).

Typically the cost for the liquid propellant is three times more than solid propellant. The engine test rig cost is very high due to the high power set-up required to do these tests, similar to the set-up used in DLR (German Aerospace Center) to test the Vulcain engine installed on the Ariane 5 launcher. The Ariane 5 was used as a baseline for the number of hours required to develop such an engine, and an empirical method was used to estimate other values (ecoSpace 2010).

| Task | Rate |
|---|---|
| Hourly rate per man for the design of the engine and propellant | 120 €/hr |
| Liquid propellant | 100 €/kg |
| Solid propellant | 35 €/kg |
| Propellant testing (test rig) | 1000 €/day |
| Complete engine testing (test rig) | 120 000 €/day |

**Table 11**
**Rates Used to Evaluate the Total Cost of the Project**

The preliminary estimates by ecoSpace showed that the overall cost for the design and acceptance of a rocket using a new propellant is around EUR 7.7B, which is about 20 percent more than the cost to design Ariane 5. A total of 8 years would be required for this development work.

One of the most important results of this analysis is the number of jobs created. For the design of the main stage more than 1,500 technical jobs would be created, suggesting more than EUR 2.3B in revenues for the contracted companies. In contrast, the required investment to design a new propellant is low, less than EUR 10M, compared to the final value of the rocket and the detailed design of the engine itself.

## 2.5 PRELIMINARY PUBLIC OPINION SURVEY AND PUBLIC OUTREACH

The subject of greener space is relatively new. As a result, the data on the knowledge of the general public about the subject is quite limited, so Team *eco*Space developed public outreach products to develop an interest in eco-friendly space technologies and to promote environmentally friendly space launch activities. These included development of web sites, use of social media and networking, as well as development of educational packages. Proposed educational packages are targeted at various age groups, from youth to university level, and towards various audiences including decision makers and journalists.

During the nine weeks of the Space Studies Program *eco*Space developed a web-based survey open to the general public, using the framework of the Kwik survey system with the goal to assess the public's knowledge about environmental impact of space launch activities. The survey was distributed to the ISU community of space enthusiasts via e-mail, and to the general public of individuals not directly associated with the space sector using Facebook. In total, 230 persons responded.

It is clear that from academic point of view *eco*Space survey performed during SSP'10 is not able to reach a level of a serious sociological research, but it did reveal some trends in the opinion of general public about environmental impact of launch activities. Four key conclusions emerged (*eco*Space, 2010):

1. There is a need for the development of an educational program about the environmental impact of space activities, and even more importantly, on possible solutions;
2. The current trend in society to buy eco-friendly/green products and the willingness to pay more for such products can be extended to space based services;
3. Space tourism companies must take into account the environmental impact of their activities. Ignoring these impacts could lead to decreased profits, based on the fact that the general opinion is that space tourism will have negative effects on the environment;
4. The public has no clear understanding of the effects of the launch industry on the environment and human health. This must be clarified to mitigate the fear of launch failures, and help promote green propellants.

## CONCLUDING REMARKS

As mentioned above, the term "sustainability" originated from studies conducted in the 1987 that suggest the strong link between economic development and the environment. When sustainability is included in an activity, the economic, environmental, and social impacts must be considered in all phases of the project cycle.

Team *eco*Space proposed a set of initiatives to accelerate the introduction of green technology into the space launch industry. These initiatives include the creation of the Sustainable Space Development Board (SSDB) to address environmental concerns, outreach programs to increase public awareness, a monitoring system to measure the environmental impact of launch activities, and an international agreement for space launch sustainability.

Team *eco*Space hopes that suggestions and initiatives produced in the report will increase public awareness about the environmental impact of the launch industry and move space agencies to adopt sustainable environmental management systems and international environmental launch regulations, while stimulating research in the development of green propellants, thus making space launch activities more sustainable and contributing to the long term success of the space endeavour.

•••

## OLGA ZHDANOVICH, MSc

Olga Zhdanovich works in the secretariat of the European Cooperation for Space Standardization at ESTEC/ESA in the Netherlands. In 2006-2010 she worked as Payload Integration Manager for the International Space Station via RheaTech, responsible for the integration of various European research payloads for the ISS, as well as University level educational projects and programs.

She graduated with honors from the Moscow Institute of Engineers in Geodesy, Aerial Surveying and Cartography, Russia as engineer of cartography in 1990. Seven years later she received an MSc in Environmental Science and Policy from Central European University,

## REFERENCES

*EcoSpace. Initiatives for environmentally sustainable Launch Activities"*
    executive summary, International Space University, 2010
*"EcoSpace. Initiatives for environmentally sustainable Launch Activities"*
    report, International Space University, 2010
ADUSHKIN V. 2000. *Environmental problems and risks of exposure to rocket
    and space technology on the environment.* Moscow: Ankil.
AGENCIA ESPACIAL BRASILEIRA (AEB; Brazilian Space Agency). 2010.
    Legislation. Available from:
    www.aeb.gov.br/indexx.php?secao=legislacao
*Air Pollution Control Act of People's Republic of China.* Available from:
    baike.baidu.com/view/84079.htm?fr=ala0_1
ANFLO K. 2004. Development Testing of 1-Newton ADN-Based Rocket
    Engines, *2nd International Conference On Green Propellants For Space
    Propulsion.*
ARIANESPACE. *Launch Log.* Available from: www.arianespace.com/news-
    launch-logs/2000-2010.asp
BAIDU. 2008. ISO14000 in China. Available from:

http://tieba.baidu.com/f?kz=309329186.

BAIDU. 2010a. Environmental Noise Pollution Control Act of People's Republic of China. Baidu Baike. Available from: http://baike.baidu.com/view/250720.htm?fr=ala0_1

BAIDU. 2010b. Marine Environmental Protection Act of the People's Republic of China. BaiduBaike. Available from: http://baike.baidu.com/view/250729.htm?fr=ala0_1

BAIDU. 2010c. Solid Waste Pollution Prevention Act of People's Republic of China. Baidu Baike. Available from: http://baike.baidu.com/view/414737.htm?fr=ala0_1

CARBON FOOTPRINT LTD. 2010. *What Is A Carbon Footprint?* Available from: www.carbonfootprint.com/carbonfootprint.html

CHINA.ORG.CA, 2010. Environment Law. Available from: http://www.china.org.cn

*China's Space Activities in 2006.* Available from: www.gov.cn/ztzl/zghk50/content_419652.htm

CHINESE PETROLEUM AND CHEMICAL CORPORATION. 2008. *Sinopec & China Space Program.* Available from: english.sinopec.com/environment_society/sinopec_china_space_program/

ESA. 2001. *Green Propellant for Space Propulsion.* Available from: www.esa.int/esaEO/ESAM1TPZ9NC_index_0.html

ESA. 2010. *Sustainable Development Report. April 2010.* Available from : http://www.esa.org

FAA, 2010 *2010 Commercial Space Transportation Forecasts*, USA: FAA.

FRENCH MINISTRY OF ECOLOGY, ENERGY, SUSTAINABLE DEVEOPLMENT AND SEA. 2010. REACH – contexte et mise en oeuvre [online, unofficial translation]. Available from: http://www.developpement-durable.gouv.fr/REACH-contexte-et-mise-en-oeuvre.html

FUTRON CORPORATION, 2010. *Satellite Industry Association.* USA, June 2010.

IAASS. 2009. An ICAO for Space? [online]. www.iaass.org

ICF KAISER CONSULTING GROUP. 1999. Environmental Assessment for the Sea Launch Project.

ICF KAISER CONSULTING GROUP. 1999. *Final environmental assessment for the sea.* pp. 31-49.

ISU 2009, editors A. Houston and M.Rycroft. *Keys to Space*, McGraw-Hill,

KISELEV A. I. 2003. *Astronautics: summary and prospects.*Springer-Verlag Wien, New York.

KUBOTA N. 2007. *Propellants and Explosives.* Wiley-VCH

LAVELLE, M. 2009. *The Climate Change Lobby Explosion.* The Center for Public Integrity. Available from: www.publicintegrity.org/investigations/climate_change/articles/entry/1171/

LEWIS, J. (1985). Lead Poisoning: A Historical Perspective. *EPA.*

LI, Juqian. (2009). Progressing Towards New National Space Law: Current

Status and Recent Developments in Chinese Space Law and its Relevance to Pacific Rim Space Law and Activities. *Journal of Space Law*. 35 (2).

LI, S. (2009). *The Role of International Law in Chinese Space Law and Its Relevance to Pacific Rim Space Law and Activities. Journal of Space Law.* 35 (2).

MEUSY, N., 2010. *European Space Agency vie of Sustainability Develop* [presentation] (Personal communication, August 12th 2010).

MINISTRY OF ECOLOGY. *Energy, Sustainable Development and the Sea.* Available from: www.developpement-durable.gouv.fr/REACH-contexte-et-mise-en-oeuvre.html

MSDS. 2010. *Material Safety Data Sheet (MSDS).* Methylhydrazine, 98%. Nitrogen Dioxide. Aluminum. Hydrazine. Polybutabiene. Kerosene. UDMH.

QUEENSLAND GOVERNMENT. 2010, Environmental Management [online]. http://www.derm.qld.gov.au/

Rocket and Space Technology. (2008). *Rocket Propellants.* Available: http://www.braeunig.us/space/propel.htm

ROSCOSMOS (2010a). *Space Program - Russian Federal Space Program.* Available from: www.federalspace.ru/main.php?id=24

RUSSIAN SPACE WEB. 2006. *The Dnepr launcher - Dnepr fails during launch with multiple payloads.* Available from: www.russianspaceweb.com/dnepr_007_belka.html

SERGEEVA A. 2004. *Analysis of the impact of rocket and space activities on the environment.*

SPACELAUNCHREPORT. 2010, Available from: www.spacelaunchreport.com

STARSEM (2001) *Soyuz Users Manual* (ST-GTD-SUM01) Russia: Roscosmos.

STATE COUNCIL INFORMATION OFFICE OF THE PEOPLE'S REPUBLIC OF CHINA (SCIOPRC). 2006. China's Space Activities in 2006. Available from: http://www.china.org.cn/english/features/book/183672.htm

THE REGIONAL ENVIRONMENTAL CENTRE FOR THE CAUCASUS. 2008. *Fuel Quality and Vehicle Emission Standards Overview for the Azerbaijan Republic, Georgia, the Kyrgyz Republic, the Republic of Armenia, the Republic of Kazakhstan, the Republic of Moldova, the Republic of Turkmenistan, the Republic of Uzbekistan and the Russian Federation.* Available from: www.unep.org/pcfv/PDF/FuelQuality_en.pdf

TRINCHERO. 2006. *Mercredis de l'espace*, Service Sauvegarde et Environnement. CNES, pp. 1-26

U.S. ENVIRONMENTAL PROTECTION AGENCY. 1996. *EPA Takes Final Step in Phaseout of Leaded Gasoline.* Available from: www.epa.gov/history/topics/lead/02.htm

USDOT-PHMSA. 2008. *PHMSA-Regulations, HMR; 49 CFR Parts 171 and 173.*

USA CODE OF FEDERAL REGULATIONS. Aug. 2010. *Title 40 Protection of Environment Part 1500-1508. Title 14 Commercial Space Transportation Part 400.*

USA CODE. 2007. *Title 42 The Public Health and Welfare Chapter 55, Section 4321.*

VALENCIA -BELL. F. 2010. *Personal communication* 11 august 2010

Valencia-Bell,F Neil Murray .2011 *Environmental Aspects of Launchers Fuel Production and Combustion Processes,* Space Access International Conference 2011

Valencia-Bell. F *Personal communication* 22 July 2014

VIRGIN GALACTIC. 2009. *Overview.* Available from: www.virgingalactic.com

WASIK, JF. (1996). *Green marketing & management: a global perspective .* USA: Blackwell Publishers Ltd.

*Water Pollution Control Act of People's Republic of China.* Available from: baike.baidu.com/view/250547.htm?fr=ala0_1

ZHAO, H. 2010. *Current Legal Status and Recent Developments of APSCO and its Relevance to Pacific Rim Space Law and Activities.* Journal of Space Law, 35(2), p

ZHAO, Y. 2010. *Current Legal Status and Recent Developments in Hong Kong Law and its Relevance to Pacific Rim Space Law and Activities.* Journal of Space Law, 35(2).

ZHDANOVICH, O. 2010. *Russian National Space Safety Standards and Related Laws,* J. N. Pelton and R. S. Jakhu, eds. Space Safety Regulation and Standards, Elsevier.

# PART IV

# SUSTAINING AND EXPANDING SPACE OPERATIONS AND COMMERCE

## INTRODUCTION TO PART IV

While the Space Age began as an effort of the world's largest nations, it has shifted over the decades to a thriving commercial domain in which dozens of companies compete successfully to provide products and services for use on Earth, including essential satellite-based communications and scientific observations. In the near future, however, we may see significant changes, as the commercial space sector increases its efforts and its impact. In Part IV we hear from many leaders in these efforts.

CHAPTER 14

# INNOVATIVE MODELS FOR PRIVATE FINANCE INITIATIVES FOR SPACE SCIENCE MISSIONS

JEFFREY NOSANOV,
NORAH PATTEN,
MICHAEL POTTER,
AND
CHRISTOPHER STOTT
INTERNATIONAL INSTITUTE OF SPACE COMMERCE

> *"No bucks, no Buck Rogers"*
> Tom Wolfe, *The Right Stuff*

## I. INTRODUCTION

We are currently witnessing the confluence of several powerful economic and policy forces in the global space sector. Due to governmental financial constraints, NASA budgets are contracting and commercialization of space markets is increasing rapidly. The lack of a clear civil space vision, matched with an unpredictable long term science and engineering prioritization mechanism, is perpetuating a cycle of budgetary, financial, and mission uncertainties. However, uncertainty in government creates private opportunity. Public-private partnerships, among other models, have been demonstrated as viable methods to meet government satellite communications requirements. Can these models also provide a new private financing model for government needs in space science missions?

> The "business as usual" approach to conducting space based science
> is not working.

Traditional mega space projects are unfortunately characterized by cost plus contracts, schedule slips, and broken budgets. The symbolic punching bag for critics are NASA's Mars Science Lab (MSL) which was $400 million over budget[1] and James Webb Space Telescope (JWST) mission over $3 billion over budget and both behind schedule.[2] A multitude of factors contribute to these issues, from changing congressional input, NASA's inability to set long term budgets for itself, and a contracting mechanism that makes it difficult to incentivize and enforce strict adherence to project milestones.

If you accept that the next frontier of human evolution involves not only protecting humanity from possible extinction from a large asteroid, but also ensuring that humanity becomes a multiplanetary species – then the subject of space finance is both exalted, and directly tied to the evolution of the human species. In this chapter we present innovative models and approaches for the private financing of space science missions that have succeeded in other government space sectors.

Government, space agencies and private businesses around the world are now scrambling to adapt to the "new normal." All parties must adapt or face the threat of irrelevance. The authors believe that new financing models can enable agencies and stakeholders to meet their project goals in these difficult economic times, and we specifically address whether these finance models can be applied to space science missions lacking a clear commercial return.

Space science is not the priority of national governments and as such receives limited funding. However, the private sector has large amounts of available capital that could be applied to space projects if governments would minimize or eliminate "market risk." In this chapter we argue for

---

[1]   http://en.wikipedia.org/wiki/Mars_Science_Laboratory
[2]   http://en.wikipedia.org/wiki/James_Webb_Space_Telescope

new thinking and innovative frameworks that can be utilized for financing future space missions.

## II. A Private Finance Model

It has been said that space missions run on money and coffee. The majority of funding for communication systems comes from the private capital markets, following the proliferation of commercial satellite entities in the 1980s and the worldwide distribution of the enabling technology.

Government, also known as "civil" and defense mission funding continues to come from government capital: taxation. However, it is rapidly becoming apparent that the breadth and depth of current space sciences can no longer be supported in the current government funding environment. Yet, where is it said that only government capital must be used to fund government missions? The authors believe that this is not a binary choice. What is the goal, space science results or the spending of government capital?

> Space science missions by their very nature are designed and executed to acquire new information and often have a large amount of uncertainty in expected return.[3] One thing must be acknowledged up front: one of the policy objectives of using government capital to fund space missions has been job creation and to further the maintenance of a highly skilled work force in the space industry.

As is stated in the National Space Policy of the United States of America, principles the United States will adhere to are 'to encouraging and facilitating the growth of a U.S. commercial space sector that supports U.S. needs, is globally competitive, and advances U.S. leadership in the generation of new markets and innovation-driven entrepreneurship.[4] However, the reduction of government capital available to fund space missions is leading to a rapid and alarming reduction in the space workforce. This is especially alarming given the highly specialized nature of this work force and the significant investment of time and money required by a young person wishing to become a useful contributor in this field. Can the replacement of government capital with private capital revitalize and strengthen the industry and protect this workforce?

---

[3]   A. Gray et al., 2005, 'A Real Options Framework for Space Missions Design', *IEEE Aerospace Conference*, March 5 – 12, 2005.

[4]   National Space Policy of the United States of America, 2010.
     http://www.whitehouse.gov/sites/default/files/national_space_policy_628-10.pdf

Since the first space mission of Sputnik in 1957, much has changed in the world of commercial space, spin-off and spin-in space based technologies and indeed financing of space initiatives. Access to space and space exploration continues to grow and change throughout the decades. Consider the shift since the 'Space Race' and the era when President John F. Kennedy announced that America would send a man to the moon and return him safely to the Earth. In the 1960's NASA was spending up to and above 4% of US budget, compared to less than 1% in 2012 and further decreasing today.[5]

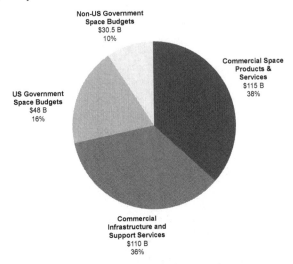

**Figure 1**
**Breakdown of Global Space Activity, 2012**
(source: www.sciencefoundation.org)

Space markets have experienced significant growth over the past two decades as outlined in *The Space Report, 2012*.[6] The breakdown of global space activity in 2011; notably, commercial space products and services account for 38% of spending. The total global space economy in 2011 was almost $290 billion, an increase of 41 percent from 2006. It is necessary now to evaluate and explore different possibilities for funding new space initiatives, enabling a continuing growth in commercial space activity. One model that could be used for financing space science missions would be for a prime contractor to bid a fixed cost mission to a governmental customer.

---

5   http://www.guardian.co.uk/news/datablog/2010/feb/01/nasa-budgets-us-spending-space-travel#data
6   *The Space Report, 2012*, 'The Authoritative Guide to Global Space Activity,' Space Foundation (www.sciencefoundation.org)

The governmental customer could provide a guaranteed contract for a multiyear purchase of data, from the contractor. The prime contractor with the data contract in hand can then shop the private financial markets for financing. There are two immediate advantages, the elimination of cost plus contracting and placing the focus predominately on the acquisition of valuable data with less focus on hardware and technology. This approach is likely to be controversial among the traditional government contractor community, but must be considered as the current path does not appear to be sustainable. The future will likely accommodate both fixed price contracting, as well as private financing models.

Recently FlightGlobal highlighted that while Private Finance Initiative (PFI) funding schemes 'may have been found to be poor financial value in hospital and school construction projects, the role of private funding in space has a better record' in the UK.[7] The Paradigm PFI funding of the UK Skynet satellite communications contract is an excellent example. Paradigm, a company owned by the European giant EADS, supplies military satellite communications primarily to the UK armed forces. Paradigm is the prime contractor for the Skynet 5 contract with the UK Ministry of Defense (MoD) with services including radio signal types and bandwidths for the UK armed forces, as well as a range of other governments and agencies. As a separate corporate entity, Paradigm operates the Skynet 5 satellite system and is therefore responsible for delivering the service, managing the network, controlling the satellites and implementing the upgrades throughout its life.[8] The service contract between Paradigm and the MoD contains service level agreements; the price varies with the service performance achieved and underpinning the contract are government guarantees for overall usage.[9]

Any extra communications capacity revenues are shared under a profit sharing agreement. The risk is transferred to the operator in return for guaranteed revenues from the government side. This model has succeeded for the Skynet constellation. Jose Del Rosario, Senior Analyst for Northern Sky Research (NSR) stated: "The budget crunch is becoming more pronounced has taken centre stage in procurement decisions. As such,

---

[7]  London Satellite Finance and Industry Conference: PPP Seed funding liked but Export Credit funding may skew market, 2012(www.FlightGlobal.com)

[8]  Website for "Government Communications The world's most advanced, resilient, secure and flexible military satellite communications service" http://www.paradigmsecure.com

[9]  D. Iron and K. Davidian, 2008, "Applying the UK's PPP Lessons to NASA's Commercial Development Policy," AIAA Space Conference & Exposition, September 9 – 12, 2008, San Diego, California

including hosted payloads and public–private partnerships (PPP) like the Paradigm-UK MoD model may be some of the ways to move forward to address fiscal constraints. Recent developments likewise signal increased non-U.S. participation in overall spending where cash-strapped allies are finding solutions within the commercial satellite industry that fit their own budget limitations."[10]

In November 2012 the British Ministry of Defense announced that it would take ownership of the Skynet system in 2022 at the end of the long-term bandwidth supply contract with Astrium Services. This transfer is planned to occur without payment and appears to be a strategic move by the government to enhance its negotiating leverage during the next communications acquisition process.[11]

The Skynet/Paradigm model could be modified as follows: Consider the example in which a federal government is willing to spend $100,000,000 to accomplish a certain data gathering mission. Typically this budget would, after selection of the implementing organization, be distributed on an annual basis in varying amounts, which may or may not actually add up to the original $100,000,000.

As an alternative financing model, assume the federal government will enter a contract with a space service provider for $13 million per year for ten years. This gives the contract a present value cost to the federal government of approximatley $100 million using a governmental discount rate of five percent (5%). This amount also assumes that both hard and soft costs and appropriate return on capital are provided to the contractor. Assuming a blended cost of funds (debt and equity) of 12% for the contractor the value of the contract to the private sector firm would be $73 million. For such a contract to be viable, the contractor must be granted operational control of how the service is to be provided. This includes a certain level of flexibility or discretion in the areas of technical solutions, programming, and management. Unfortunately, the current CCDEV demonstrates NASA's resistance to such a level of discretion to the private sector.

Let's assume the private sector is willing to place its capital at risk on this project for 15%, a typical return on a low risk subsidized rent affordable housing project. That results in a project budget of about

[10]  http://www.satellite-evolution.com/group/site/?p=8154
[11]  "British Military To Take Ownership of Skynet Satellites in 2022," Nov. 27, 2012. http://www.spacenews.com/article/british-military-to-take-ownership-of-skynet-satellites-in-2022#.ULodbqXk4co

$65,000,000 to accomplish the mission as compared to the traditional cost plus government contract.

Or, alternatively, the US Federal Government may be willing to underwrite the private sector's return requirements resulting in an annual contract of about $20,000,000/yr@10/yr or $200,000,000. This new "return" variable will make projects more challenging.

Naturally, the implementing organization will be trying to lower cost of capital, lower project costs and push the government to pay for the required returns. The implementing organization will further explore other entrepreneurial inputs such as finding additional customers for the data to enhance the revenue stream, partnering with no-cost partners (universities), leveraged financial structures, naming rights, or other monetization strategies such as selling project derived technology developments.

The Skynet example shows that this financial flexibility can work, and is worth exploring on larger scales. It appears reasonable to assume that capital is available at a price that will buy a stream of payments from the government for a return of their capital and a healthy return on it. This will drive the state of the industry to greater reliability, modularity and new financing structures probably comparable to early trade missions during colonial times, high risk real estate, or media production financing models.

It will really help when the industry matures and there are institutional sources of syndicated/pooled equity money to spread risk across multiple investors and debt that will help leverage returns for investors. When the capital markets understand the market, technology and the risks and the total project can be syndicated or spread in some form, the costs of capital will drop, making project financing easier.

The Sea-viewing Wide Field-of-view Sensor (SeaWiFs) project office was initiated at the NASA Goddard Space Flight Center in 1990 and seven years later the sensor was launched by Orbital Sciences Corporation (OSC) under a data-buy contract to provide 5 years of science quality data for global ocean-biogeochemistry research.[12] NASA was to have insight, but not oversight, into the spacecraft and sensor design, construction, and testing. Under the contract, OSC was responsible for building, launching, and operating the spacecraft. The payment schedule was front-end loaded so that most of the fixed costs were paid at the completion of specific

[12] C. McClain, G. Feldman and S. Hooker, (2004) 'An Overview of the Sea WiFS Project and Strategies for Producing a Climate Research Quality Global Ocean Bio-Optical Time Series', *Deep Sea Research II Tropical Studies in Oceanography*, Vol. 51, Issues 1-3

milestones during the prelaunch system development and post launch data acceptance phases.

After acceptance, fixed monthly payments were made, subject to penalties if the data quality was less than nominal. In McClain et al. (2004) it is highlighted that by having scientists with vested interests in the data as element leaders: a commitment to data quality and data access, and a focus on research and development in every aspect of the mission was ensured. In addition, in the post-launch phase approximately 80-90% of the SeaWiFs Project Office (SPO) personnel were onsite contractors provided under support services contracts defined, completed and negotiated by the SPO management, helping to ensure SPO received the best support available. Many of the aspects of this program were implemented as a result of the lessons learned from the proof-of-concept Costal Zone Colour Scanner (CZCS), which provided an understanding of what needed to be done differently in an operational setting.

With this in mind, it needs to be further investigated how the successful financial models used to fund non-space science missions can be used as examples of lessons learned, in transferring the successful models into funding of new space-science missions in the world of ever contracting budgets. In these times of global change, the world is striving to face and adapt to inevitable, possibly profound, alteration. But the question in relation to private financing of space science missions is, How?

Smaller spacecraft could also mean smaller rockets, and therefore lower launch costs. Sending large satellites into orbit on the workhorse Atlas 5 rocket can cost from $102 million to $334 million per trip. Some examples of small spacecraft are the suitcase-sized "Canadian Hubble telescope," and any number of cubesat (loaf of bread-sized) science platforms currently orbiting the Earth. So-called "Fractionated spacecraft," in which the various subsystems are free-flying, without a containing structure, could further reduce costs and increase launch flexibility. Satellite designers at the U.S. Defense Advanced Research Projects Agency (DARPA) are pushing forward with a project to develop fractionated satellites - or small, networked clusters or orbiting satellite modules. The System F6 program seeks to develop the enabling technologies for fractionated spacecraft, which consist of small, individually launched, wirelessly networked, and cluster-flown spacecraft modules designed to replace large, monolithic satellites.[13]

---

[13]  DARPA moves ahead with fractionated-satellite System F6 program with solicitation for affordable satellite bus May 20, 2012.
http://www.militaryaerospace.com/articles/2012/05/darpa-moves-ahead-with-

Such architecture enhances the adaptability and survivability of space systems, while shortening development timelines and reducing the barrier-to-entry for participation in the national security space industry.[14] The general philosophy that underlies the technical approach and structure of the System F6 program is to arrive at on-orbit functional demonstrations through a disaggregated series of efforts.

DARPA has also begun a program to demonstrate that small satellites produced and launched on demand can provide imagery on request directly to individual soldiers, named the Space Enabled Effects for Military Engagements (SeeMe). DARPA's goal is to show that a constellation of 24 satellites, each weighing less than 100 lb., can be launched into Low Earth Orbit (LEO) at a fraction of the cost of acquiring additional unmanned aircraft to provide the same imagery.[15] Raytheon has received the first contract of $1.5 million for the nine-month first phase to design a small imaging satellite. DARPA says other contracts will be awarded as well. With support from the National Science Foundation (NSF) and NASA's Educational Launch of Nanosatellites (ELaNa) Initiative, researchers from SRI International and the University of Michigan obtained the first-ever measurement of naturally occurring auroral turbulence recorded using a nanosatellite radar receiver.[16] The National Science Foundation (NSF) is an independent federal agency created by Congress in 1950 and is a major source of federal backing in many fields such as mathematics, computer science and the social sciences. NASA's Educational Launch of Nanosatellites (ELaNa), managed by the Launch Services Program (LSP) at NASA's Kennedy Space Center, is an initiative created by NASA to attract and retain students in the science, technology, engineering and mathematics disciplines. ELaNa missions were the first educational cargo to be carried on expendable launch vehicles for LSP.[17]

As outlined in the USAF Posture Statement 2012, there is an ongoing

fractionated-satellite-system-f6-program-with-solicitationfor-affordable-satellite-bus.html

[14] System F6: http://www.darpa.mil/Our_Work/tto/Programs/System_F6.aspx

[15] DARPA Plans Smallsat Imaging Constellation; http://www.aviationweek.com/Article.aspx?id=/article-xml/awx_12_18_2012_p0-530082.xml

[16] "Space Weather Researchers from SRI International and the University of Michigan Take First-Ever Measurement of Auroral Turbulence Using a Nanosatellite Radar Receiver." March 22, 2012 http://www.sri.com/newsroom/press-releases/first-measurement-auroral-turbulence-nanosatellite-radar-receiver

[17] Project ELaNa: Launching Education into Space. http://www.nasa.gov/mission_pages/smallsats/elana/index.html

need to plan, design and implement space advancements as a result of the rapid technology advancements and the long-lead time associated with new space technologies. The acquisition strategy for the Efficient Space Procurement (ESP) of complex space systems is designed to identify efficiencies, to provide enduring capability and help provide stability to the space industrial base.[18] The current fiscal environment demands exploration of alternate paths to provide resilient solutions. As discussed in the Posture Statement, the USAF is incorporating tenets of the National Space Policy and National Security Space Strategy and are "actively developing architectures that take into consideration the advantages of leveraging international partnerships and commercial space capabilities."

It is clear that budget constraints are driving a shift in contracting mechanisms.[19] While operationally responsive space didn't originally have "low cost" as an imperative, it is now a selling point. Is that the answer to the budget predicament? Hosted payloads are one option, in which military sensor, communications or other technologies piggyback on commercial satellites.[20] The US Air Force Research Laboratory (AFRL) has awarded $34 million to Orbital Sciences to develop ESPA Augmented Geostationary Laboratory Equipment (Eagle), a system for boosting to and keeping small payloads functioning at geostationary orbit.[21]

In 2011, the launch of CHIRP marked the first ever commercially hosted payload for the US Air Force.[22] This arrangement illustrates how government customers can gain affordable access to space with the help of a commercial satellite operator. The CHIRP team is a government-industry collaboration led by the Air Force's Space and Missile System Center Development Planning Directorate. NASA also has a program in which science instruments piggyback on commercial GEO satellites. A $90 million NASA instrument mounted on a commercial communications

---

[18]   United States Posture Statement, February 28, 2012.
[19]   Military Space Quarterly | USAF Eyes Purchase of Commercial Satellite Platforms Jul. 11, 2012
       http://www.spacenews.com/article/usaf-eyes-purchase-commercial-satellite-platforms#.ULof5qXk4co
[20]   "Air Force Embraces Small Satellites As Budget Outlook Grows Dim," *National Defense*, July 2011.
       http://www.nationaldefensemagazine.org/archive/2011/July/Pages/AirForce EmbracesSmallSatellitesAsBudgetOutlookGrowsDim.aspx
[21]   "US Air Force Funds Small Satellite Platform." October 24, 2012.
       http://www.flightglobal.com/news/articles/us-air-force-funds-smallsatellite-platform-378045/
[22]   Air Force Commercially Hosted Infrared Payload Launches Successfully from Guiana Space Center, Kourou, French Guiana, September, 2011.
       http://www.afspc.af.mil/news/story.asp?id=123273037

satellite in geostationary orbit will monitor air pollutants over North America beginning in 2017.[23] Another option is leveraging the services these private companies offer or carrying out joint programs with allies. Smaller satellites are also an option: "Can we distribute sensors and network them together so we have smaller satellites that give you adequate capability when they are networked together?"

Not only is the emergence of commercial space creating new technologies, it is leading to innovative models for funding such projects. Today, very small satellites are used primarily for educational and technology demonstrations. However, defense and civil agencies are demonstrating new applications such as remote sensing, communications and fractioned satellite architectures to support future systems.[24] The US Army is in the process of developing a family of satellites under 25kg, Kestral Eye imaging satellites and SNAP communication satellites. Changing technology impacts the way business is done, impacts the cost of missions, the key players involved in addition to financing initiatives.

## III. THE PFI EXPERIENCE IN THE UK

Public Private Partnerships (PPPs) or P3s are partnerships between public sector organizations and private sector investors and businesses for the purpose of designing, building, planning, operating and financing projects normally provided through traditional procurement procedures by the government. P3 is not just about the private sector financing capital projects in return for an income stream, but also makes use of the private sector skills and management expertise to deliver and operate public projects more efficiently over their lifetime.[25] As outlined in the Policy Framework for P3s in Ireland, the benefits of a P3 are that the public and private sectors will have certain advantages relative to each other and these advantages can be exploited to achieve the most economically efficient agreement.

In addition, the risks are identified at the outset and should be placed

---

[23] "NASA Instrument On Commercial Sat To Track U.S. Air Pollution," November 2012.
http://www.aviationweek.com/Article.aspx?id=/article-xml/asd_11_12_2012_p04-01-515368.xml

[24] "Leveraging small satellites and commercial space for NASA technology infusion."
http://lunarscience.nasa.gov/articles/leveraging-small-satellites-and-commercial-space-nasa-technology-infusion/

[25] "The Policy Framework for Public Private Partnership in Ireland: Evolution of PPP in Ireland," November 2003.

with the party best able to manage them. As the private sector is paid according to their performance, the incentive to perform is heightened. In the UK, since the initial launch of PFI in 1992, PFI has matured and the emphasis has shifted. It is now clear that the impact on government borrowing is much less significant than at first supposed, and P3/PFI are now being seen as essentially a new approach to risk allocation in public infrastructure projects.[26]

It is estimated that the UK's experience of P3/PFI includes over 700 contracts that altogether raised over $130Bn of private investment (Iron and Davidian, 2008).

According to a report issued in 2009 by Professor Mustafa Alshawi, of Private Finance Initiative (PFI) is 'a public service delivery type of P3 where the responsibility for providing public services is transferred from the public to the private sector for a considerable period of time.'[27] In developing countries, the high demand for infrastructure development coupled with pressure on national budgets is making governments move toward encouraging the private sector to invest in infrastructure projects. The aim of PFI is to bring the private sector's finance, management skills and expertise into the provision of public sector facilities and services.[28] Separation of responsibilities and work teams involved in the P3 is a key aspects that make these types of partnerships successful and efficient, and essentially is a mechanism for mitigating risk.

In PFI projects, the private sector develops, finances and maintains an asset used in delivery of public services. This means that the government departments purchase services from the private sector rather than being the owner and operator of the asset. As discussed in Bing et al. (2005), only 15% of construction cost and 13.2% of the operation Net Present Value (NPV) cost of the fifty-three PFI projects they surveyed were less than £10 million, which demonstrates some of the difficulties for smaller companies to engage in the larger scale P3s. In addition, this type of contract can add significant complexity and requirements between the concerned parties, which the smaller companies are often not capable of dealing with. In these types of projects, design and construction become integrated up-front, and the ongoing service delivery, operation,

---

[26]  Bing et al. (2005) "The Allocation of Risk in PPP/PFI Construction Projects in the UK." *International Journal of Project Management*, Vol. 23, pp. 25-35.

[27]  M. Alshawi, 2009, 'Concept and Background to Public Private Partnership (PPP) / Private Finance Initiative (PFI), UK Experience', I.I.E.R.

[28]  P. Carrillo, H. Robinson, P. Foale, C. Anumba and D. Bouchlaghem, 2008, 'Participation, Barriers and Opportunities in PFI: The United Kingdom Experience,' *Journal of Management in Engineering*, ASCE.

maintenance costs become a single party's responsibility for the duration of the concession period. This makes the bidding costs in these projects considerably higher.

According to Iron and Davidian (2008), there are two examples in the space sector that stand out, Skynet and Galileo. Galileo failed to obtain significant PFI due to political and institutional difficulties and because there was no guaranteed revenue.[29] In a press release issued in 2007 by the BBC News, the UK transport minister told the House of Commons that "Galileo is considered a key [European] Community project, but we are clear that it cannot be carried out at any price; it has to be affordable and has to be value for money."

> "It needs better governance and risk management, open competition and a firm focus on the opportunities for getting the private sector to share the costs and risks."[30] This statement reinforces the role of P3 in the space industry, and others in general, that clear and defined risks to the project must be addressed at an early stage in the project, and the financing of these projects must be viable for industry as well as government.

In the US, financial constraints have forced government agencies to outsource many aspects of its projects to private companies. For example, NASA established a two-phased approach to stimulate the commercial space transportation industry.

The first stage is the development and demonstration phase, in which NASA is investing finances within the private sector to develop and demonstrate cost-effective space transportation capabilities. The program is investing financial and technical resources to stimulate efforts within the private sector to develop and demonstrate safe, reliable, and cost-effective space transportation capabilities.[31] The COTS program was designed to facilitate companies in the development and demonstration of cargo vehicles and systems, driving the involvement and participation of the private sector.

Phase 2 is the ISS Commercial Resupply Services (CRS) which are managed by the ISS Transportation Office.[32] In 2008, two contracts were

---

[29] London Satellite Finance and Industry Conference: PPP Seed funding liked but Export Credit funding may skew market, 2012

[30] UK Presses Private Galileo Role, July 3, 2007.
http://news.bbc.co.uk/2/hi/science/nature/6266264.stm

[31] NASA's Commercial Crew and Cargo Program Website.
http://www.nasa.gov/offices/c3po/home/index.html

[32] http://www.nasa.gov/offices/c3po/about/c3po.html

awarded to SpaceX and Orbital, demonstrating a break from the traditional contracting mechanisms at NASA, a shift from the traditional NASA approach of creating government owned and operated systems and working in a government/contractor relationship with industry. These shrinking budget constraints have opened the doors in driving innovation and creativity, allowing new player's access to the space community. In October 2012, SpaceX's Dragon capsule completed NASA's first commercial resupply mission to the ISS. Orbital was awarded the other contract for the ISS CRS, and will carry out the design, manufacturing and testing of the Antares launch vehicle and the development, production and integration of the Cygnus spacecraft and cargo modules. The ISS CRS will provide a US-produced and operated automated cargo delivery service in an innovative partnership reflective of the changing dynamics of funding space initiatives. For space science missions, the focus becomes a market risk issue with regards the financing of the mission.

According to NASA, a key feature of the COTS Space Act Agreement is that NASA will make fixed-price payments only when performance based milestones are completed according to objective success criteria. NASA's investment will not result in the government assuming any equity in the participating companies and as a result will benefit from lower costs to the ISS. If a similar rationale were to be applied to the financing of space science missions, the issue of market risk needs to assessed. The reduced market risk associated with the COTS program can be related back to the fact that as the ISS needs cargo delivery, there is a real market and customer.

With the retirement of the Space Shuttle, there was an opportunity to deviate from the usual business case and allow more commercial involvement in not only accessing space, but also working with key government players. This partnership between government and industry, in some cases, can reduce the barriers to entry into certain markets, for example the System F6 program. In a Futron report issued in 2011, the Isle of Man's competitive position was benchmarked against five peer jurisdictions, Bermuda, Gibraltar, Hong Kong, Singapore and the United Kingdom. The report found that the Isle of Man possesses a unique strategy and approach that has resulted in significant economic benefits to the people and business on the Island.

One of the key findings highlighted that as part of the government structure and policy through innovative government industry relationships, such as the unique use of a Public Private Partnership to process ITU orbital slot filings (ManSat), the Isle of Man appears to be at the leading edge of a new industry model in space finance and services.[33]

In a similar manner, funding of space science missions needs to be addressed, and the development of such relationships between industry and government examined in order to provide an infrastructure for innovative models for funding.

Recently Brazil's Embraer and Telebras signed a shareholder agreement to form Visiona in which Embraer will hold 51% of the share capital and Telebras will hold the rest. Visiona will focus on the Brazilian Geosynchronous Satellite, which will serve the satellite communications needs of the Federal Government. Moreover, it will support the National Broadband Program as well as strategic defense transmissions. With this type of arrangement, the market risk is mitigated because the government has direct responsibility and involvement. For example, in the case of P3 arrangement, there is a certain element of market risk associated with the contracts. Government structures change, government plans and priorities change and most importantly government finance allocation changes. In forming a partnership between government and industry to minimize the market risk associated with space missions, huge barriers can be overcome. For the financing of space science missions, the market risk associated with the mission needs to be analyzed in order to overcome this problem. If the market issue can be addressed, confidence could be instilled in the private sector to pursue space missions.

The National Geospatial-Intelligence Agency (NGA) provides timely, relevant and accurate geospatial intelligence in support of national security. ClearView and NextView are NGA's primary image purchase vehicles and provide for purchase of U.S high resolution imagery and limited 'Value Added' products/services.[34] It is highlighted in this report that the NGA was 'aggressively' integrating commercially remote sensed imagery into its business plans, processes and architecture.

Enhanced View is a 10-year NGA contract, signed in 2010, with GeoEye and DigitalGlobe to purchase imagery on a service level agreement basis[35] In June 2012, GeoEye received $111 million from the NGA in a cost-share payment for the construction of GeoEye-2. The

[33] Innovative Strategies for Space Competitiveness: Assessing the Space Isle's Policy and Results, February 1, 2011.
http://www.futron.com/upload/wysiwyg/Resources/Whitepapers/Innovative_Strateg ies_for_Space_Competitiveness_0111.pdf

[34] D. McGovern, "NGA Commercial Imagery Acquisition Programs," US Chamber of Commerce Public Workshop, October 19th 2006

[35] "Enhanced View News not so Rosy for GeoEye," June 2012
http://www.spacepolicyonline.com/news/enhancedview-news-not-so-rosy-for-geoeye

development of this satellite is part of NGA's EnhancedView program.[36] However, in that same month an article published in *SatelliteToday* expressed the concerns for GeoEye as a result of NGA not renewing the EnhancedView contract it awarded them.

GeoEye shares fell 22 percent as a result of the announcement[37] and with the EnhancedView program accounting for 41 percent of GeoEye's 2011 full-year revenue, the survival of GeoEye relies on extensions and renewals of contracts. However, the loss of EnhancedView is not only potentially fatal to the future of GeoEye, the loss of the contract would make GeoEye an ironic potential take-over target as the company offered to buy DigitalGlobe in May 2012.[38] This again highlights the problems associated with market risk in space missions; with no long-term guarantee of sources of funding the consequences to the commercial sector can be significant.

There needs to be real downside risk for contract cancellation for the government, and incentives structured to mitigate the market risk associated with the missions. The long term planning and contract type is essential in sustaining space science missions, and the possibility of including clear incentives and penalties in the agreements with regards the financing of the project should be addressed.

Without a real downside for a government in canceling a contract, it becomes much more difficult to mitigate this issue associated with the market risk and therefore much more difficult to finance space science missions.

As outlined in Iron and Davidian (2008) (Applying the UK's P3 Lessons to NASA's Commercial Development Policy), the Exploration Commercial Development Policy (ECDP) encourages funding of commercial companies with capability demonstrations of technologies of a mid-level Technical Readiness Level. With this approach, the commercial company has the "opportunity to license pre-existing technology to the government in exchange for a royalty, or permit the fixed price acquisition of the eventual operational capability by NASA."

---

[36]  "GeoEye Receives $111 Million Cost-Share Payment from the NGA" http://geoeye.mediaroom.com/2012-06-28-GeoEye-Receives-111-Million-Cost-Share-Payment-from-the-NGA

[37]  "GeoEyes Shares Drop with latest Enhanced View Twist," June 2012. http://defensesystems.com/articles/2012/06/28/agg-geoeye-shares-nga-satellite-development.aspx

[38]  "GeoEye Mulls Survival Options After Losing NGA AdvancedView Contract." June 26, 2012. http://www.satellitetoday.com/publications/st/2012/06/26/geoeye-mulls-survival-options-after-losing-nga-enhancedview-contract/

## IV. PRIVATIZATION OF A SPACE TELESCOPE

NASA recently signed a Space Act agreement which turns the nine-year-old, $150-million Galaxy Evolution Explorer (Galex) ultraviolet space telescope over to the California Institute of Technology. Caltech will take the responsibility to raise money to put the telescope back into operation.

The Keck Institute of Space Studies, a consortium of Israeli universities led by the Weizmann Institute of Science near Tel Aviv, Cornell University, and an international consortium named GAMA/Herschel-Atlas/DINGO has committed resources. Caltech will continue to pursue additional funding from private donors, foundations, as well as from corporate sponsors.

"If this Space Act is successful, it might provide a model for other space telescopes such as Hubble, which NASA plans to decommission in a few years. NASA plans to replace Hubble with the James Webb Space Telescope, but Webb is designed primarily for observations in the infrared, rather than visible wavelengths. There is likely to be some unhappiness when the public realizes that the visible-light images, which have made Hubble so popular, will be coming to an end. With the advent of low-cost launch and space servicing, it might be possible to keep space observatories like Hubble operational indefinitely. The question is, will the low-cost revolution come soon enough, or will Hubble be deorbited before it arrives, much as Skylab came down while waiting for the Shuttle to become operational?"[39]

NASA uses Space Act Agreements as the primary vehicle for partnering with the external community. Space Act Agreements enable us to enter into partnerships with organizations that give us access to a wider range of technologies and capabilities that are not part of NASA's core competency.[40]

## V. INNOVATIVE PARTNERSHIPS & COMMERCIAL INFUSION

NASA's Innovative Partnerships Program (IPP) provides needed technology and capabilities for NASA's Mission Directorates, Programs, and Projects through investments and partnerships with Industry,

[39] DARPA unveiled a new program to engage amateur astronomers in helping to protect satellites. http://www.citizensinspace.org/category/citizen-science-blog/astronomy/
[40] The NASA Space Act Agreement: Partnering with NASA http://www.nasa.gov/open/plan/space-act.html

Academia, Government. Agencies, and National Laboratories. As one of NASA's Mission Support Offices, IPP supports all four Mission Directorates and has Program Offices at each of the NASA Centers. In addition to leveraged technology investments, dual-use technology-related partnerships, and technology solutions for NASA, IPP enables cost avoidance, and accelerates technology maturation.

IPP consists of the following program elements: Technology Infusion which includes the Small Business Innovative Research (SBIR)/ Small Business Technology Transfer (STTR) Programs and the IPP Seed Fund; Innovation Incubator which includes Centennial Challenges and new efforts such as facilitating the purchase of services from the emerging commercial space sector; and Partnership Development which includes Intellectual Property Management and Technology Transfer, and new innovative partnerships.[41]

## VI. OTHER FUNDING INITIATIVES

According to the Solar System Research Institute (formerly NASA Lunar Science Institute) discussion of small satellites and commercial space has great potential:

"A wide range of science can be accomplished by leveraging current small satellite technology. For example, by maintaining organisms in a dry state for months and then rehydrating them on command, the effects of significant exposure to space radiation can be assessed. These experiments could also resolve fundamental biological phenomena related to human health & safety, measure dust toxicity and the effects of long-term exposure on the lunar surface, and help determine effectiveness of radiation shielding. Small satellite payloads such as GeneSat and PharmaSat could gather critical information to provide in-situ validation of long-term radiation effects in support of extended missions to the Moon, Mars or asteroids.

Tomorrow's explorers are eager to launch their payloads and scientific experiments to shed light on the most pressing questions for a sustained human presence in space. In some ways, the technology is like a cosmic hitchhiker—all it needs is a ride. Given today's budget constraints and the high price of access to space, researchers are looking more and more to the ride-along concept— with someone else providing the vehicle and fuel, with the payload going to wherever the driver is already heading. But where are these rides?

---

[41] "Technology Development and Infusion from NASA's Innovative Partnerships Program." Douglas A. Comstock

The Google Lunar X Prize (GLXP) is igniting a new era of lunar exploration by offering the largest international incentive prize of all time. These teams must be at least 90% privately funded, although sales to government customers are allowed without limit, and thus GLXP teams could offer excellent platforms for NASA science and technology payloads.

A growing number of commercial, government, military, and civilian space launches now carry small "secondary" science payloads at far lower cost than dedicated missions; the number of opportunities is particularly large for so-called cube-sat and multicube satellites in the 1 – 10 kg range. Small payload instruments may also fit into unused mass margins of larger lunar or planetary lander, orbiter, or even impactor missions. The tools of bio- and micro-technology combined with automation and integration enable a range of experiments in small space platforms.

Dr. Antonio Ricco of NASA Ames Research Center recently gave an overview of technologies for low-cost autonomous small satellites & payload instruments as part of the Lunar Commerce Virtual Seminar Series.

During even the most fiscally challenging times, NASA is still making big strides towards discovery with its small instruments and payloads. Leveraging commercial space may allow the Agency to advance science and technology while, at the same time, furthering its broader goals for future human exploration."[42]

The commercial space industry has evolved as part of the global space activity and is driving innovations and creation of new technologies and space based applications. In order to access this commercial space, it is essential that companies generate the necessary funding to realise their potential.

KickStarter, a funding platform launched in 2009, provides the necessary infrastructure that 'creative projects' utilise in an effort to generate funding for the project. Since its launch it has provided the infrastructure that has enabled funding for more than 30,000 projects, with over $35 million coming from pledges by more than 2.5 million people (www.kickstarter.com). For example, 'KickSat', a personalised miniature satellite project, raised almost 250% of the funding needed through KickStarter.

Another example of a space based project called SkyCube was successfully funded recently. SkyCube is a nano-satellite that allows bidders to take Earth images and 'tweet' from space. Through the use of

---

[42] "Leveraging small satellites and commercial space for NASA technology infusion." http://lunarscience.nasa.gov/articles/leveraging-small-satellites-and-commercial-space-nasa-technology-infusion/

such a platform like Kick-Starter, not only are necessary funds generated but in addition the project generates support from many different people and the profile of the project is raised. 'Backing a project is more than just giving someone money, its supporting their dream to create something that they want to see exist in the world' (www.kickstarter.com).

Prize competitions have been used since the 18th century to accelerate the development of a many different technologies. Prizes have stimulated technological advances and enhanced various sectors in society including industry, military, public health and safety and tourism/adventure.[43] There were numerous airship and aeronautical prizes offered in the early 20th century including the Deutsch Prize, the Daily Mail English Channel Crossing Prize and Orteig Prize. With the evolution of space exploration, in more recent years the popularity of prize competition has increased and some landmark events, including the Ansari X Prize, have highlighted the success of utilizing this approach. The most successful competitions produce excitement among the public, media, and educators, and this in turn encourages young people to pursue careers in that area. Government research grants typically require that the funding agency determines who will receive the funds to achieve a certain goal and determine the different approaches for achieving that goal. In contrast, public inducement prizes allow the government to establish a goal without being prescriptive as to how the goal should be met or who is in the best position to achieve that goal.[44] This was vividly illustrated by the outcome of the Orteig Prize in the early 20th century. Taking the ten million dollar Ansari X Prize, which was financed by a one million dollar insurance policy as an example, the X Prize Foundation reported that the prize stimulated at least one hundred million dollars in private sector investment. Kalil (2006) highlights that with the passage of NASA's 2005 authorisation legislation, NASA can sponsor a prize of any dollar amount, and can also accept matching funds from the private sector.

As noted in Kalil (2006) prizes can also address some of the problems that are associated with government support for applied R&D – "researchers funded on the basis of an outsider's assessment of potential rather than actual product delivery have incentives to exaggerate the prospects that their approach will succeed, and once they are funded, may even have incentives to divert resources away from the search for the

---

[43]  K. Davidian (2005), "Prize Competitions and NASA's Centennial Challenges Program," International Lunar Conference.

[44]  T. Kalil, (2006), "Prizes for Technological Innovation, The HamiltonProject – Advancing Opportunity, Prosperity and Growth."

desired product." The X-Prize Foundation's "Revolution through Competition" provides a large scale monetary award for the achievement of a specific goal set out by the Foundation. This method aims to incite innovation by tapping into the competitive and entrepreneurial characteristics, rather than directly funding research. The Foundation looks to accelerate the real world impact of science, technology and information and focuses on competitions to stimulate innovative breakthroughs to positively impact humanity.[45]

The use of prizes in advancing space science missions should not be neglected as a viable source of funding as this method has proved to be both popular and successful. Not only can this method attract teams with fresh ideas, it mitigates the situation where the government provides a grant or contract and pays even if the recipient is unsuccessful.

## VII. OPENSOURCE SPACE

Crowdsourcing is when a person or company puts out a call to the world for ideas relating to a particular job that person or company wants done.[46] As highlighted in a recent report, some agency programs are falling victim to the sweeping cuts as NASA struggles to balance budgets.[47] The start-up company called Uwingu was founded by a team of noted astronomers, educators, scientists and other industry officials with the hope of privately raising millions of dollars to fund scientific research and space exploration projects. The company launched a crowdsourcing campaign to raise a minimum of $75,000 to officially launch the company and funs its operation.

A new method emerging: 'selective crowdsourcing' or 'smartsourcing' to foster collaboration between experts and identify a particular community of skilled persons who may not otherwise communicate with each other. Take for example ResearchGate, which provides a platform for scientists and researchers to network and connect with a global scientific community.

---

[45] Prize Development http://www.xprize.org/prize-development
[46] Jason Schreier. "How To Selective Crowdsourcing: The More Experts, the Merrier." InnovationNews Daily, May 3, 2011.
http://www.technewsdaily.com/2517-selective-crowdsourcing-science.html
[47] Denise Chow. "New Start-Up Aims to Fill Funding Gap for Space Projects," SPACE.com, August 3, 2012.
http://www.space.com/16908-space-projects-private-funding-uwingu.html

## VIII. CONCLUSION: RISKS AND GUARANTEES

Traditional financing of space exploration and development models and projects are being severely tested. "The real risk is all 'customer/market' related, which drives up the cost of capital and lowers access to capital at the same time. The government is an awful customer. It changes its mind. You can get paid this year, but next year they will not put funding in the budget. Even if you have a deal with the current government, and that government leadership is truly committed to the deal, governments change. The next government can decide to not keep the contract commitment made by the previous government."[48]

On the other hand, "anything is financeable if a credible revenue stream, market validated income stream, contractual payments, or governmental commitment/guarantee is available."[49]

As we have shown in this chapter, there are emerging frameworks for innovative financing future commercial space and scientific missions. While there are still further areas of further research required to determine how best to optimize P3s for space science missions, promising areas to examine include the recent lessons learned in more mature industries such as utility scale energy.

Key questions remain: What are the key lessons, and how can these lessons benefit private partnerships in the area of space science? How can legal and regulatory barriers to P3 be minimized through the legislative process, to facilitate multiyear purchase agreements from private parties? Innovative finance cannot remedy inadequate strategy, poor policy, or ill conceived efforts to create artificial industrial base, or other sub-optimal political outcomes, but in the right applications they can enable tremendous accomplishments in space, and innovative structures may allow humanity to pursue critical space activities for the benefit of all humanity in a more financially efficient and results oriented fashion. Rather then saying farewell to the "golden era" of the Space Age, we must encourage creativity, flexibility and new approaches to financial models.

Adapt or perish!

•••

---

[48]   Charles Miller, President NextGen Space LLC
[49]   Justin McCarthy, Economic Development Executive

Special thank you to the following for reviewing earlier drafts of this chapter: Justin McCarthy, Economic Development Executive, Charles Miller, President of NexGen Space LLC, and Ken Davidian of the FAA. A version of this paper was presented at the AIAA Space2013 Space Systems Engineering and Economics Track on the subject of Innovative Finance, San Diego, California September 12, 2013.

## Jeffrey Nosanov

Jeffrey Nosanov is the Program Planning and Assessment Lead, Nuclear Space Power Office at JPL, a NIAC Fellow at NASA, Visiting Research Associate UCLA, Visiting Associate, California Institute of Technology, University of Nebraska-Lincoln LLM Space & Telecommunications Law, New York Law School JD Law.

## Norah Patten

Norah Patten, Chair - Space Management and Business Department at International Space University. She worked as part of the Space Studies Program at ISU since 2011 (Teaching Associate in 2011 and Emerging Chair in 2012). She studied Aeronautical Engineering at the University of Limerick, where she spent 8-months working at The Boeing Company. She completed her Ph.D. in aerodynamics at the University of Limerick and during this time, completed an internship in the Thermal Management Research Group in Bell Labs Alcatel Lucent, Dublin, Ireland. Norah has worked at the University of Limerick, Ireland as a lecturer and project manager in the Mechanical, Aeronautical and Biomedical Engineering Department. Norah is keenly interested in the prospects of the establishment of an Irish space agency.

## Michael Potter

Michael Potter is a Senior Fellow at the International Institute of Space Commerce, Isle of Man. He serves as Director, of Paradigm Ventures a family investment firm focused on high technology ventures. Potter has served as faculty at the Singularity University. Previously Potter was Vice Chairman, founder and President of Esprit Telecom plc., a pan-European competitive telecommunications services provider. He was formerly an international telecommunications analyst at the Center for Strategic & International Studies (CSIS) in Washington, D.C. Potter was also Vice Chairman of the founding Board of the European Competitive Telecommunications Association (ECTA). Potter is an Advisory to Odyssey Moon and Space IL, and served as a member of the Board of Trustees of the International Space University. Potter is a member of the TED community.


## CHRISTOPHER STOTT

Chris Stott is Chairman and Chief Executive Officer of ManSat LLC. Originator of the ManSat concept, Chris is also ManSat's co-founder. Mr. Stott also serves as a Main Board Director of Odyssey Moon Ltd.

Serving on the Faculty of the International Space University (ISU) since 2003, Chris is also a Main Board Trustee of the University and a former Co-Chair of its School of Management & Business. He is a published Director of the International Institute of Space Commerce (IISC). Chris has served as faculty at the Singularity University.

Chris left his position as Director of International Commercialization & Sales with Lockheed Martin Space Operations in Houston to found ManSat Limited. Chris came to Lockheed Martin from the Boeing Space & Communications Company in Huntington Beach, California, where he worked International Business Development for the Delta Launch Vehicle program.

Mr. Stott was co-author of Europe's first work on space privatization, "A Space For Enterprise," via the Adam Smith Institute of London in 1994.

Mr. Stott is also the Chairman of the Manna Energy Foundation, and is active in space the space education community through his service on the Boards of the Challenger Centers, Conrad Foundation, the Foundation for International Space Education, and the Society of Satellite Professionals International. Mr. Stott is a donor and attendee of the TED conference series.

Chapter 15

# AN ALTERNATIVE MODEL FOR SPACE COMMERCE SUSTAINABILITY

Thomas E. Diegelman
NASA

## Introduction

"Commerce" is the act of making transactions between humans, and thus "space commerce" is these exchanges that involve or occur in outer space.

There is much talk about the emergence of space commerce, but after 50 years of being a spacefaring civilization we are yet to sustain a viable vision or program for commerce in outer space that is global, and in fact have used space for just about everything other than trading goods.

Humanity has traded weapons, threats, prestige, electronic signals, and some economic benefit, but little else.

Hence, this chapter addresses the question, "What is the necessary combination of people, resources, events and cultural ingredients that would constitute success in space commerce sufficient to attaining and then sustaining space commerce?" Clearly, the answer lies not in technology issues and concepts of spacefaring, as it is a human question, not a technology question.

This contrasts with a concept of space commerce defined around building rockets to loft things and people into orbit, or observational and communications satellites. While those activities are today considered to be space oriented commerce, the position taken here is that these are the prerequisites, but not true space commerce. Space commerce should mean creation and movement of goods for commercial purposes, and movement of people for commercial purposes as well.


## THE EXAMPLE: SALT

We gain insight into the prerequisites for the sustainability of Space Commerce by examining the history of last 3000 years as seen in the impact that salt, common sodium chloride, has had on human commerce, culture, and human quality of life.[1] Our intent here is to forge a model for sustainable space commerce that will continue to evolve, although hopefully more quickly than the 3000 years required for the commerce of salt to develop during human history.

This discussion draws from the story told in *Salt: A World History*, which traces written the history of the world with the commodity salt as the epicenter, and which explains why the world's trade routes were established, and why ships were designed as they were.[2]

Salt, the book explains, was valued as a sound currency in commerce, as a god in the coinage of many nations, and until roughly 100 years ago its possession meant power and control in commerce. It financed wars, caused wars, and inspired revolutions.

---

[1]   *Space Commerce*, ATWG 2010, ISBN 978-0-578-06578-6;
      *International Cooperation for the Development of Space*, ATWG, 2011. ISBN 9-781478-186235
[2]   Mark Kurlansky. *Salt: A World History*. Penguin Books, ISBN 0-802701373-4, 2002.

A few selected comments that highlight the points that will be made in our model development.[3] Our words "salary" and "salad" are both derived from Latin connoting salt. Roman soldiers were sometimes paid in salt, hence "salary;" and a "salad" is a collection of vegetables that is doused in a brine sauce before eating.

In Catholic Europe, the church forbade meat on Fridays and church holidays, which created an enormous market for fish. Before refrigeration, the only way to package and transport fish was through salting. Cod was the perfect fish for salting, and was discovered around this time and became an enormous culinary hit throughout Europe. The massive scale of the cod trade was enabled by salt, and it was the search for cod that eventually led explorers to North America.

"…..Until about 100 years ago, when modern chemistry and geology revealed how prevalent it is, salt was one of the most sought-after commodities, and no wonder, for without it humans and animals could not live. Salt has often been considered so valuable that it served as currency, and it is still exchanged as such in places today…. Demand for salt established the earliest trade routes, across unknown oceans and the remotest of deserts: the city of Jericho was founded almost 10,000 years ago as a salt trading center."

By comparison with the global extent of the salt trade, the extent today's space commerce occurs in a very limited scale trade within a single space faring community of technical experts, a miniscule community within the much larger human community. Hence, we can speculate that we are just emerging from the "flags and footprints" era of space endeavors, where the bulk of the commercial activity is based satellite technology that all still exists within the Earth's own gravity well, toward something quite different.

What we're looking for, then, is the next salt that is capable of opening space to commerce. How do we reconfigure the space commerce endeavor in the sophisticated, high tech, communications-intensive world of today to accommodate and exploit commerical goods, like salt, that would enable what is is currently a very limited, high tech industry, to become global in scale? For space will really be "for all mankind" only upon the discovery of the "next salt " from space.

The proposal put forth here, which cannot be proved or disproved, is a model that articulates an alternative approach to creating a sustainable space commerce system. In previous writings I have discussed the concept

---

[3]    http://www.bookrags.com/studyguide-salt/#gsc.tab=0
     https://www.bookbrowse.com/reviews/index.cfm/book_number/960/salt

of spaceports and the need to develop them, why they are functionally akin to terrestrial ports, and where they will likely be located.[4]   Here we will examine the nature of the commerce that is likely to emerge and which will make use of these ports.   And via these ports, we can speculate that the very interesting history of space commerce lies in the future, yet to be created or discovered.

Here are the steps we will follow:

- Recognize the reality of where we are in the space endeavor, and in particular, assess space commerce today with respect to achieving sustainable space commerce as a goal:
  - Specifically, we are just emerging from "flags and footprints" era that has left most of the population of the world without any direct benefit from space, only the spin-offs.
  - Where the spin-offs exist, while significant, they are not commerce as defined here, wherein people who are geographically separated are trading goods and services.
- Develop the checklist of pertinent questions that will grow more focused, robust and cohesive over time through use of the model and analysis of results.
- And lastly, suggest some terrestrial problems that, if solved, would directly benefit millions and billions of people, such that space and space commerce will be seen as investments in humanity.
  - This is backed up by data on what it would take to support a small city (roughly 5000 people) as a first approximation to off the planet needs of a trading commerce outpost.

## FOR ALL MANKIND

As with all endeavors of humanity as it reaches out on its learning quest, ironies are present.   Even the moon race of the 1960's has left the ultimate irony on the moon:

> *"Here men from the planet Earth first set foot upon the Moon, July 1969 AD.   We came in peace for all mankind."*
>
> Richard M. Nixon
> President of the United States

---

[4]   *Space Commerce*, ATWG 2010, ISBN 978-0-578-06578-6;
*International Cooperation for the Development of Space.*  ATWG 2011, ISBN 9-781478-186235

While this reflects well upon the intent of the foundation of NASA, for which it was stated that, "The Congress hereby declares that it is the policy of the United States that activities in space should be devoted to peaceful purposes for the benefit of all mankind."

However, Apollo did not involve the entire world then and still doesn't now, despite many spin-offs and benefits. As with any new technology, especially transportation, the first participants, called "early adopters" in technology circles, are more daredevil than entrepreneur.

And while the first participants were indeed taking risks "For All Mankind," there are currently more global billionaires than there are people who have been to space. In 2014, Forbes counted 1,645 billionaires with an aggregate net worth of $6.4 trillion,[5] while there have been fewer than 700 people in total in space since the dawn of the space age.

Space faring has been opened up to all races, with no gender barriers and limited age barriers, so we could make the case humanity is "maturing" in spacefaring, but in a larger global sense it is still seen as the domain of elite nations or "tribes," but not an open frontier with opportunity for all.

This does not however, render the progress in technology and partnerships such as on the International Space Station (ISS) meaningless, but it does serve to illustrate the difficulty of opening up an intra-tribal activity to other tribes.

As commerce is the linking mechanism that builds tribes into nations and nation-states, the Roman Empire grew through trade and cultural assimilation,[6] and while the world has certainly moved away from acceptance of "cultural assimilation," trade has much greater power today than it did in Roman times: The higher the standard of living achieved, the greater is the need for trade and commerce.

The question in focus, then, is, "How do we develop a sustainable space commerce system?"

The answer comes in two parts:

First, defining the system: What is it, how does it work, and what do we wish to control? Once that is understood, we can then approximate the system in a loose mathematical sense.

Second, What inputs are required, given initial conditions and definable system dynamics that give us the desired outputs?

---

[5] http://www.forbes.com/sites/luisakroll/2014/03/03/inside-the-2014-forbes-billionaires-list-facts-and-figures/

[6] http://en.wikipedia.org/wiki/List_of_languages_by_number_of_native_ speakers

The concept presented here is that the problems of space-faring and non-space faring nations (the distinction being at the moment and is not permanent) tend to get solved wherever the largest group of benefactors are. That would be terrestrially. Therefore, if we focus the solutions upon the terrestrial problems at hand, with the added resources of the space faring community, the terrestrial solution will be in general a richer, savvier solution.

Developing terrestrial products with spin-in to space and space commerce might yield the corporate infrastructure necessary to assure the funding stream in the commercial sector to sustain space commerce. But the current model of space development relies on governments that buy services and systems from industry and commercial vendors, resulting in associated technology spin-offs (i.e., spin out of space technology), but which has not lead to sufficient infrastructure in the space commerce endeavor.

Increasing the rate of technology and idea sharing will accelerate the march toward sustainable space commerce, which requires developing markets and customers, as well as providing value to as much of the human race as practical.

The salt story tells us that salt benefitted everyone by preserving their food. The story of wireless technology shows the benefits for nations and peoples who lack the financial means to build the infrastructure required for landline communication. The quest for sustainable space commerce, then, is the search for space based commodities and activities that might similarly make life better for the entire human community, whether they live on Earth, in a moon colony, or are space workers working in space shuttling between these ports much as merchant mariners did hundreds of years ago.

Simply put by the late motivational speaker Zig Ziglar:

"You can have everything in life you want, if you will just help enough other people get what they want."[7]

The question raised above was, "How do we develop a sustainable space commerce system?" The answer, then, is collaboration for the benefit of all.

---

[7]   http://www.ziglar.com/quotes/you-can-have-everything-life-you-want

## FRAMING THE STEPS

This is an oversimplification for illustration only, but if we take the combined American, Russian, and Chinese space budgets, and doubled that figure per year, we would have about $300+ billion per year of funding. Let us assume that the entire human race was indeed interested in space commerce and the derived benefits, and decided to levy the cost equitably across the entire population of the world. Given 7 billion people alive today, that would work out to less than $50 per person/year. Further, assume half could not pay because of poverty in their countries. That comes still to less than $100 per person per year, if we assume the "progressive tax" approach is applied.

Could we create benefits that almost every human being might be convinced was worthwhile? Could we derive benefit commensurate with this level of individual support, across international interests, and all the divergent needs of the planet's inhabitants? Theoretically, the answer is yes; practically speaking, an understanding of the needs of the planet's inhabitants would be necessary, and a program would be required to visibly address these needs and produce results in a timeframe considered "reasonable."

If space commerce is to be sustained and sustainable, it must be launched from Earth as a commercial endeavor, and therefore must be rooted terrestrially in its technology development. In addition, and more importantly, it must seek to solve social, economic and fiscal issues. Ethereal prognostications and lofty academic claims will not satisfy these needs; only results that solve terrestrial problems will connote success.


## WHAT ARE THE PROCESS STEPS AND INFORMATION TO BE GATHERED THAT ARE REQUIRED TO INITIATE A SOLUTION FOR SUSTAINABLE SPACE COMMERCE?

Thus, the relevant questions to be addressed in the search for a compelling purpose around which genuine space commerce may be organized, include:

- What is the terrestrial problem being addressed?
- Which nations will speak to this problem and which nations are vitally interested in solution of it?
- Which nations will lead it?
- Which nations will own the terrestrial technology, and which will own it in space?

- Is the ownership perhaps an international company?
- What segment or size of Earth's population will benefit initially, and ultimately, from this solution?
- Does this solution have technical, and social, and religious, and ethnic components to it?
  - o  Will those delicate issues be successfully addressed within the participating nations or tribes with respect to acceptance?
  - o  Are there insurmountable issues even formulating the team to address the problem?
- What issues of sustainability are there terrestrially and in space regarding the solution?
- Are there perhaps multiple solutions and an associated "trade space" for selection of a more "acceptable" answer?
- Can a nation that is a member of the team assimilate the solution into the culture of its population?
- Can a given problem be solved without the geopolitical implications of the "one world government" ideology, and the results being labeled as destabilizing to the world order as a potential result?
- Can adaption to and adoption of a global solution to terrestrial problem, while a successful solution in space, be perceived and actually be supportive of cultural diversity, at least terrestrially?
  - o  Are we as the human family willing to trade survival for the potential annihilation of some cultural groups?
  - o  Which nation(s) would make such a choice and enforce it?
  - o  Is this an option at all?
- Are nations that team well, contribute, and lift the fortunes of other nations, cultures, peoples, and tribes rewarded for the best of human behavior, or are we doomed to the moral frailty of human history? (Humanity's report card is far from a passing grade over 5000 years of recorded history. Even the last 200 years, post the Age of Enlightenment, is bloody and tragic.)
- Is there any chance that these endeavors can be the transformative events that, by paving the path off the planet, humanity assures that those "left behind" are not left wanting, but in fact are left in a better state because of the off ramp to space has been created?

## FOUR SPACE COMMERCE ISSUES APPROACH:
## SOLVE THE TERRESTRIAL PROBLEM FIRST

Here we will explore four broad and also widely understood issues and problems as examples: water, food, cancer, and genetics. The purpose is to illustrate the method of defining the terrestrial problem, formulating the team, and identifying its applicability to space commerce.

### 1. THE OCEAN'S DILEMMA:
### WATER FOR HUMANITY IS THE NUMBER ONE TERRESTRIAL NEED

We all recall the geography lesson that three-fourths of the world is covered in water. It is the world's commercial highway ...

Who has and who needs water is a very different situation than one might envision. While rivers and dams create lakes, without rivers, created lakes or natural lakes, the options are few and difficult to obtain. Many countries today rely on quite antiquated technologies, wells and desalination of ocean water, to create the potable water they need. Desalination is the default solution where wells are not possible or cannot supply the needed quantities, and may even be required for brackish or brine water commonly found.

A few observations:

*   The US and Egypt have about the same reliance on water from the ocean – even though Egypt is a desert country.
*   Qatar – tiny country that it is – generates 8 times more potable water from the ocean than China does, and twice what Australia does.
*   Not all large desalination producers are in arid regions of the world.
*   California is likely to require desalination plants on a massive scare because of the unparalleled drought over the last 5 years, showing it shares a common problem with much of the Third World.

Perhaps more striking, the most prevalent desalination technology – osmotic pressure cells – is essentially unchanged since 1961 when President Kennedy endorsed a worldwide push to eradicate water shortages in the name of peace between neighboring countries.

Hence, the disappointing progress in desalination technology is an excellent example of insufficient analysis and unguided inputs resulting in an excellent opportunity, with the funding being essentially squandered because the dynamics of the total system were not understood. Since around 60% of the cost per gallon of water is the energy invested in making it, often coming as electricity generated from relatively high impact sources. Using coal to power electric plants to create potable water ... think about the carbon footprint! It's enough to turn a "green" person red with anger.

And rightfully so; had the investment been made and a viable team formed starting with President Kennedy's initiative in 1961, the technology would certainly be much further along. If all this water was produced using 20%, 30% or maybe 50% less energy, how much more would that help the poor, and simultaneously reduce the human carbon footprint?

Might this be the problem that creates a niche technology, purifying water from not only from salty sea water but any brine or reuse water? Is not this technology, especially if energy efficiency is the secondary target, ready for launching into the space commerce / human space endeavors?

But our recurring question, the litmus test is the Third World; it will remain not connected with the space race, space today, or space commerce tomorrow if nothing changes. Would clean, cheap, abundant water that was Earth friendly at its source and footprint be sellable to the Goroka tribe in Papua New Guiana, as well as to Qatar and orchard owners in California? Or to Mars explorers or moon base inhabitants processing crater-shaded ice? Most likely.

## 2. THE PERFECT PARTNERSHIPS EXAMPLES: SOLVING HUNGER, DISEASE AND GENETICS - FOR ALL MANKIND - INCLUDING SPACE WORKERS

No part of the world today is immune from hunger, as it is present even in the wealthiest nations.

To venture off the planet, food will by necessity be required to be efficiently produced, efficiently processed and by-products repurposed and recycled.

Even terrestrially, this is being noticed. Jerry Glover, the subject of a recent Discover Magazine article, commented, "We have the urgent need of increasing food productivity. It's a perfect storm of opportunity and challenge, really."[8]

This nicely matches our thesis: go terrestrial in solution and extrapolate to space and space commerce, for all of humanity.

While there are concerns about genetically modified organisms (GMOs), given the choice between GMO issues and mass starvation,

---

[8]    Elzia Barclay, "Feed The World." *Discover Magazine*, May 2014.

terrestrials will not pick starvation.    And surely, long term space adventures, space commerce and the like will require many travelers who must eat nutritionally balanced foodstuffs.    Clearly, their food supply cannot be part of the "bring along" gear.  Clearly too, this is a terrestrial / space issue needing closure.

On Earth, the issues of water runoff, nitrogen depletion of soil, relatively low production yield of annual crops, crop disease, and general microbial soil health are issues that plague humanity.  Whether soil or hydroponically based in space, the issues remain the same.

Work to date has yielded some significant results.  It appears that the annual crop approach, which has less to recycle and is kinder on the growth media (soil or hydroponic cell), has potential that is largely unexplored.  It does require some GMO potentially, which makes it a great problem for an integrated team searching for an integrated solution.

Coupling the innovations, superior results have been achieved with the promise of more to come.  However all this is tailored to terrestrial results; there is no participation by space faring partners or infusion of technology from the space faring community of practice.

When new, high-yield plant technology was introduced over the last 50 years across the globe, cultural resistance to the required changes faded as food became more plentiful.

Adaptations were made to accommodate and pay homage to cultural mores and traditions, and yet the United Nations Food and Agricultural Organization (FAO) has clearly shown that success in agriculture does affect culture positively.

### 3.  CURING CANCER:  THE CURSE OF MANKIND AND CURSE OF SPACE FARING

Medical science has been aware of the hazards of space travel on the bodies spacer travelers since the beginning of the space era five decades ago.

The last 30 years has provided to medical science a diverse, large and robust database of people who have had exposure to space courtesy of the Space Shuttle program and its large astronaut contingent.  As space does not have a positive effect upon human tissue and the human organism as a system, for nearly 20 years NASA conducted a Longitudinal Study of Astronaut Health (LSAH), wherein an astronaut was paired up with a non-spacefaring NASA employee of similar build, activity level, and age. Health issues identified in the space travelers included vision problems due to macular degeneration and cataracts, bone degeneration, and nerve degradation. Principally these arose due to lack of gravity, which can be overcome, and from radiation which at some level cannot.

The reason that radiation cannot be eliminated is because of the very nature of the universe. In the near-vacuum of space, once a particle is released especially high energy particles from the galactic background of the universe itself, their high energy and the sheer quantity of them can overcome any form of mechanical protection, especially during long missions. To beat radiation, a magnetosphere like that of the Earth is required, which is very energy intensive and complicated, and while certainly achievable, quite difficult.[9] (Again, earth bound applications? Fusion reaction for energy?)

All spacecraft designs require tradeoffs between risk and performance. What this means for humans in space is that there will a relatively short career that can be spent before its time for "medical transfer" or "retirement." However, as the NASA data and other sources have made clear, the risk is not finished when spacefaring is over, as there are residual effects. Like smoking, these changes are often subtle but traceable. Genetic changes, even more subtle and difficult to establish, have also occurred. (This is addressed below.)

This situation seems to be a perfect opportunity for cancer investigations, theory testing, and therapy testing. To conduct space commerce, one encounters exposure, and if one is exposed one must be in long-term care and monitoring not only for personal attention, but to continue to grow the database on the human species and its diseases. Doubtless prophylactic measures designed for space will find favor terrestrially, and vice versa. It would therefore seem that this perfect storm should lead to sharing of data, funds, and technology across many years and nearly universally across cultures. With rare exceptions in the world today, diseases affect humans universally (although there are pockets in the world where certain diseases are nearly unheard of due to lifestyle, diet, and genetics).

An excellent example of time-dependent decision is captured in this discussion. 50 years ago, even if the desire were there for expansion of space travel and commerce, the supporting data on the health issues related to space were not available. Nor was the understanding of the sources and lethality of the cosmic background radiation.

In fact, we have only recently begun to cope with what high speed particles can do to spacecraft and spacefarers. Micrometeorites, like radiation, are also known threats, but the defenses are imperfect, as pictured in Figure 1, which shows damage to a cable on the ISS that was

---

9 "Cancer and Human Space Flight," *Aerospace America Journal*, American institute of Aeronautics and Astronautics (AIAA), May 2014, page 30-35.

struck by a micrometeorite. The Earth's atmosphere, while a very effective shield against meteorites and radiation is not perfect, and does not prevent 100% of the penetration by either. Therefore, we clearly have a case for terrestrial + spacefaring collaboration.

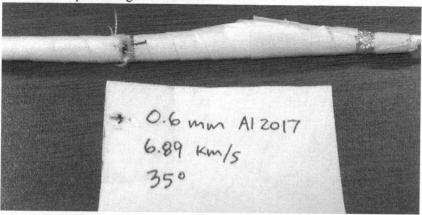

### Figure 1
### Micrometeorite Damage from An ISS External Cable
Photo by T. Diegelman

## 4. GENETICS: CULTURALLY SENSITIVE, POLITICALLY UNTOUCHABLE, MORALLY CONTROVERSIAL, AND ABSOLUTELY ESSENTIAL TO UNDERSTAND WHEN SPACEFARING FOR SUSTAINABLE SPACE COMMERCE

Molecular evolutionary biologist Masatoshi Nei is working on his theory that all members of the human race are mutants with connected, but extremely diverse and unpredictable genetic programming within. He postulates that Darwin's thesis of natural selection has not been proven according to the scientific principle to be the forcing function behind evolution of all species. He may have a point. Despite being a secular gospel for modern education, it is far from complete with respect to conclusions above scientific debate.

Darwin married his first cousin Emma Wedgwood, an act certainly counter-intuitive for a man given to the theory that diversity in the species produces robust offspring and strengthens the species as a whole.

If Nei is to be believed, he can genetically connect the races that currently inhabit the earth and discuss with solid data, the strengths and weaknesses of each.[10]

---

[10] Gemma Tarlach, "We are All Mutants." *Discover Magazine*, March 2014.

Nei raised more than a few eyebrows in his 2013 book, *Mutation-Driven Evolution*.[11] His thesis is that to have natural selection, if that even is a workable selection process, you must first have an unpredictable, not anticipated or controllable mutation that provides another workable genetic solution. That has no counterpart or connection to natural selection as Darwin explained genetics. He further postulates that a mutation may be a positive, negative, completely neutral, or even an arbitrary change. Therefore, for humanity to have evolved, we are all mutants, or descendants of mutants. Starting at some point in the past, humanity is continually mutating toward a destiny that is not known and is not, as Darwin would have us believe, predictable. And Darwin's model of natural selection did not include the factors we have discussed in this section, the effects of diet, meteorological events, food availability, radiation, or the possibility of gene alteration from external sources.

For purposes of this discussion concerning the impacts of this theory on space faring, let us assume that Nei is correct and Darwin is not. We then see that phenomena such as low or no gravity, deep space radiation, space dust, or microbes from another world (we will explore that shortly) will contribute to the probability of a "random mutation" in the human genetic material. Which is to say that given sufficient exposure of enough people to enough space environments, we may spawn genetic mutations that may fare well in space or terrestrially, or in neither place. More remote in probability is the spawning is a superior genetic strain.

Think for a moment what that might do to the social and cultural aspects of life. From the work of Gregor Johann Mendel (1822 – 1884), we understand the role of variation in breeding of existing traits. What we do not know from his work is the effect of breeding in traits that are mutations not a result of controlled adaptation. We will focus on that topic through an activity which has been practiced since Biblical times: cousin marriage.

Prevalent in Europe in the time of royal families, it resulted in hemophilia in the English royalty, while the French monarchy had issues of dementia from inbreeding before its demise. Egyptian rulers also promoted inbreeding in the pharaohs such that after several hundred years they were mentally unfit as rulers.

Why is this important to the sustainability of space commerce? Because without alternatives or a plan to avoid these risks, the spacefaring community may similarly be required to accept breeding of humans that are

---

[11]   Masatoshi Nei. *Mutation-Driven Evolution*. Oxford University Press, 2014.

"at risk" of mutating in a very unpredictable and undesirable manner. [Editor's note: Please see Chapter 10 for more detailed discussion of evolution and genetic engineering as it pertains to space and space travelers.]

Relatively few of humankind will actually leave the planet, so must we "ask" them not to breed once they have been "exposed" to space environmental factors "sufficiently" to have high probability of genetic mutation? And who defines the boundaries on the quotation-marked terms? What data are to be used to determine these parameters, and what theory of evolution of the species is truly the mechanism – are we naturally selected or all mutations?

This issue is target rich for both space and terrestrial application of insights developed.

We already have some data on the topic. Before NASA probe Curiosity was launched toward Mars it was scrubbed with alcohol and baked at 203 degrees F. A final check revealed the stunning fact that 56,400 organisms were still present on the rover, including 377 different bacterial strains. The design team then tried UV radiation, ultra high ph levels, and even hydration to sterilize the spacecraft. Still, a majority of organisms survived, including Gracillbacillus, Pseudomonas, Staphylococcus, Moraxella, and Streptomyces![12]

Curiosity was launched anyway, and thus the question raised: by exploring the cosmos, are we also altering its own evolutionary trajectory? Are we now a participant in *War of the Worlds* as the perpetrator? Will a return sample from another world be our just payback? As pointed out in this section, nothing in the work of Mendel, Darwin, or religion prepares us for this event. This issue opens up more dimensionality of the randomness of the universe, as perhaps the naturally occurring radiation shielding on the comets (dust) might be a radiation shield of the future? Alternatively, perhaps we get into the genetics of bacteria and determine how they cope with radiation without dying? Clearly, mankind has few of these answers, even to the terrestrial part of the question.

## CONCLUSION

In this chapter we have examined a set of questions intended to engage the diversity of the human race in a supportive role for the development of sustainable space commerce through solutions to related terrestrial

---

[12] *Popular Science* Magazine, "Points of Interest: Four Ways Spacefaring Microbes Could Muck Up the Solar System." January 2015, Page 36.

problems. Further, it has been demonstrated that spinning the terrestrial solutions into the space commerce arena is a unifying, enriching and sustainable approach to space commerce.

The work that remains is delicate, difficult and far more challenging than the technical problems of spacefaring and space commerce. It is working with, through and within the cultures and peoples of the human race. Happily, the human race has been wildly successful in terrestrial commerce, and the model is historically documented and established.

The story that is the kernel for the model here is *Salt: A World History*.[13] As noted, commerce related to salt defined trade routes, blended cultures, changed diets, enabled long journeys on the seas, and much more. It largely defined the structure of our current geopolitical world. Through an epic story took humanity more than 3000 years to complete, while never leaving the planet, and without ever having to worry about the many and ponderously enormous trappings that burden the developed world of today, without ever having addressed inequities, and basic human dignity as requirements, and with no concerns for radiation or microbial contamination. No, it was just the collection and trading of salt.

And from this historical concept we have a proposed model for successfully and sustainably reaching out into the infinitely more complicated sector called "space," in which to establish sustainable commerce. And of course there is more work to do.

As the vanguard team on the precipice of this new extraterrestrial commerce challenge, it is ours as humanity to discover and define the "space salt." What is it we must have? Who are the members of the team that is going out there to get those space salts, bring those commodities back to Earth, pushing forever the sphere of commerce irrevocably beyond the Earth's atmosphere? It is our destiny to do this as humanity, or we risk extinction from the unforgiving and hostile cosmos should fail to rise sustainably beyond the vulnerability of the Earth's imperfect incubation and protection.

•••

Editor's note: Please see the following chapter, which suggests that H3 on the moon could precisely be the "space salt" that Mr. Diegelman is seeking.

---

[13]   Mark Kurlansky. *Salt: A World History*. Penguin Books, ISBN 0-802701373-4, 2002.

## THIS CHAPTER IS DEDICATED TO
## DR. KEN COX

What can you say except "Thank you, and I love you like a brother" to a man who, because of an abiding belief in the integrity, importance, and perhaps most importantly, the fragility of the American Space Program, took you aside and said: "Help me keep NASA vibrant, and in return, I will teach you how to do that"?

Well, thank you, my dear friend, Ken! Thank you for the lessons on doing what is right. Thank you for inspiring me to keep the head high when I wanted to hang my head in desperation. Thank you for teaching me your political savvy and wisdom. Thank you for your hilariously delightful wit, and for showing me how to balance wit and wisdom. Thank you for inviting me into the ATWG "writers guild," leaving a message for future generations. Thank you for encouraging my foray into elected politics, and igniting the fire to forge together technical and political wisdom. Thank you for the opportunity to collaborate with some of the greatest minds in the world while writing these ATWG books. These books and their collective wisdom will endure way beyond our limited years.

Lastly, thank you for being my friend for the last 25 years. Because you and I are friends till the cosmos ceases to exist, I dedicate this contribution to *Space for the 21st Century* to you. God bless you my friend!

## THOMAS E. DIEGELMAN

 Tom Diegelman has been in the aerospace community for over 40 years, involved in the research, development and operation of training simulators, ground based flight control installations and facility operations. Tom started his career with Cornell Aeronautical Laboratory as a research engineer, working on early versions of shuttle handling quality study simulations and shuttle shock tunnel testing.

Tom moved to Houston in the late 70's to join Singer / Link Flight Simulation and worked in the Shuttle Mission Training Facility (SMTF) as a model developer, and later a manager of simulation projects. In 1988, Tom joined NASA to lead the $170M redesign of the SMTF. Assignments at NASA / JSC include projects in advanced mission control technology, technology development, and project manager for Mission Control technology projects. He served as Facility Manager for Mission Control for 3 years before accepting an account manager position in the Technology Transfer Office, developing partnerships and Space Act Agreements.

The design of the training facility for the Constellation Program culminated his nearly 30 years of experience at JSC in the Jake Garn Astronaut Training Facility. Tom became the ISS Vehicle Safety Engineer for Communications and Tracking Subsystem. Most recently, Tom accepted the position as the JSC Safety and Mission Assurance (S&MA) representative to the Office of JSC Chief Engineer Joint Technology Working Group that has the responsibility for identifying and funding game changing technologies in human spaceflight.

Tom was elected to Seabrook City Council in 2006, and served two terms, during which he worked closely with the Port of Houston Authority on the Seabrook / Bayport Terminal Facility issues. Tom is a member of Baytran, a non-profit organization promoting inter-modal transportation solutions in the Houston / Bay Area, and continues to be involved in local, state and federal government on behalf of the space technology community.

A dedicated writer, he is the coauthor of numerous chapters in the previous volumes in this ATWG series.

Chapter 16

# HELIUM 3: THE FIRST TRUE COMMERCIAL SPACE ACTIVITY
## AND
# RECOVERY OF LUNAR RESOURCES BY MINING AND SPACE TRANSPORTATION INDUSTRIES

Thomas C. Taylor
President, Lunar Transportation Systems
Senior Member, AIAA

Note:  The intent in this chapter is to stimulate interest in mining off-planet, beginning with mining of the moon for He3 to support Lunar Transportation Systems, Inc. moving forward in raising money from investors and completing a commercial-cargo-only space vehicle design. This chapter was to be presented as a paper at the 2015 AIAA Annual Conference, Disney World, Florida, July 2015. A section on nomenclature may be found at the end of the chapter.


## LUNAR COMMERCE

Commercial lunar development has not progressed during the 46 years since the Apollo program, other than the mapping of helium3 (He3) locations and quantities by Chinese satellites. Their research revealed that He3 is available on the moon in significant quantities that are currently worth approximately $14 billion per ton ($16,000 per gram) on Earth.

This is significant because He3 is used in the $100 million per year market for nuclear detection equipment that is used at US international borders to inspect for nuclear weapons in shipping containers. He3 causes radioactive waves from any nuclear material that may be located inside shipping containers to be altered and detected by instruments without opening them, making it an important resource in the current fight against the threat of terror.

However, He3 is not naturally occurring on Earth, and therefore must be made at great expense by the nuclear industry. Consequently, the possibility of obtaining He3 from the moon is potentially attractive as a commercial activity, and as a pioneering activity in developing sustained lunar-Earth commerce.

He3 in larger quantities may also be mined and transported back to Earth for use in future fusion power solutions for Earth's energy grid. In either or both instances, it would thus constitute humanity's first off-planet trade route. Further, some of the oxygen, hydrogen and other elements derived from Helium3 mining and processing would be used for propellant and for supplies for those living on the moon's surface, providing further economic advantage and reducing costs.

In this chapter, we examine Lunar Transportation Systems (LTS) proposed round trip cargo transportation and mining system between Earth orbit and the moon.

## The Future Source of Earth's Needed Resources, and Trade Route Transportation

The need for resources helped push early humanity across the Earth in search of food and shelter, and the same motivation will inevitably push us to explore space for non-renewable resources.  Our company, Lunar Transportation Systems, intends to create a trade route of transportation from Earth and Low Earth Orbit (LEO) to the moon, and returning to Low Earth Orbit, at affordable commercial rates, by using commercial launch vehicles from Earth to LEO, and a Space Plane or Space Cruiser to deliver He3 to customers on Earth at locations near international airports and spaceports capable of receiving Space Cruiser deliveries.

Mining is to be accomplished by robotic excavator/processors in a two-step temperature process to recover He3.  However, lunar mining camps will be different than Earth's remote resource camps because of the lack of atmosphere, increased radiation levels, and the danger of impacting rocks from off-planet. Normal lunar radiation levels are higher than Earth, and periodic solar flare eruptions increase radiation peaks to five times greater than normal.  With only 8 minutes of warning before arrival, adequate protection requires better radiation-shielded habitats, and/or better future pressurized underground living.  Limited work durations may force underground-pressurized protective living for crews, permitting longer stays.

The commercial lunar transportation systems vehicles to service this trade will be expendable initially during the early development period. LTS has expressed the intent to design, build and operate vehicles to perfect the design, communications, automation systems and other technical details, leading to development of a robust cargo-only vehicle. Expended vehicles would remain on the lunar surface to then be available to other companies and governments, which would use the discarded hardware, tanks, water, communications equipment, etc. as part of a concerted effort to develop and utilize the moon for commercial purposes, the first celestial body in our near solar system to be so developed.

The company anticipates no plans for human-rated passenger vehicles, as the LTS vehicle is completely automated and controlled from either Earth, the moon, or from orbital stations.

Eleven US patents have been issued, and more may be awarded, which define and explain the inner workings of our vehicle and its use for cargo, machinery, robotics and supplies for delivery to lunar mining camps, and for transportation of He3 product to Earth orbit. [Editor's note:  The

patents pertinent to these vehicles are listed following the chapter.]   A
seven minute video is available on YouTube at:
http://www.youtube.com/watch?v=26Y5w0vqtIU


## THE POTENTIAL VALUE OF HELIUM 3, AND WHERE IS IT?

As noted above, He3 currently sells for $16,000 per gram, which translates
to $14.5 billion per ton, with a current market demand of about $100
million per year.

However, He3 is difficult to mine or retrieve from the surface of the
moon, and requires excavation and processing.   Approximately 55,000
cubic yards of regolith yields one ton of He3.

Helium-3 is Rare on Earth but Abundant on Moon
Map from Chinese Chang'e-1 spacecraft, South at top

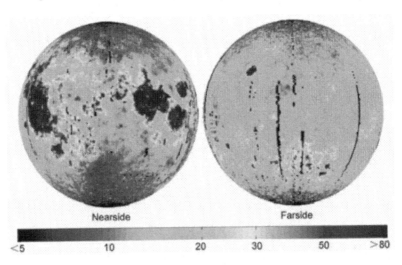

### Figure 1
### Chinese Satellite Maps of the Moon for He3
These maps, showing south at the top, indicate the location of He3 on
the moon, both near and far side, as identified by a Chinese satellite.

Helium3 can be mined in place on the moon by a moving Bucket
Wheel Excavator (see below) and Two Step Temperature process, digging
the regolith continuously, first by taking out the big rocks seen in Figure 3

and moving the rest of the regolith into a two-step temperature process to drive off and recover the volutes seen in Figure 2. The whole process would be done as the excavator moves across the flat bottom of a crater, and after removal of Helium3 and other valuable products, the regolith may be discarded or used as insulating material over habitats and workshops. Figure 2 also shows the resources that would separated and stored from the mining process, including hydrogen for fuel and, water in some form, nitrogen, carbon dioxide, methane, and carbon monoxide, all helpful for life support, plus Helium4 for cryogenics, and one ton of Helium3.

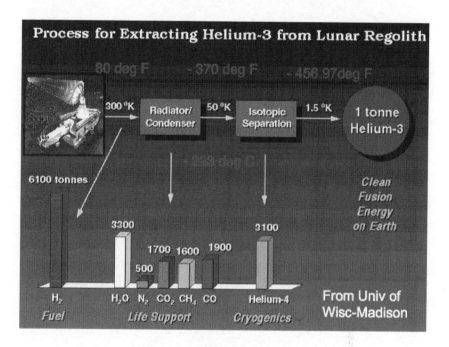

**Figure 2**
**The Bucket Wheel Excavator**
Extracts the above materials from each 55,000 cubic yards of regolith processed. The concentration, of course, varies as shown in the He3 maps in Figure 1.

## THE BUCKET WHEEL EXCAVATOR (BWE)

The Bucket Wheeled Excavator, as shown in Figures 3 and 4, is currently in its fourth design and prototyping iteration at the University of Wisconsin at Madison, and continues its evolution and refinement under the excellent leadership of former astronaut Harrison "Jack" Schmitt.

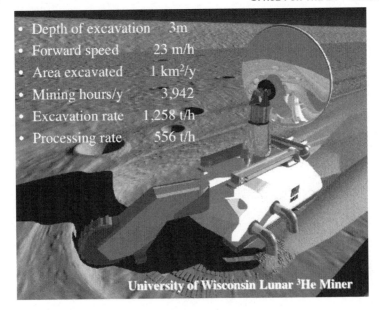

**Figure 3**
Version 1 of the Bucket Wheel Excavator (BWE) processing regolith
and discarding small rocks, using power beamed through the antenna.

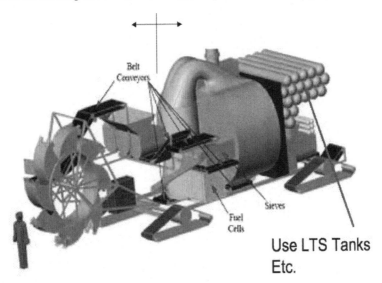

Use LTS Tanks
Etc.

**Figure 4**
Version 3 of the Bucket Wheel Excavator scoops up regolith, discards
big rocks, uses 2-step thermal process to drive off and collect useful
elements and He3.

The Bucket Wheeled Excavator and Mining Machine in operation is similar to a wheat combine, which harvests by moving forward and reaping the rows of wheat without disturbing the adjacent rows. The BWE requires continuous power and communications, is probably unmanned and controlled from a distance, and is repairable in underground or at breakdown locations in the open, all the while monitored from a distance by lunar, LEO, or Earth locations. Like a wheat combine, the BWE must have an onboard storage capability for all mined materials, in this case including He3, He4, CO2, and perhaps 5 to 10 other elements needed to supply a lunar surface economy with oxygen, nitrogen and some limited hydrogen, all of which are probably in liquid or gas forms. The BWE probably should not disturb by driving over future mining areas and use wheat harvesting techniques that try not drive over or disturb uncut wheat areas or knocking the wheat from the uncut stalks. Helium3 is likely smaller than wheat and it appears to have electrostatic qualities.

Because these activities are difficult to automate, Bucket Wheeled Excavators will likely be assembled, possibly operated, and maintained by humans on the moon. They will be powered by beamed power, or possibly individual nuclear reactors, or a combination of both. A large reactor positioned on a crater rim with an elongated beam envelope could power more than one BWE machine simultaneously.

Mining staff would also handle He3 products, including change-out of storage containers on the mining machines, and shipping He3 to Earth using standard containers.

## LTS Transportation for Cargo.

Figure 5 shows the proposed unmanned cargo vehicle for operations from LEO to the lunar surface and return to LEO. The system would be launched from Earth on a commercial or government rocket via either a disposable or reusable system.

Payload tanks are mounted in a system that can scale from 13.8 feet in diameter to 33 feet or ~10 meters in diameter.

The vehicle would be tested with early payloads and developed to become a two way vehicle transporting heavy difficult mining equipment payloads, water and food payloads.

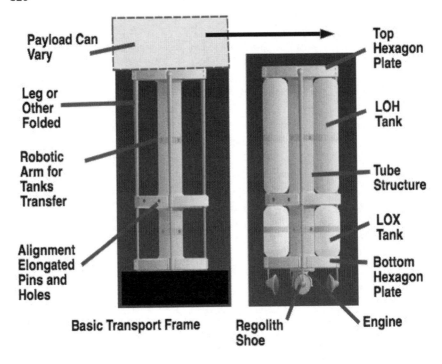

Payload Can Vary

Leg or Other Folded

Robotic Arm for Tanks Transfer

Alignment Elongated Pins and Holes

Basic Transport Frame

Top Hexagon Plate

LOH Tank

Tube Structure

LOX Tank

Bottom Hexagon Plate

Regolith Shoe

Engine

**Figure 5**
The LTS Transportation hardware is eventually reusable and acquires
propellant and payloads at LEO and the lunar surface.

## THE GROUND SERVICE VEHICLE (GSV)

This unmanned LTS vehicle can transport heavy mining hardware called Ground Service Vehicles (GSVs) which are transportation and crew vehicles used on the moon to service, load, unload and transport cargo and propellant.

Lift beds install payloads on launch vehicles, and hexagonal pallets can be moved on and off the vehicle level using a track slot embedded in the top level. These GSVs use lift platforms to hold and rotate the propellant 12-tank rack of cartridges on hexagon pallet levels, to lower and raise remove empties and place loaded propellant tanks and vehicle payloads. The launch vehicle is unmanned.

The early living and equipment repair workshops can be remote or located underground. Mining operations require assembly; test, repair and salvage of a variety of mine related equipment on Earth. Beyond Earth in a very different risk environment, where there is really "no air pressure" and literally no actual environment to breath.   Bucket Wheel Excavator equipment design will require remote diagnostic and even remote software and electronically delivered repairs, plus on site plug-in solutions for wear parts and potential problems that designers will find as we proceed into equipment design under the very different constraints of another celestial body.

# Reloading LTS Vehicles

## Multi-Use Equipment

**Hydrogen Tank**           **Payload**

**Utility Vehicle with Lift Capabilty**

**Figure 6**
**Two General Service Vehicles (GSVs)**
Loading a launch vehicle with propellant tanks, one on top and 12 in the racks.

## LIVING ON THE MOON

Habitation on the lunar surface will probably start by using small structures transported from Earth, as shown to the far right in Figure 7, until humans are able to utilize local lunar materials and "Living Off the Land" techniques.   Similarly, oil companies operating at Prudhoe Bay, Alaska and other resource recovery locations regularly use "Living off the Land" techniques, but they don't call this a technique, they just utilize local materials to eliminate commercial transportation costs.   At Prudhoe Bay local materials include water, gravel for pads for buildings to protect the fragile permafrost top layer, and air.

The same benefits will be available on the moon, significantly reducing the investment that would otherwise be required to send materials there.

Lunar residents will likely face lifetime space radiation limits, as well as object impact problems that are significantly more challenging than on Earth due to the lack of atmosphere and magnetic fields, but space architects and human factors designers can help humans adapt and accelerate or jump start the lunar resource development cycle and develop longer range space habitation design parameters and designs.

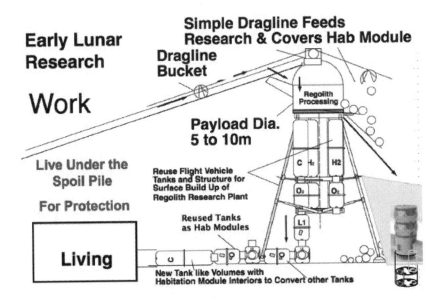

**Figure 7**

After several months on the lunar surface, the LTS NASA-JSC Minimum Functional Habitat Element (MFHE) can be buried by waste regolith from the early Exploration Mining Crew Feasibility Research Lab, providing additional radiation and object impact protection for the human crew. The MFHE (lower right) was developing during the MFHE Study managed by Boeing and funded by NASA-JSC.

Given the higher amounts of potential radiation exposure, the maximum time allowable for living on the moon may be limited to protect the health of the miners. Figure 7 shows a research lab on the moon sitting on its delivery vehicle, and capable of testing He3 mining equipment and recovery methods using draglines to gather He3 samples. The living quarters are buried under regolith at the far left, to protect the humans inside.

Space habitats must provide radiation protection, object impact protection, air, water, heat, and agriculture to sustain the mining effort, while providing mining crews with support while working in 700-mile diameter polar craters that never see sunlight, or on lunar equatorial sites with 14 days of darkness.

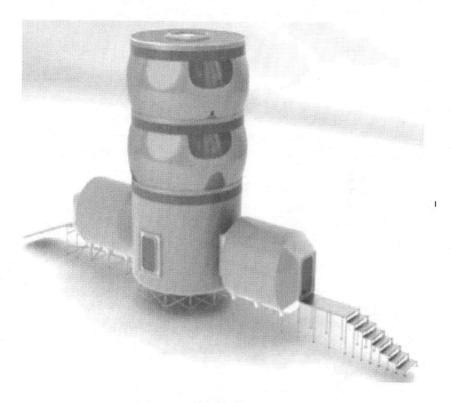

**Figure 8**
Artist Conception1 of hard shell 1st floor and inflatable upper floors plus salvaged SPACEHAB and Spacelab hardware for spacesuit storage. Habs are transported from Earth initially, but may evolve into underground habitation for protection from hazards.

The Functional Habitat Element (MFHE) Study, which was managed by Boeing and funded by NASA-JSC, was an exploratory early study on lunar habitats, conducted mostly by technical aerospace experts and one commercial space entrepreneur. Each team submitted early concepts and ideas from their applied technical fields in aerospace.

Figure 8 shows the Minimum Functional Habitat Element (MFHE) solution submitted by the Lunar Transportation Systems, which could be launched from Low Earth Orbit to the moon's surface after arriving in LEO as a cargo-only payload.

Indeed, many believe that LEO is a natural location from which space vehicles will transit to and from the moon's surface, similar to Earth's shoreline, where surface vehicles transit from land transportation to ocean transportation vessels including cargo ships and ferries.

Reaching Low Earth Orbit, and the trip beyond LEO to the moon are quite different, because reaching LEO from Earth requires 92% of the total rocket mass as propellant to overcome Earth's gravity and the atmosphere. At LEO and beyond vehicles discard the propellant stages and face a new class of problems reflecting the change to a no-environment, micro-gravity conditions. Once in LEO, a space-only vehicle starts a rocket engine, the vehicle accelerates to speed, and then the engine is shut off and the vehicle continues at that speed indefinitely.

It can then rotate 180 degrees as it arrives near its destination, slow down with the same engine to enter lunar orbit, and use rocket thrust to transit to a gentle landing on the lunar (or other) surface. Hence, great vehicle flexibility is possible once Earth's gravity and atmosphere are overcome.

## CONCLUSIONS

In our present biological form, humanity on Earth would not survive a life extinction event similar to the any of the previous six known events that have already occurred. In time of need, people living on the moon could thus offer Earth a work force capable of supporting evacuation should a seventh major extinction event occur. Further, enabling humans to live on the moon might also be our only way to survive as a species.

From a more commercial perspective, the current global economy cannot function without nonrenewable resources, and now that exotic resources and rare Earth metals are important for cell phones and other high technology, they are in increasing demand. This need for resources is likely to be the reason that humans settle the solar system, and non-renewable resource mining is likely to become a bigger industry than aerospace, although aerospace will certainly help these resource mining industries get started elsewhere in the solar system.

Mining the moon's resources will occur when humans need its resources, and buyers probably won't care much where the resources come from; they'll just want them, and if they're willing to pay the requisite price, then commercial contracts will expedite the development of a lunar resource extraction industry.

We anticipate that the vehicle designs, development process, patents and financing will then follow the typical pathway of entrepreneurship, aerospace development, fabrication, space business development and financing, and that consortia of companies and possibly groups of countries will likely cooperate to develop humanity's first celestial body for commercial purposes.

The author's earlier work was with SPACEHAB, Kistler Aerospace, and Lunar Transportation Systems, Inc. all started by Bob Citron, Walter Kistler and Tom Taylor, which raised $1.2B in private equity capital and hired major aerospace contractors to perform work. They launched hardware on more than 30 Space Shuttle missions, including 21 SPACEHAB Module missions.

Early on, SPACEHAB shares sold for five cents each, and eventually rose to $15 per share. Our family of commercial aerospace companies has been proud to call Mitsubishi and Alenia Spazio as investors, customers and contractors, as well as most American aerospace companies, NASA, and the US Air Force as customers.

•••

## Acknowledgment

The author wishes to acknowledge the significant entrepreneurial help, financing and learning experiences over 34 years from Walter Kistler and the late Bob Citron in the development of SPACEHAB (SPAB), the Kistler Aerospace Corporation (KAC), and Lunar Transportation Systems, Inc. (LTS).

## THOMAS C. TAYLOR

Thomas C. Taylor is President of Lunar Transportation Systems. He is an entrepreneur, inventor and a Professional Civil Engineer in the commercial aerospace industry. His goal is building commercial space projects including an unmanned transportation cargo service to and from the moon's surface with Lunar Transportation Systems, Inc.

Since 1979, Tom has helped to form 22 entrepreneurial aerospace startup companies with four successful commercial space startup companies raising a total of $1.2B in private equity financing. These four actually completed the commercial space startup process and evolved into meaningful private commercial space companies, but each took almost a decade to unfold.

Tom enjoyed working in the trenches for 4 to 12 years on each of these commercial space successes with Walter Kistler and Bob Citron, the founders of most of the successful startups, as described in this chapter, including SPACEHAB, Inc., Kistler Aerospace Corporation, and Lunar Transportation Systems, Inc., an unmanned logistics service anticipating commercial cargo to the moon's surface at commercial rates with scalable hardware. Started in 2005, LTS proposes a privately financed logistics service for commercial lunar development. The goal is a sustainable commercial transportation system for the moon to support government and commercial efforts.

He is also a senior member of the AIAA.

## Nomenclature

BWE    Bucket Wheel Excavator + 2-Stage Temperature Process mining $He_3$ on the Lunar Surface

EVA    Extra Vehicular Activity requiring addition human protection in a Space Vacuum

GSV    General Service Vehicle providing transport, PL & propellant loading & repairs to equip. on moon

$He_3$    Isotope of helium from sun on solar wind, impacts upper inches of regolith worth \$451B/ton on Earth

JSC    NASA Johnson Space Center, Clear Lake, Texas

KAC    Kistler Aerospace Corp., Founded by Walter Kistler & Bob Citron, ~1985

LDEF    Long Duration Exposure Rack, in LEO ~5+ yrs. recording object impacts on various test materials

LRL    Lifetime Radiation Limit for the Human Body

LTS    Lunar Transportation Systems, Inc., Founded by Walter Kistler & Bob Citron, owned by author

MFHE    Minimum Functional Habitat Element Study[1], lunar living shelter study \$ NASA-JSC, led by Boeing

PL    Payload or cargo on a space vehicle

SPE    Solar Proton Event, 5 times normal lunar radiation w/ 8 min warning require more protection shelters

## Patent References

For patent information see:
http://www.pat2pdf.org
http://patft.uspto.gov/netahtml/PTO/patimg.htm

Patent: SPACEHAB Module, Taylor, T.C., Citron, R.A., both of SPACEHAB, Inc., U.S. Patent 4,867,395 issued Sept. 19, 1989 "Flat End Cap Module in the Space Shuttle," Assignee: SPACEHAB, Inc. (Washington, DC), later Astrotech, Inc. Titusville, FL.

Patent: Space Trans Node w/ Tether Sys, Kistler, W. P., Taylor, T.C., Citron, R.A., All of Lunar Transportation Sys, Inc., U.S. Patent 7,681,840 B1, issued Mar. 23, 2010, "Space Transportation Node Including Tether Sys," Assignee: Taylor, T.C., LTS, Inc., Las Cruces, NM

Patent: Sys & Method for Trans & Storage of Cargo in Space, Kistler, W. P., Taylor, T.C., Citron, R.A., All of Lunar Transportation Sys, Inc., U.S. Patent 7,114,682 B1, issued Oct. 10, 2006, "Sys & Method for Trans & Storage of Cargo in Space," Assignee: Taylor, T.C., LTS, Inc., Las Cruces, NM

Patent: Plat & Sys for Mass Storage & Transfer in Space, Kistler, W. P., Taylor, T.C., Citron, R.A., All of Lunar Transportation Sys, Inc., U.S. Patent 7,118,077 B1, issued Mar. 10, 2006, "Sys & Method for Trans &

Storage of Cargo in Space," Assignee: Taylor, T.C., LTS, Inc., Las Cruces, NM

Patent: Plat & Sys for Mass Storage & Transfer in Space, Kistler, W. P., Taylor, T.C., Citron, R.A., All of Lunar Transportation Sys, Inc., U.S. Patent 7,370,835 B2, issued Oct. 13, 2008, "Plat & Sys for Mass Storage & Transfer in Space," Assignee: Taylor, T.C., LTS, Inc., Las Cruces, NM

Patent: Large Cryo Tank Logistics for In-Space Vehicles, Kistler, W. P., Taylor, T.C., Citron, R.A., All of Lunar Transport Sys, Inc., U.S. Patent 7,681,840 B1, issued JUL. 14, 2009, "Large Cryogenic Tank Logistics for In-Space Vehicles," Assignee: Taylor, T.C., LTS, Inc., Las Cruces, NM

Patent: Platform & Sys for Propellant Tank Storage & Transfer in Space, Kistler, W. P., Taylor, T.C., Citron, R.A., All of Lunar Transportation Sys, Inc., U.S. Pat 7,156,348 B1, issued Mar. 23, 2010, "Space Transportation Node Including Tether Sys," Assignee: Taylor, T.C., LTS, Inc., Las Cruces, NM

Patent: Flat Valve for Orbital Applications, Jones, H. S., Kistler, W. P., Taylor, T.C., Citron, R.A., All of Lunar Transportation Sys, Inc., U.S. Patent 7,562,670 B1, issued Mar. 23, 2010, "Flat Valve for Orbital Applications," Assignee: Taylor, T.C., LTS, Inc. Las Cruces, NM

Patent: Space Transportation Node including Tether Sys, Kistler, W. P., Taylor, T.C., Citron, R.A., All of Lunar Transportation Sys, Inc., U.S. Patent 7,503,526 B1, issued Mar. 17, 2009, "Space Transportation Node including Tether Sys," Assignee: Taylor, T.C., LTS, Inc., Las Cruces, NM

Patent: Inflatable Habitation Volumes in Space, Taylor, Lunar Transportation Sys, Inc., U.S. Patent 6,439,508 B1, issued Aug. 27, 2002, "Inflatable Hab Volumes in Space," Assignee: Taylor, T.C., LTS, Inc. Las Cruces, NM

Patent: Large Cryogenic Tank Logistics for In-Space Vehicles, Kistler, W. P., Taylor, T.C., Citron, R.A., All of Lunar Transportation Sys, Inc., U.S. Patent 7,559,509, B1, issue Jul. 14, 2009, "Large Cryogenic Tank Logistics for In-Space Vehicles," Assignee: Taylor, T.C., LTS, Inc. Las Cruces, NM

Chapter 17

# AN EMERGING BUSINESS MODEL FOR COMMERCIAL SPACE

## Michael Wiskerchen, Ph.D.
### Founder, Space Venturers Holdings, LLC

During the past ten years the barriers constraining commercial space projects and investments have been reduced through key factors such as the emergence of a robust international commercial space transportation system and non-government management of the International Space Station. The commercial space transportation field is being expanded each year with new commercial players from various countries. Government designed, built, and managed space laboratories have made space R&D to be overly expensive, inefficient, and a management nightmare for commercial users. So when the US Congress made the decision to make the US part of the International Space Station (ISS) a National Laboratory, the doors were opened to encourage commercial ventures in space by establishing a non-profit entity to manage a significant portion of the National Laboratory. This was also a message to private investors that this new frontier was now open for business.

The unique microgravity environment of space has dramatically altered R&D in a number of fields, including space life sciences research for both plant and animal research, microgravity materials processing, and protein crystal processing. These are emerging as significant commercial domains that are currently attracting increasing investment and are likely to attract much more in the coming years, and we will examine each briefly here.

## LIFE SCIENCE RESEARCH

### PROTEIN CRYSTAL EXPERIMENTS

Proteins, a major component of living organisms, have been attracting the attention of researchers who seek to clarify their functions and structures. Compared to protein crystals grown in Earth's gravity (1g), protein crystals grown in space microgravity are much larger in size and purity, which permits detailed structural analysis using x-ray and neutron diffraction techniques. The biotech industry can utilize this structural information to produce new drugs on Earth for many diseases. One example is the recent research of Dr. Ng (University of Alabama, Huntsville) on Protein Crystals (enzyme inorganic pyrophosphatase (IPPase) for Neutron Crystallography studies. IPPase plays a critical role in DNA replication, gene expression processes, fatty acid synthesis and other critical biological reactions. The microgravity environment of the ISS is proving to be essential to Dr. Ng's neutron diffraction study of IPPase to determine how it functions in cells.

### PLANT AND ANIMAL RESEARCH

When exposed to microgravity environments, the unique responses of living organisms can provide deeper understanding of cellular structures and processes, and this knowledge can lead to novel biotech applications on Earth.

For example, about 50% of the energy expended by terrestrial-bound plants is dedicated to structural support needed to overcome gravity. By removing gravity as an environmental factor, plant cells whose genetic heritage have evolved in a gravity environment suddenly have "extra" energy available to them, causing the plant's genetic structures to express differently than in a gravity environment. With normally dormant genes within the cell now expressing, some plants are able to adapt quickly to additional changes and stresses that researchers introduce into the environment, such as disease-causing organisms. This results in the development of new varieties of plants with commercially beneficial attributes through a development process that is an order of magnitude faster than traditional Earth-based methods. By accelerating the learning process by a factor of ten, of course, the overall pace of R&D is significantly accelerated, with potential benefits for the researching companies and end users.

## MICROGRAVITY MATERIAL PROCESSING

### SPACE MANUFACTURED SILICON CARBIDE

Gravity frequently causes defects in materials manufactured on Earth, limiting the quality and size of crystals, increasing manufacturing cost, and restricting applications. One example is space manufactured Silicon Carbide (SiC) wafers, which can operate at 1000°C as opposed to 300°C for Earth manufactured silicon based devices. They're also able to withstand ten times greater electric fields than silicon, and offer high radiation resistance, high thermal conductivity, high maximum current density and more.

## BARRIERS

With an increasing portfolio of successful past projects to refer to, and the promise of increasing commercial value through future projects, the existing portfolio of space ventures nevertheless remains very limited if we measure by the number of projects being pursued or the total amount of invested capital, indicating that barriers to full exploitation of the potential that space science offers remain significant. What, then, is constraining the commercial growth of microgravity research and development?

One significant reason is that commercial space projects are highly complicated. To successfully carry out space-based R&D projects, the following key elements are required:

- It's necessary to identify a novel science or engineering experiment that can benefit from a microgravity environment and could likely lead to commercial applications, and this is not necessarily easy to do.

- Having done so, one then needs to understand and develop the experimental equipment required to carry out this experiment in space, which will, at this early stage of our experience with space science, probably be unique. Hence, the researcher/experiment designer may also have to be a skilled equipment designer, or have one on the research team.

- The next requirement is to understand the international launch services and payload integration environment that will transport an experiment to space and return the results to Earth.

- Researchers must also have knowledge of the unique operational environment of the space laboratory facility, and how it can be used (and misused, or used ineffectively).

- Next they must develop partnership arrangements with international groups including governments, universities, and industry, for use of needed personnel and facilities. In particular, the scientists on the ISS will inevitably be involved in setting up, running, and monitoring experiments, which means that their capabilities and availability must be factored into the experiment design. As crews rotate in and out, continuity is essential for commercial results, which means multiple cycles of trainings may be needed to acquaint new ISS crews with ongoing experiments.

- All of this inevitably requires that researchers must also negotiate intellectual property (IP) agreements between the various partners, private and public.

- All of this may not be inexpensive, which requires that prospective space scientists must also become at least somewhat proficient in finance in order to raise sufficient investment capital, which may come from private or public sources, each of which will have its own unique standards, goals, methods, and requirements.

- And of course scientific progress proceeds through progressive learning as a result of iteration, which means that researchers must plan for subsequent generations of experiments over time that clarify, amplify, and build on prior results.

This list may not be entirely complete, but it does give us a good sense as to why the management of space experiments is indeed complex, and of course that very complexity is itself a barrier. Hence, an entrepreneur with even a spectacular scientific idea may find these requirements to be quite daunting. And then comes the business part, as once their money is in, investors will be monitoring progress to verify that the entrepreneur is successfully navigating through the barriers and executing a cohesive business plan.

## SYSTEMATIC SUPPORT: SPACE VENTURE INCUBATION

The massive complexity of commercially-oriented space science does also, however, in and of itself, define a new entrepreneurial opportunity, that of facilitating scientists with great ideas to progress quickly by helping them to overcome the barriers to the commercialization of space.

Based on the premise that we have to date only scratched the surface of what can be done in the life and materials sciences in space, and that projects of great value are still waiting to be created, our new organizational model consists of an incubator/facilitator and linked holding company to find, support, and promote promising space ventures.

We have established such a company in San Diego, Space Venturers Holdings, LLC (SVA). As we already have the scientific, managerial, and technical staff with the necessary experience, expertise, and contacts in all of the key areas mentioned above, we're able to significantly expedite the implementation of great research ideas in the expectation that great companies will emerge from these efforts.

Of course the model itself is not necessarily new, as it has already been implemented in many parts of the world. We know it well as the incubator that links venture capital with high technology ventures, and these exist in all developed and many developing countries as an essential part of the growing economy. The model works so well because most entrepreneurs and researchers with big ideas have minimal resources, and thus it can be a very productive partnership.

## SPACE VENTURERS HOLDINGS, LLC (SVH) ORGANIZATIONAL CONSTRUCT

### THE PARENT COMPANY

SVH will operate at two distinct levels. The parent company, SVH, will be a technical, scientific and managerial overseer of qualifying child C Corp companies, individually either wholly or partly owned by SVH and its individual investor community. SVH's brief will be to profit from its unique set of talents and connections in space science, technical, regulatory, international relations, financial investment, business management and global marketing. It will do this by identifying competent candidate technologies, and entrepreneurial teams and companies that could benefit by allying themselves with the SVH group. In exchange for equity or other positioning in successful candidates, SVH will provide access through its growing investor community. This includes investment partners who might not have invested in the individual company without the spectrum of capabilities and skill-sets that SVH brings to the business arrangement.

## THE SUBSIDIARY CHILDREN C CORPORATIONS

Each qualifying technology, management team, entrepreneur or existing company will be considered a candidate for resource allocation from within SVH's asset stable. Resources would include investment dollars by offering our network of high net worth investor's access to new technologies and cutting edge companies in which to invest that have been prequalified by SVH. SVH would also bring technical and scientific prowess through its network of key highly qualified personnel who have agreed to assist in our vision as described above. In exchange for these resources being made available, SVH will take an equity position in the child company along with the individual investor's equity stake. SVH might also make available technical skills in exchange for equity, or might bill for services rendered, on a case-by-case basis. Investment in each child corporation will be dictated by the individual risk appetite of each high net worth individual or group to which the potential project is offered as an investment opportunity.

## EXIT STRATEGY

SVH might have a lofty vision, but its choices of investments are firmly rooted in logical and diligent business management norms. Each child company will be viewed with an eventual exit strategy either through IPO, licensing of IP or outright sale.

The formula for execution is well proven, as each idea becomes the basis of a new venture, owned jointly by the entrepreneur/researcher and the incubator/investor, and each then succeeds or fails on it own merits, but with the benefits of the incubator and expert advisors in support of it. Using this model, great ideas originating anywhere in the world can be identified and nurtured using resources from international entities, public and private, leveraging international and off-Earth resources to significantly expand our commercial presence in space, and to bring the benefits of space science and space manufacturing to the broadest market in the shortest time.

•••

## MICHAEL WISKERCHEN, PH.D.

Founder, Space Venturers Holdings, LLC
Emeritus Faculty, Mechanical & Aerospace Engineering
University of California, San Diego

Dr. Michael J. Wiskerchen received his Ph.D. degree in Physics from the University of Denver in 1973. He has had a diverse academic and research career in space-related science and engineering at the University of Arizona, NASA, Stanford University, the University of California at San Diego, and several commercial companies. For more than forty years he has managed and participated in space missions involving deep space satellites, the Space Shuttle, and the International Space Station. During much of this time his efforts have been focused on the development, application, and operations of space related projects that involve alliances between industry, university, and government partners. From 2005 to the present, Dr. Wiskerchen has devoted considerable effort in developing private – public partnerships involving both space-related research and commercial programs. His recent efforts have concentrated on the founding of an incubator/facilitator company, Space Venturers Holdings, LLC, that stimulates and promotes commercial space enterprises.

CHAPTER 18

# FIREFLY

MICHAEL BLUM
CO-FOUNDER & CHIEF FINANCIAL OFFICER
FIREFLY SPACE SYSTEMS

*"When you're in low earth orbit, you're halfway to everywhere."*
Robert Heinlein

As data collection and transmission increasingly moves into space, the need for a dedicated small satellite launcher is becoming clearer.

What used to be the job of large, multi-hundred million dollar geosynchronous satellites, is more frequently being done by small satellites – some as big as your coffee table weighing several hundred pounds, others fitting into the palm of your hand weighing a mere few pounds. Today, these spacecraft operate using automotive grade components that cost a fraction of what traditional space grade componentry bills out for. A small satellite now often costs a few million dollars and some manufacturers are pushing satellite manufacturing costs down to the tens of thousands of dollars.

The applications of this technology are vast and will have profound implications for our society. Earth sensing and earth imaging missions in low earth orbit (LEO) that can observe the same spot on our planet frequently - sometimes multiple times per day - will allow for near real time monitoring of coastal erosion, illegal mining and logging operations, infrared sensing of port activity, the macro economy by measuring car movements in our metropolitan areas, etc., etc. Today, all of this big data can be accumulated at a far lower cost than ever before.

Networks of small satellites in LEO will soon be broadcasting high-speed broadband internet data across the globe, offering backhaul services previously restricted to terrestrial fiber-optic cable. Terrestrial solutions have proven to be unjustifiably expensive, resulting in vast areas of low broadband internet penetration. The democratization of information will take a giant leap forward as every person on Earth will be a simple receiver away from free flowing ideas, data, news and more.

But only, if these small satellites can get to space frequently, quickly, reliably, accurately and on a budget. Today, most small satellites have to ride-share – in the industry, we call this multi-manifesting. And it's a big problem for small sat operators. Think of it as hitch-hiking across the country: one doesn't get to pick when and where the ride goes and one has to put up with any of the driver's idiosyncrasies. In reality, this often means ending up in the wrong orbit and not being able to completely fulfil a mission. It often means getting delayed.

To get a full picture of Planet Earth, small satellites typically want to fly in polar orbit, circling the earth as it "raster-scans" beneath the satellite. Most large rockets do not fly to polar orbit as it is out of the way of their primary destinations: the ISS, geosynchronous orbit, etc.

Firefly Space Systems is solving this problem. We are developing and building a family of launch vehicles, employing best practices from mass manufacturing industries to dramatically reduce the cost of access to space. Our core philosophy is driven by a "simplest / soonest" approach to designing and engineering our technology: Firefly Alpha, our first vehicle, will be a pressure fed, all-carbon composite rocket, burning LOx / hydrocarbons and manufactured almost entirely by robots. Alpha will be able to deliver a ~400kg payload to a 400km equatorial orbit or ~200kg to a 500km sun-synchronous orbit.

Demand for launch is growing and is trending towards becoming a significantly constrained commodity. Currently, there are over 2,000 satellites in orbit. With the advent of large LEO constellations that number will grow enormously. Several well-funded groups have announced plans to launch constellations of many hundreds to even thousands of small satellites, each weighing a few hundred kilograms. Existing launch capacity is neither well suited nor available to deploy these types of projects.

Our focus has been on simplicity, relying on well documented and researched technologies at high Technical Readiness Levels, and the aggressive reduction in the number of parts that make up the vehicle to bring down cost. In time, we will add reusability to our product line-up further leveraging improvements in technology to decrease the cost of space launch.

**Figure 1**
**Firefly's In-House Designed Engine Control Board**
A fraction of the cost and size of commercially available alternatives.

To accomplish all of this, Firefly has built a team of ~100 propulsion, structural, avionics and test & launch engineers since early 2014. Many of our senior leaders have had extensive New Space experience working for companies such as SpaceX, Blue Origin or Bigelow Aerospace. We have aggressively hired young talent with undergraduate and graduate degrees from top universities and given these young leaders responsibility over and ownership in projects they are working on.

Our company believes in vertically integrating as much of our operation as possible. While we do work with outside partners to accelerate the development of our know how, we aim to bring all of the critical technical and engineering components in house as our infrastructure expands. This is an important strategy to control and reduce the cost of each vehicle. In the long run, our aim is to build each Alpha vehicle for around $1M.

By late 2017, Firefly expects to be conducting a sub-orbital test flight program. Our maiden launch to orbit will take place around March 2018 – a NASA Venture Class Launch Services mission. Over the course of several years, we will ramp up our launch cadence. By 2022, Firefly Alphas should be flying weekly. And our advanced vehicles, currently in early stages of design, should be taking to the skies. These vehicles will incorporate reusability and other cost saving features, in addition to allowing Firefly to service the entire sub-one metric ton payload segment.

**Figure 2**
**Firefly's 200 Acre Test Site in Briggs, TX**

As a business, our primary driver is to make money. Yes, the owners of the company want to earn a return on their investment and become rich. But only a financially successful space company will have a chance to innovate and make ever more challenging products to design missions that stretch the imagination of what is possible today.

This is the embodiment of Firefly's motto:

*Making Space For Everyone.*

•••

# MICHAEL A. BLUM
## CO-FOUNDER & CHIEF FINANCIAL OFFICER
## FIREFLY SPACE SYSTEMS

Michael is an entrepreneur at heart. At Firefly, he has been responsible for overall strategy and fund raising activities. He is also the Co-Founder & President of Hedgeye Risk Management, a leading online financial media company. Born in Hong Kong, he has spent many years in the region and is Managing Director of Asia Leisure Capital, a Macao based casino and hotel management and financing firm. Michael was formerly Senior Manager at PayPal, where he started PayPal Germany and later ran payments in South East Asia. Always passionate about space, Michael has been a frequent speaker on commercial access to space and space entrepreneurs. Among others, he has given lectures at Columbia Business School, the Yale School of Architecture and the Institute for Tourism Studies in Macao. He earned his Bachelor's Degree in Economics and International Studies from Yale University.

•••

Editor's Note:

Firefly Space Systems announced a successful test firing of its first rocket engine, Firefly Rocket Engine Research 1 ("FRE-R1") on September 10, 2015.

Chapter 19

# ARE SOLAR POWER SATELLITES SITTING DUCKS FOR ORBITAL DEBRIS?

Alfred Anzaldua,
Brad Blair,
and
David Dunlop

## Background

Even conservative estimates of the energy needed in coming years to meet surging world demand are staggering. Energy use, currently over 120,000 billion kWh annually, is forecast to double by 2030–2040 and to quadruple by 2090–2100.[1] To meet this growing demand, many space enthusiasts are

---

[1] Mankins, John C.; *The Case for Space Solar Power*, 2014, p. 21.

promoting the idea of space solar power for terrestrial use. Unfortunately, a fierce and growing spoiler lies in wait: orbital debris.

Individuals and governments around the globe are becoming aware of the danger that orbital debris presents both to our modern life and to future plans for the utilization of space. According to NASA, there are over 21,000 Earth-orbiting objects larger than a softball (10 cm) and 500,000 shrapnel fragments between 1 and 10 cm. The number of shrapnel smaller than 1 cm exceeds 100 million.[2]

Because of their high relative velocity on impact, as much as 10 km per second or 22,000 miles per hour in Low Earth Orbit (LEO), orbiting shrapnel as small as 5 mm can disable a spacecraft.[3] The debris is an ever-growing hazard to the International Space Station, future space flights, and the approximately 1,200 operational satellites.[4] Indeed, Jer-Chyi Liou, Chief Scientist for NASA's Orbital Debris Program Office, using estimates drawn from six space agencies, recently declared that even without a new catastrophic collision or explosion in orbit, and with 90% compliance with the 25-year deorbiting-after-use guideline, debris will continue to grow over the next 200 years.[5] Moreover, it seems reasonable to expect that the increase in debris, by knocking out stationkeeping capabilities of impacted satellites, will worsen Liou's current estimate[6] that there will likely be a major catastrophic collision every 5 – 9 years.

Within two LEO altitude bands, the density needed to initiate the "Kessler Syndrome," i.e., a cascading chain-reaction of collisions leading to uncontrollable growth of debris, *may have already been reached*.[7] Although most of the debris is in LEO, with the greatest concentration

---

[2] NASA Orbital Debris Program Office; "Orbital Debris Frequently Asked Questions." (http://orbitaldebris.jsc.nasa.gov/faqs.html#7), 12 November 2013.

[3] Jer-Chyi Liou speaking at the 2014 NewSpace Conference Orbital Debris Panel on July 26.

[4] Jer-Chyi Liou; Ibid., stated that working satellites represent only about 7% of the large objects in orbit around the Earth. Other have put the fraction at 7.6%.

[5] Jer-Chyi Liou; Op. Cit.

[6] Jer-Chyi Liou; Op. Cit.

[7] For first reference to what became known as the "Kessler Syndrome," see Kessler, D. J. and B. J. Cour-Palais (1978), "Collision frequency of artificial satellites: The creation of a debris belt," J. Geophys. Res., 83(A6), 2637–2646, doi:10.1029/JA083iA06p02637. For a reference to the Kessler Syndrome being approached, see Grossman, Lisa, "NASA considers Shooting Space Junk with Laser," *Wired*, 15 March 2011; National Research Council, "Orbital Debris: A Technical Assessment," *The National Academies Press*, 1995.

found near 750–1000 km altitude, there is also a considerable amount in or near Geostationary Orbit (GEO). Even at slower-than-in-LEO velocities, the large debris objects now tumbling out of control within or near this latter altitude pose a significant collision threat.[8] In sum, debris in LEO and GEO should be the first targets for orbital debris mitigation and remediation.

## PRECIOUS TARGETS IN THE SKY

### THE RISK TO PRESENT ASSETS

Orbital debris, by threatening our satellites and related spacecraft, is also threatening to shred the very fabric of modern life. Satellites are intimately involved with our everyday activities. Anyone using GPS, checking the weather forecast, watching TV, listening to the radio, flying on a plane, using an ATM while traveling, accessing many Internet sites, taking a cruise, or calling long-distance on a phone makes use of satellite technology.

### THE RISK TO FUTURE DEVELOPMENTS

Worse yet, *future* space technologies and missions are threatened. For example, Solar Power Satellites (SPS) for terrestrial use, an energy technology with enormous potential to improve lives, is also at stake. In 2009, retired astrophysicist Donald Kessler, who started NASA's work on orbital debris more than 30 years ago, stated, "large structures such as those considered … for building solar power stations in Earth orbit could set up a situation where a single satellite failure could lead to cascading failures of many satellites."[9]  Solar power satellites are not the only future spacecraft that will be threatened.  Bigelow Aerospace plans to have its BA 330 habitats serve as tourism hotels in orbit starting as early as 2016.[10]  Add to this the thousands of satellites and other spacecraft scheduled to go into Earth orbits until well into the future, and the available orbits become crowded.

---

[8]   NASA; Op. Cit.
[9]   Kessler, Donald; "The Kessler Syndrome."
      (http://webpages.charter.net/dkessler/files/KesSym. html) webpages.charter.net, 8 March 2009.
[10]  See lasvegascitylife.com/sections/opinion/knappster/George-knapp-infinity----and-beyond.html, April 14, 2013.

## RISK REDUCTION STRATEGIES

But would a hyper-modular space-based solar power system (SPS), such as proposed by John C. Mankins also be vulnerable? Mankins admits that micrometeoroids and orbital debris might impact the SPS and cause damage, but then he argues, "Fortunately, with a hyper-modular architecture such as SPS-ALPHA[11] there are no 'single' points of failure. Impacts will cause damage, but it will be mostly inconsequential and will only occasionally require repairs."[12]

This statement bears examination. Much shrapnel debris exists below current detection limits, so quantification of risk remains problematic. Further studies of risk and greater detection capacity are needed to reduce uncertainty and to encourage potential investors that the risks to capital invested in solar power satellites (SPS) are acceptable.

Admittedly, the hyper-modularity of the SPS-ALPHA system would mitigate damage from orbital debris. But Mankins proposes multiple SPS-ALPHAs to solve our energy concerns, each measuring approximately 3 km x 5 km.[13] These structures would be very large targets, "sitting ducks," in the case of a Kessler-type runaway debris growth at the 35,700 km, and the damage would likely go beyond "inconsequential." Even if the satellite remained structurally intact, maintenance costs would sharply rise. Keep in mind also that to build such a large SPS in the first place, many SPS module-carrying spacecraft would have first to pass through shrapnel-cluttered LEO bands before carrying modules to GEO for construction by telerobotically operated spacecraft.[14]

Perhaps SPS-ALPHAs require not only hyper-modularity, but hyper-*permeability*, such that the modular elements can each separately move to avoid debris. Ideally, the modules would describe an array of SPS-ALPHA elements flying in precise formation and with the ability to self-adjust to avoid danger, reminiscent of a school of fish adaptively avoiding the lunge of a predator.

## LARGE DEBRIS COLLISIONS MAKE SPACECRAFT-KILLING SHRAPNEL

Large debris, i.e. larger than 10 cm in diameter and 1 kg in mass, can range in size all the way up to 9-ton rocket bodies and 5-ton satellites. These

---

[11]  SPS-ALPHA stands for "Solar Power Satellite via Arbitrarily Large Phased Array."
[12]  Mankins, John C.; Op. Cit. pps. 461.
[13]  Mankins, Op.Cit.; pps. 8 and 424.
[14]  Mankins; Op. Cit. p. 9.

multi-ton bodies make up the mass of approximately 6300 tons of orbital debris, with approximately 2200 tons in Low Earth Orbit (LEO) alone, and collisions among them are the source of millions of shrapnel fragments.[15] For example, in 2007 China intentionally destroyed its Fengyun-1C weather satellite, and in 2009 a non-functioning Russian Cosmos 2251 satellite collided with an U.S. Iridium 33 satellite. At least one-third of all orbital shrapnel can be traced to just these two collisions.[16] Worse yet, orbital shrapnel smaller than 10 cm/1 kg is currently untrackable, and because of the high collisional velocity in LEO of around 22,000 mph, *even shrapnel as small as 5 mm* can take out a spacecraft.[17]

## POTENTIAL REMEDIES

### THE LARGE-OBJECTS-FIRST STRATEGY

A consensus is building among persons studying the orbital debris problem that the greatest danger will come from *inevitable* catastrophic collisions between large debris objects, which will produce immediate and *subsequent* financial loss due to untrackable shrapnel. Because the subsequent financial loss will dwarf the immediate loss, Jerome Pearson and his colleagues Joe Carroll and Eugene Levin in a recent article argued strenuously for dealing with such large objects as soon as possible.[18]

But which large debris objects should be the priority? Launching countries are naturally sensitive about the nature of their satellites. Therefore, to induce international cooperation to remove, recycle, or rehabilitate large debris objects, it is best to start with the much less sensitive, but still dangerous, upper stages (i.e. basically aluminum tanks).

---

[15] Joe Carroll, President of Tether Applications, Inc., speaking at the 2014 NewSpace Conference Orbital Debris Panel on July 26. Also, Jer-Chyi Liou, Op. Cit.

[16] Multiple publications speak of this disaster, including NASA, Op. Cit.

[17] Jer-Chyi Liou stated at the 2014 NewSpace Conference that shrapnel from 5 mm to 1 cm is the most dangerous because of its ubiquity and relative velocity. It is the collision of car-sized objects, from 1-9 MT, however, which has contributed and will continue to contribute to the vast growth of these objects.

[18] $200 million subsequent cost versus $30 million immediate cost per catastrophic collision. But this conservative estimate does not take into consideration all downstream costs due to loss of communication and electronic services on the ground, which could run into the billions, especially with multiple collisions. See Pearson, Jerome;Levin, Eugene; Carroll, Joseph; "The Long-Term Cost of Debris Removal from LEO," 64th International Astronautical Congress, Beijing, China, 2013.

At around 1100 tons, they make up about half of the debris mass in low earth orbit. Capturing aluminum tanks would also be a lot less complicated than grabbing satellites with solar arrays, antennas, and nuclear reactors. Because most of the large debris is of Russian origin, a bilateral treaty with Russia would be a good place to start. (See below.)

## LARGE DEBRIS REMEDIATION IN GEO

### CELLULARIZATION

The DoD's Defense Advanced Research Projects Agency (DARPA), under a demonstration project called Phoenix, is teaming up with the private sector to harvest and "repurpose" still functional components of nonworking satellites in GEO to create new space systems at greatly reduced cost. Beginning in 2016, the project proposes to attach nano-satellites to parts of retired U.S. government and commercial satellites, making the debris a resource. In a process called, "cellularization," nanospacecraft separately carrying out functions such as power, communications, attitude control would be launched into orbit as secondary payloads. A service-tender spacecraft would then be teleroboticially directed to attach such miniature devices to large antennas or other large parts of dead satellites to produce working satellites at a fraction of the cost of new ones launched from Earth.[19]

### REFUELING

Another way that defunct satellites in GEO can be rehabilitated if they are not already too damaged by orbital debris is through refueling. The Canadian company MacDonald, Detwiller, and Associates' (MDA) 2010 Space Infrastructure Services (SIS) project envisioned both refueling and otherwise servicing satellites in orbit teleroboticially. Although MDA and Intellsat in 2012 cancelled their collaborative agreement in which MDA was to develop a satellite capable of servicing Intelsat's 50 operating satellites, MDA remains interested in the concept and is waiting for a possible DARPA contract.[20]

---

[19] Dykewicz, Paul; "DARPA Advances Plans to Salvage Antennas of Retired, In-Orbit Satellites," www.onorbitwatch.com/program/darpa-advances-plans-salvage-antennas-retired-orbitsatellites, 24 November 2013. The cellularization process was also described by David Barnhart, Program Manager for DARPA Phoenix, during the Orbital Debris Panel on July 26 at the 2014 NewSpace Conference.
[20] "Intelsat Picks MacDonald, Dettwiler, and Associates Ltd. For Satellite Servicing," www.mdacorporation.com/corporate /news/pr/pr2011031501.cfm, 15 March 2011;

In this connection, it is important to note that in May 2013, NASA carried out a series of telerobotically operated "propellant transfer experiments" on an exposed platform of the International Space Station (ISS).[21] Although the ISS is in LEO, the refueling technology being developed is intended for use in GEO.

## Remediation of Large Debris in LEO

### "EDDE," The ElectroDynamic Debris Eliminator Vehicle

Various ideas and technologies are being developed potentially to remove, recycle the metal component of, or reuse through rehabilitation or repurposing large debris objects in LEO. For example, three companies, Star Technology and Research, Inc., Tether Applications, Inc., and Electrodynamic Technologies, LLC have been developing a technology called ElectroDynamic Debris Eliminator (EDDE), wherein a long conductor is energized using solar energy to thrust against the Earth's magnetic field. Operating without propellant, EDDE can repeatedly change its altitude by hundreds of kilometers per day and its orbital plane by degrees per day.[22]

Assuming effective EDDE or other non-propellant debris remediation technologies[23] are developed, which LEO orbits are ripe for remediation? About half of the mass of orbital debris in LEO is at inclinations of 71° – 74°, 81° – 83° and sun-synch orbit. According to Jerome Pearson, President of Star Technology and Research, Inc., and Joe Carroll, President of Tether Applications, Inc., disposing of upper rocket stages in these inclinations

---

Spark, Joel; "MDA, Intelsat Cancel On-Orbit Servicing Deal," *Space Safety Magazine*, 20 January 2012; Foust, Jeff; "The Space Industry Grapples with Satellite Servicing," *The Space Review*, 25 June 2012.

[21] Adrienne Alessandro for Goddard Space Flight Center, "NASA's Robotic Refueling Mission Practices New Satellite-Servicing Tasks," *Space Daily,* 13 May 2013.

[22] Pearson, Jerome; Levin, Eugene; Carroll, Joseph; "Affordable Debris Removal and Collection in LEO," 63rd International Astronautical Congress, Naples, Italy, 2012. http://www.star-tech-inc.com/papers/Affordable_Debris_Removal_IAC_2012.pdf.

[23] Hungarian J. Szentesi has designed a non-propellant space device, using what he calls an Electro-Magnetic Propulsion System (EMPS), which also thrusts against a planet's magnetic field and theoretically could be used to move orbital debris. See J. Szentesi; "Electro-Magnetic Propulsion System (EMPS) for Spacecrafts and Satellites, 43rd Lunar and Planetary Science Conference, 2012.

would remove 79% of the collision-generating debris potential is a crucial first step to stopping the growth of shrapnel.[24]


## THE ISS AS *THE DEMO SITE* FOR DEBRIS REMEDIATION TECHNOLOGIES

There are good reasons for testing and developing EDDE and other debris remediation technologies from the ISS. In the first place, the ISS generates ten tons of waste annually, and money and effort is already being spent to remove it.[25] The ISS also has features that can facilitate early demonstrations of debris removal technologies: its own electrical power supply, a redundant international supply chain, human extravehicular capabilities, robotic grappling and docking, a Ka Band microwave transmission antenna, and a potential for servicing and refueling other spacecraft. Joe Carroll suggests that EDDE vehicles could bring another 100 tons of orbital debris to the ISS for either de-orbiting or salvage.[26] Testing and developing EDDE and other technologies, such as energy-beaming and solar-electric propulsion (SEP) at the ISS could inform the space development community on techniques and technologies for capturing and handling orbital debris for subsequent de-orbiting, metal recycling, or repurposing.

Once we have learned to deal with this smaller amount of debris in connection with the ISS, we will be better prepared to deal with the estimated 2200 tons of dangerous large debris objects in LEO and elsewhere.

The ISS has occasionally to dodge space debris, and this involves moving its million-pound mass with rocket engines using chemical propellants. Perhaps the ISS-connected debris remediation demonstrations, done with free flyers operating within power beaming-distances,[27] could evolve into technologies specifically to protect the ISS and thus obviate the need to burn precious chemical propellant.

---

[24]  Pearson, et al; both publications Op. Cit. Joe Carroll noted more specifically at the 2014 NewSpace Conference that 3/8 of the mass of orbital debris is at 81° - 83° inclination and 3/16 is in sun-synch orbit.

[25]  Joe Carroll; Op. Cit.

[26]  Joe Carroll; Op. Cit.

[27]  Although EDDE vehicles use solar energy to create an electrical flux, other spacecraft potentially involved in debris mediation may very well benefit from energy beamed from the ISS.

## Considerations for a Debris-Remediation Economic Model

### Transportation and Removal Costs

Assuming that Space X does indeed manage to get the payload price to LEO down to $2200/kg[28] using the Falcon Heavy, and eventually half that cost with routine first-stage reuse, debris remediation at LEO and higher using only rockets would remain prohibitively expensive. Fortunately, using EDDE and other non-propellant using vehicles to carry out the actual removal of at least a thousand tons of large debris from LEO will make a noticeable difference at a more reasonable cost. In this regard, Jerome Pearson, et al., in considering an orbital debris removal campaign removing only upper stages from LEO, estimate that in seven years of operation "1000 tons of upper stages and 79% of the collision-generated debris potential can be removed at an average cost of less than $500 per kg and an average annual cost of about $70 million."[29]

### Salvaged Raw Materials & Their Market Value

To the above considerations we must add the salvage value of 2000 metric tons of refined metal. Aluminum scrap on Earth is currently valued at around $1730 per mt.[30] Thus, at a minimum the large pieces of debris in LEO represent at least $3,460,000 in raw material value. Finished products would have many multiples of that value in orbit. However, as shown below, manufacturing finished products in orbit from raw metal would involve heavy production costs.

## Production Cost for Value-Added End Products

### The Market Model

Salvaged metal can only be worth something to a company ready and able to process it into new tools, devices, or spacecraft – for a profit. To get that profit, the potential buying company will have to figure in capitalization costs necessary to transform the metal into final products. Then the buyer must either sell the new tools, devices, or spacecraft, or use them to provide

---

28 "Upgraded Space X Falcon 9.1.1 will launch 25% more than old Falcon 9 and bring the price down to $4209 per kilogram to LEO," http://nextbigfuture.com2013/03/upgraded-spacex-falcon-911-will-launch.html

29 Pearson, et al.; "Affordable Debris Removal....."; Op. Cit.

30 See http://www.cinelli-iron-metal.com/market.php.

a service for which there is demand. All these actions within the cislunar market will determine the actual value of the salvaged metal to the first buyer. It is unlikely, however, that *all* those tons of salvaged metal will be bought for space construction in the foreseeable future, as a good number of the smaller upper stages and "zombie" satellites will likely be deorbited.

Beyond these preliminary market figures and considerations, on-orbit recycling of materials for construction and manufacturing would counteract the throwaway culture that has made space operations, with the exception of commercial communications and GPS satellites, largely beyond the reach of the commercial economy.

## THE RISK-REDUCTION & QUANTIFICATION MODEL

Lowering or eliminating the likelihood of large debris collisions that threaten a satellite industry grossing over $200 billion annually, by whatever means, is a valuable service that must be quantified. The community of satellite users must therefore collaborate to remove large debris objects safely and thus lower the risk of catastrophic collisions, or face operating losses, customer anger and loss, coupled with much higher costs for satellite replacement. However, retiring this risk of collision will avoid subsequent much larger losses. Market-based insurance and salvage quantification-models could be used to provide economic incentives to remove, reuse, or recycle space debris, and thus significantly help this industry.

## SPACE DEBRIS AND ORBITAL MECHANICS

Serious thought should be given to where orbiting scrapyards would best be located, and what sorts of vehicles should emplace them. Most orbital debris resides within 1500 km of the Earth's surface, although there is a significant band of debris around GEO. The orbits of scrapyards below 600 km would degrade, within a few years or months because of atmospheric drag and de-orbit, depending on the particular altitude

At around an altitude 650 km, however, orbital debris is relatively sparse, and scrapyards there would need only infrequent boosting to maintain altitude. EDDE vehicles could therefore carry large debris objects to cross-truss scrapyards at that altitude. Carrying defunct upper stages to 650 km for collection would make the raw aluminum more accessible for subsequent construction in LEO[31] and would be quicker than carrying them to deorbiting altitudes.

Orbiting scrapyards could also be located within other sparse debris bands in Medium Earth Orbit (MEO), High Earth Orbit (HEO), *or even around Earth-Moon Lagrange points*. Scrapyards embedded in cross-frames in meta-stable halo orbits near Earth-Moon L1/L2 (E-M L1/ L2), with a little stationkeeping, could serve as a metal-resource sites and a nexus for cis-lunar infrastructure, facilitating the later growth of staging sites, fuel depots, spacecraft construction sites, comsats, and habitats with telerobotic capabilities.

Note that it takes a bit *less* chemical propellant to reach Earth-Moon L1&2 from LEO than to reach and circularize on orbit in GEO.[32] On the other hand, in comparison to going to GEO, reaching Earth-Moon *L4 or L5* from LEO would take a little *more* chemical propellant. Scrapyards in these latter locations, however, could remain in stable-bean shaped orbits without stationkeeping for many years. When dealing with low-thrust Solar Electric Power (SEP) from LEO to Lagrange orbits in comparison to GEO, the propellant cost is not as favorable.[33] With SEP, however, we would be dealing with much less propellant in the first place.

Of course, every proposed salvage operation should entail reducing the risk of orbital-debris collisions, not increasing it. Moreover, the act of grappling, controlling, or moving debris should not generate more of the material. Any international system monitoring such salvage operations should operate transparently and give notice of voluntary space "clean up" activities by sovereign nations or parties registered with those countries to do business. Opportunities for third-party review, comments, filing of objections, and unilateral "holds" should all be part of the process. Finally, liability assignment under various scenarios will have to be agreed upon by all parties before orbital remediation can begin.

[31]   Pearson, et al.; "Affordable Debris....";  Op. Cit., http://www.star-tech-inc.com/papers/Affordable_Debris_Removal_IAC_2012.pdf
[32]   http://space.stackexchange.com/questions/2046/delta-v-chart-mathematics; http://www.strout.net/info/science/delta-v/; http://en.wikipedia.org/wiki/Delta-v_budget.
[33]   Ibid.

## The Russian Stakes for Cooperation to Remove Orbital Debris

More than 70% of the total mass of debris in LEO is Russian. Relations between the United States and Russia have fallen to another nadir, and the relationship is filled with the growing suspicion and mutual hostility harkening back to the Cold War. So why would Russia cooperate with the United States (and others) to deal with orbital debris?

First, those Russian debris pieces, large and small, carry with them liability under the 1972 Liability Convention. Second, the debris represents enormous value in an emerging economic model for space debris cleanup, both as already emplaced highly refined metal, and as buses for enabling nanosats. Third, the U.S. Department of State, in coordination with the UN Committee on the Peaceful Uses of Outer Space (COPUOS) could play a crucial role in putting together an agreement between Russia and the United States to remove the 79% of the orbital debris with the greatest shrapnel-generating potential. Fourth, through public-private partnerships, addressing this heritage of Russian space debris may provide a strategic opportunity for Russia to enhance its international position in the commercial development of space.

## Sweating the Small Stuff: What About the Shrapnel?

### Ground-Based Lasers

To remove small debris, "laser nudging" from ground-based lasers appears to be the most viable option. In this concept, a powerful ground-based laser would ablate the front surface off a debris target to slow and thereby deorbit it. To remove the political-military element, the system would have to be ground-based, transparent, and under international civilian control. In this regard, several observers have proposed an international civilian consortium to manage, in transparent fashion, ground-based lasers targeting debris shrapnel smaller than 10 cm for ablation to induce deorbiting.[34] Under such a regimen, consensus would be needed to select targets and timing for deorbiting. A bounty system could also be used to facilitate commercial participation so that the consortium members need not pay for

---

[34]  For general information about how ground-based laser could be used to deorbit space debris see work by Dr. Claude Phipps, http://photonicassociates.com/.

complex and expensive development project, or for failures, but only for results.[35] A system of paying bounties for results could also be used to facilitate the removal of large debris.

Along with property rights in space, international laws and treaties touching on the delicate issue of lasers to remove orbital debris will eventually have to be modified. The prestigious Institute of Air and Space Law of McGill University has recently proposed a meeting to carry out just such reconsiderations.[36]

## SHRAPNEL-TRACKING IMPROVEMENTS ARE CRUCIAL

When it comes to using lasers to deorbit debris, beyond having to organize a large group of space-involved countries and getting their consensus, there is another major problem standing in the way: small orbital debris is currently not being tracked, and even larger debris is not tracked in real-time.[37]

Doug Beason, Senior Vice President for Special Programs at the Universities Space Research Association, decries the lack of tracking of the most immediately dangerous shrapnel, i.e. debris larger than 5 mm but smaller than 10 cm. He suggests public/private efforts to "find, fix, track, and target" orbital debris objects, so they can be engaged and that engagement later be assessed.[38] Also, because the Joint Space Operations Center is part of U.S. Strategic Command, much of its tracking technology is secret. According to Beason, an international station with optical, radar, and polarization debris-tracking technologies under civilian control is urgently needed, not to find and track sensitive satellites, but to find and track in real-time, orbital debris, including shrapnel.

Instead of depending on military debris detection and tracking wrapped in secret capabilities and protocols, international investment in a transparent non-military commercial tracking system is needed. We have transitioned from a military-only GPS system to commercial dependence on GPS tracking that now includes Russian GLONASS and European

---

[35] Robertson, Donald F.; "A Commercial Approach to Debris Control," *Ad Astra*, pps. 20 – 23.

[36] Co-author Dunlop's personal communication with Andrea DiPaolo of McGill University Institute of Air and Space Law

[37] Doug Beason, speaking at the 2014 NewSpace Conference Orbital Debris Panel on July 26, noted that Space Command does not track even larger debris in real-time and can only predict location of large debris seven days in advance and with error-bars of 1.5 – 10 km.

[38] Doug Beason; Ibid.

Galileo systems, in addition to Chinese and Indian positioning systems. We can make the same type of transition with orbital debris tracking systems.

## "FREE PARKING" ORBITS AS COMMERCIAL DISINCENTIVES TO RISK REDUCTION

Beyond remediation, steps can be taken to mitigate orbital debris as well. Joe Carroll decries the 25-year "free parking," which results from international guidelines for satellite companies to deorbit their satellites after 25 years of non-use.[39] Technology already exists that can begin deorbiting a satellite the moment it stops functioning. In this regard, the company Tethers Unlimited has developed a conductive "terminator tape," which uncoils out of a defunct satellite, causing both aerodynamic and electrodynamic drag to deorbit the spacecraft. Two of the company's terminator tapes are currently being tested as demonstrators on cubesats.[40]

The European Space Agency (ESA), in coordination with the University of Surrey, is about to test the de-orbiting capabilities of the pop-out 5 x 5 meter gossamer solar sail technology it developed.[41] Other countries and institutions are developing debris mitigation and remediation technologies based on grab-and-plunge spacecraft, solar sails, electro-static nets, and balloons.[42]

---

[39] Joe Carroll at 2014 NewSpace Conference; Op. Cit. For general background on problems caused by free parking relevant to the LEO commons, see book by Donald Shoup, *The High Cost of Free Parking*, Updated Edition, 2011. Also see article by Garret Hardin, "The Tragedy of the Commons," *Science*, 13 December 1968, http://www.sciencemag.org/content/162/3859/1243.full

[40] Robert Hoyt, CEO and Chief Scientist of Tethers Unlimited, described on July 26 this de-orbiting system during the Orbital Debris Panel at the 2014 NewSpace Conference.

[41] Schenk, Mark; "Cleaning Up Space Debris with Sailing Satellites," University of Surrey, 31 January 2014.

[42] See Quigley, J. T.; "Japan will Cast a 'Magnetic Net' for Space Junk, 16 January 2014; Rutkin, Aviva Hope; "Japan's Huge Magnetic Net Will Trawl for Space Junk," *New Scientist*, 22 January 2014. Also see Pousaz,Lionel; "The Time Has Come to Destroy Debris," http://www.s-3.ch/en/mission-goals; Barraud, Emmanuel; "Cleaning up Earth's Orbit: A Swiss Satellite to Tackle Space Debris," *Mediacom*, www.s-3.ch/en/mission-goals and actu.epfl.ch/news/cleaning-up-earth-s-orbit-a-swisss-satellite-to-tac/. S3 is itself partnering with 12 other companies and one major investor/sponsor (Breitling).

## Pay Now or Pay (Much More) Later[43]

The risk of generating hundreds of millions of dollars of damage from orbital debris collisions is currently not properly balanced against the costs of deorbiting or moving satellites. An economic model creating disincentives for extending the risk-time of defunct satellites is needed, and is a topic for consideration vis-à-vis new space laws and treaty provisions. At present, we have not developed the mix of carrots and sticks needed for an effective economic orbital debris cleanup model.

There is an old adage that "A stitch in time, saves nine." In the spirit of this saying, we should note that if no action is taken to address orbital debris until there are multi-level Kessler cascades, the tab will be much higher than now in terms direct financial costs for satellite insurance, satellite replacement, and satellite service disruption in various industries and businesses. Therefore, nearly all persons living in industrialized societies will eventually have to pay the tab one way or another if in no other way than through increased user fees.

If voluntary "seed money" plans are not instituted, it might fall to the international space-users community to empower the International Telecommunications Satellite Organization (ITSO) or some other intergovernmental organization to collect a universal and mandatory tax on satellite services to finance bounties to be paid to commercial entities for orbital-debris remediation. For such a plan to work, all satellite-service providers would have to contribute (i.e. no "free riders"), so that no competitive advantage would result from non-compliance.

## Orbital Debris: Resource Ladder to the Stars

Not only is orbital debris the "low hanging fruit" with regard to a vast supply of already refined metal and emplaced structures on which to hang empowering nanosats, the technologies that will be developed to deal with debris will also be useful for dealing with capturing and mining near-Earth asteroids (NEOs),[44] lunar mining, on-orbit assembly of spacecraft, robotic

---

[43] McNight, Darren; "Pay Me Now or Pay Me More Later: Start the Development of Active Orbital Debris Removal Now,"
http://www.amostech.com/TechnicalPapers/2010/Posters/McKnight.pdf.

[44] Robert Hoyt; Op. Cit. at the 2014 NewSpace Conference, also described company mechanisms that could be used for de-spinning orbital debris and NEOs.

transport of materials, and other technologies. Growing out of a commitment to remove, re-cycle, or rehabilitate orbital debris could come new cislunar industries, including materials reprocessing, spacecraft manufacturing, multi-purpose platform construction, propellant depots and staging site construction, cislunar transportation development, new communication networks, and navigation infrastructure.

## SUMMARY

The issue raised by the consideration of solar power satellites as sitting ducks is merely illustrative of the risks and rewards of both present and future economic activities. It also raises the issue of international legal reforms and new initiatives. Space debris could be a show stopper for the future - *but only if we let it*. It is well within our power to constructively address the risks and investments to facilitate an emergent, and eventually booming, cislunar space ecosphere.

The investment to manage and remediate space debris is first an "insurance" cost to maintain acceptable levels of risk in doing space-connected business. Remediating orbital debris is also the road to a vibrant cislunar economy. The emerging cislunar economy will include solar power satellites, GEO communications platforms, Earth-Moon Lagrange facilities, and a reusable transportation infrastructure in cislunar space, and will be orders of magnitude greater than the current space economy. Solving the challenges of orbital debris opens the door to that greater economy.

•••

## Alfred Anzaldua

Al Anzaldua is a retired U.S. State Department diplomat and 30-year veteran of space advocacy. His undergraduate educational background is in Science and International Studies, followed by a Master's degree in Latin American Studies. In the U.S. Foreign Service, he carried out diplomatic and science-related work, primarily in Latin American and Caribbean countries. Al is the National Space Society (NSS) Region 3 Board Director, the NSS International Chapters Coordinator, Vice-Chair of the NSS International Committee, Secretary of the Tucson L5 Space Society, and active with other space or astronomy organizations. He has authored a series of articles on space issues and gives frequent presentations and exhibitions on space-related subjects in English and Spanish. Al is dedicated to seeing humans become a prosperous multi-world species, but understands that we must take many difficult, practical steps to realize that goal.

## Brad Blair

Brad Blair is a geologist, mining engineer and mineral economist. After several years working in mineral exploration and engineering, he began to study and model space resource economics as a returning graduate student at Colorado School of Mines. After completing a second Master's degree, his professional aerospace experience has included consulting to NASA, DARPA, DigitalSpace, Raytheon, Bechtel and the Canadian Space Agency on lunar and asteroid mining architectures, ISRU technologies, real-time 3D engineering simulations, and cost/benefit analysis of using in-situ propellants for human planetary exploration missions and space commerce. He has subcontracted on simulant and engineering design as well as economic analysis for robotic asteroid missions, with customers including NASA-NIAC, Altius Space Systems, Excalibur Exploration and the Robert A. Heinlein Foundation. Brad has been a guest lecturer at Singularity University and is on the board of directors for the National Space Society. He is a trained and experienced cost estimator using the NASA and Air Force Cost Model (NAFCOM). He competed in two NASA centennial challenges and is currently consulting to, advising and participating in NewSpace startup companies.

# DAVID DUNLOP

Dave Dunlop is a member of the Board of Directors of the National Space Society, Chair of the NSS International Committee which supervises NSS role as an organization with permanent observer status at the United Nations Economic and Social Council and COPUOS. He is active in organizing cislunar programming at the ISDC Conferences. He is currently working as an NSS representative to the International Lunar Decade Working Group to promote an expanded international participation in the further exploration and economic development of cislunar space including the lunar surface and Lagrange points.

He is also an editor of *To The Stars International Quarterly* which is published electronically on the NSS web-site. He is a contributor to NSS' Publication Ad Astra.

He has collaborated in articles in Space Review with NSS BOD colleague Al Anzaldua, about a range of issues involving space solar power, further economic development of economic infrastructure in cislunar space, and the challenges of cleaning-up space debris.

He has a BA and Master's Degree from the University of Illinois in Chicago. Currently he lives in Green Bay, Wisconsin. He has worked in diverse settings including, social work, mental health systems administration, teaching, education and public outreach, and with NASA contractors.

Chapter 20

# Update on Commercial
## and
# Academic Space Endeavors

## The Editors

Commercial efforts to develop space-related businesses have seen significant successes during the last few years, and with increasing emphasis and investment by a wide range of private, public sector, and academic investors, it is reasonable to expect continuing expansion of both investment and operational success.

Our intention here is not to provide a comprehensive overview of these efforts, but to mention at a high level some of the more notable aspects of this ongoing and growing process.

## PRIVATE AND PUBLIC INVESTMENT

As of this writing, April 2016, there are dozens of private space efforts under way. A page maintained on Wikipedia entitled "List of private spaceflight companies" summarizes the companies engaged in development efforts already under way in these categories.[1]

| Category | Number of Companies Listed |
|---|---|
| Crew and Cargo Transport | 17 |
| Space Stations | 5 |
| Launch Vehicles | 13 |
| Landers, Rovers, and Probes | 18 |
| Mining Companies | 6 |
| Propulsion | 11 |
| Satellite Launch | 8 |
|  | 78 * |

* Four companies are included in multiple categories,
giving a total of 74 active companies.

Bloomberg reports that the total of invested capital represented by these companies is more than $10 billion,[2] and of course these do not include the major private sector suppliers to satellite, government and military space efforts, such as Boeing and Northrop Grumman in the US, Airbus and Thales in Europe, Mitsubishi in Japan, and dozens of others around the world that support a thriving satellite industry.

Their major customers are listed below, which are the largest of the national space efforts worldwide, the nations with the 23 largest budgets. As you can see, current public investment in space totals more than $40 billion, of which about 70 percent is accounted for by the US, Russia, and the ESA.

[1]  https://en.wikipedia.org/wiki/List_of_private_spaceflight_companies  –  Accessed April 12, 2016
[2]  http://www.bloomberg.com/news/articles/2015-02-05/galactic-gold-rush-private-spending-on-space-is-headed-for-a-new-record

| Country/ Region | Agency | Budget (millions of US $) | Year |
|---|---|---|---|
| United States | NASA | 17,800 | 2012 |
| Russia | Russian Federal Space Agency | 5,600 | 2013 |
| ESA | European Space Agency | 5,510 | 2013 |
| Japan | Japan Aerospace Exploration Agency | 2,460 | 2010 |
| France | French Space Agency | 2,170 | 2014 |
| Germany | German Aerospace Center | 2,000 | 2010 |
| Italy | Italian Space Agency | 1,800 | 2014 |
| China | China National Space Administration | 1,300 (Euroconsult) 500 (official) | |
| India | Indian Space Research Organisation | 1,200 | 2016 |
| Canada | Canadian Space Agency | 489 | |
| UK | UK Space Agency | 414 | |
| South Korea | Korea Aerospace Research Institute | 366 | 2007 |
| Ukraine | State Space Agency of Ukraine | 250 | |
| Argentina | Comisión Nacional de Actividades Espaciales | 180 | 2016 |
| Iran | Iranian Space Agency & Iranian Space Research Center | 72 + 67 = 139 | 2015 |
| Spain | Instituto Nacional de Técnica Aeroespacial | 135 | 2009 |
| Netherlands | Netherlands Space Office | 110 | |
| Sweden | Swedish National Space Board | 100 | 2011 |
| Brazil | Brazilian Space Agency | 100/R$ 299 | 2015 |
| Pakistan | Space and Upper Atmosphere Research Commission | 75 | 2011 |
| South Africa | South African National Space Agency | 12 | 2015 |
| Switzerland | Swiss Space Office | 10 | |
| Mexico | Mexican Space Agency | 8 | |
| World | Total of listed budgets | ~ 41,800 | |

## SUCCESS AND FAILURES

The efforts of these entrepreneurial companies generates a significant volume of news coverage from their recent successes, which is indicative of a vibrant and active community engaged in investment, engineering, and testing. The more prominent firms, including SpaceX, Blue Origin, Orbital Sciences, and Virgin Galactic have received extensive press coverage of recent supply missions to the International Space Station, launch and retrieval of launch vehicles, extensive programs of test flights, and in the case of Bigelow Aerospace, the delivery of one of its inflatable modules to the ISS for testing as of this week, April 2016.

There have also, of course, been some notable failures, including the crash during a 2014 test flight of the Virgin Galactic SpaceShip 2, and some failed launches by SpaceX. Following the release of the National Transportation Safety Board report on the fatal SpaceShip 2 crash, Virgin Galactic founder Richard Branson offered this comment: "Every new transformative technology requires risk, and we have seen the tragic and brave sacrifice of Mike Alsbury [who was killed in the crash] and the recovery of injured surviving pilot Pete Siebold. Their tremendous efforts are not in vain and will serve to strengthen our resolve to make big dreams come true."[3]

This theme is quite consistent among space entrepreneurs – the intent to dream big. Their work to fulfill those dreams has been a consistent theme throughout the Space Age, and it has provided significant motivation for all types of participants. We can anticipate that continuing successes among the various entrepreneurial and commercial firms will only accelerate those efforts.

## RISKS TO SUSTAINABILITY

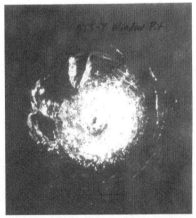

However, it should also be clear based on many of the chapters in this volume that all of that investment is at risk, and all of the benefits that space has to offer to humanity are likewise at risk, should the problem of space debris overwhelm the capacity LEO and GEO space to accommodate successful satellite and station orbits.

The NASA photo to the right shows the damage that was sustained to a window on the Space Shuttle during STS 7, 1983, and as we saw on page 306, the ISS has also sustained

---

[3]   http://www.space.com/30073-virgin-galactic-spaceshiptwo-crash-pilot-error.html#sthash.0lzbo5eN.dpuf

damage from a micrometeorite. NASA's Orbital Debris Program Office is focused exclusively on this issue, and maintains an interesting library of data on debris strikes on various spacecraft, as well as procedural standards and ongoing mitigation efforts.[4]

NASA reports that it tracks more than 21,000 pieces of orbital debris larger than 10 cm, the estimated number of particles between 1 and 10 cm in diameter is 500,000, while he number smaller than 1 cm exceeds 100 million.[5] The image below, also from NASA, shows orbital debris in GEO.

NASA is of course not the only space agency that is concerned.

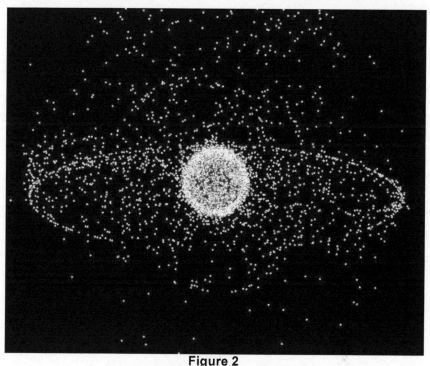

**Figure 2**
**Orbital Debris in GEO.**
Mostly orbiting at 17,500 miles per hour. Photo: NASA.

The Swiss research institute École Polytechnique Fédérale de Lausanne (EPFL) announced its plans to build and launch a spacecraft called CleanSpace One by 2018 that would grab orbital debris and then carry it back towards Earth, burning up in the atmosphere with it on its way down.

---

[4] http://www.orbitaldebris.jsc.nasa.gov/index.html
[5] http://www.orbitaldebris.jsc.nasa.gov/faqs.html#3

Originally designed using a claw-like grasping tool, EPFL's further research revealed that a folding conical net designed by students at the Western Switzerland University of Applied Sciences that would gobble up bits of space garbage is expected to be more effective.


## GROWING ENTREPRENEURISM

While some efforts are geared at attempting to collect space garbage, others are creating tools to enable more students and professionals to be active space scientists and entrepreneurs on very small budgets. For example, CubeSat Kit will be happy to sell you, for only a few thousand dollars, the materials you need to build a cube sat, a 10 cm x 10 cm x 10 cm satellite device that has become a standard tool for low budget space operations (http://www.cubesatkit.com/). And NanoRacks, the first company to operate on the International Space Station, will launch your cubesat for you at a modest price. They have already launched more than 300 successful payloads (http://nanoracks.com/).[6]

These are just two among dozens of small companies that are being developed to serve what could become a thriving industry in the coming decades. And like companies operating on Earth, which routinely buy services from dozens of suppliers that provide water, electricity, sewage disposal, waste removal, recycling, janitorial services, internet access, package delivery, and office supplies, space-based entrepreneurs will need all of these services, as well as unique services that ground-based businesses never need to think about. Like orbital debris monitoring, threat removal, and radiation monitoring.

The market research firm NewSpace Global now counts more than 800 private space companies globally (http://newspaceglobal.com/). One of them, the space imaging firm Planet Labs has raised more than $160 million in investment, which it used to launch 73 global imaging satellites. "A lot of people were very skeptical, and what we were trying to do seemed ludicrous," said founder Will Marshall, a former engineer at NASA Ames Research Center in Mountain View, California. "But we have assets in orbit, and customers who are very interested in the data. Every time we

---

[6]   For more on NanoRacks see also: Cummins, Christopher K. "An Open Source, Standardized Research Platform for the International Space Station," in *Space Commerce: The Inside Story by the People Who Are Making It Happen.* Morris and Cox, Editors. ATWG, 2010. P. 105.

take a picture, we can see how the world is changing." [7]

What was once ludicrous is now common, and what may seem to us unimaginable may one day in the not so distant future also become common. That's the potential of space commerce to further extend the human presence into and eventually perhaps throughout the Solar System. Such progress would transform the concept of space sustainability from a concern of governments and foresighted observers to an everyday business concern, and the efforts of COPOUS and all of the related organizations globally to address these issues will be fully rewarded not only with new economic opportunities, but tremendous social and cultural benefits as well.

•••

---

[7] http://www.bloomberg.com/news/articles/2015-02-05/galactic-gold-rush-private-spending-on-space-is-headed-for-a-new-record

# PART V

# SPACE AND SUSTAINABILITY ON EARTH

## INTRODUCTION TO PART V

What are the long term benefits from space, and how can humanity assure that we take the actions today that will lead to enduring success in the discovering new opportunities, innovating constantly to create new resources, and sustaining the effort to bring benefits to all humans in the 21$^{st}$ Century? These are the topics we address in Part V.

CHAPTER 21

# THE MEANING OF SPACE SUSTAINABILITY FOR DEVELOPING COUNTRIES

CIRO ARÉVALO-YEPES
FORMER COPUOS CHAIR
SECURE WORLD FOUNDATION ADVISORY COMMITTEE MEMBER

When first launched during the United Nations (UN) Committee on the Peaceful Uses of Outer Space (COPUOS) 51$^{st}$ session in 2007, the concept of space sustainability was essentially circumscribed to safety and security in the space environment. The initiative to establish a set of space sustainability guidelines was focused on addressing the risks posed by space debris, which had already become a primary concern, particularly for spacefaring nations.[1] Gradually, it was recognized that the existing international legal framework governing space activities must be considered both with regard *to legal obligations and rights* to take preventive measures that limit the risk, as well as to the *legal consequences* should such a risk materialize.[2]

The debate that followed the adoption of the proposal[3] reflected different perceptions on the sustainability issue. For some developing nations[4] the space debris problem should be handled under the principle of common but differentiated responsibility. Initial reactions were skepticism since many developing countries were not direct producers of the intentional and non-intentional debris affecting the space environment. The sense of an additional burden to the existing space technological gap was on the mind of many non-spacefaring nations. After a long effort, the set of draft long-term sustainability (LTS) guidelines – addressing policy, regulatory mechanisms, international cooperation, and management - will be open for political discussion in the upcoming meetings of COPUOS and its Scientific and Technical Subcommittee (STSC) with an aim for their final adoption in 2016.

Among the issues raised during the discussions, particularly by the Latin America and Caribbean Regional Group (GRULAC), was the perception that the LTS guidelines were not sufficiently linked to the implementation of space technologies for sustainable development, and to the equitable access of the spectrum and geostationary orbit resources. The first is related to the recommendations of the World Conference on Sustainable Development Rio+20 and the Post-2015 Development Agenda, a priority in the developing world. The second has been on the agenda of the International Telecommunication Union (ITU) and COPUOS for a

---

[1] There is no explicit international legal obligation to mitigate risks associated with space debris. The international legal principle of due regard (under Article IX of the Treaty on Principles Governing the Activities of States in the Exploration and Use of Outer Space, including the Moon and Other Celestial Bodies) may be considered to oblige spacefaring nations to take appropriate measures to prevent harm to other States.

[2] The current draft COPUOS guidelines constitute fundamental guiding principles recognizing the problem of space debris and expressing political commitment to address and mitigate it, but they are voluntary and create no legal obligations.

[3] Under the chairmanship of the author of this paper, COPUOS in its 52ⁿᵈ session in 2009 (3-12 June) agreed that, starting from its 47ᵗʰ session, a new agenda item entitled "Long-term sustainability of outer space activities" under the multi-year workplan should be included. Document A/64/20

[4] More than 65 percent of COPUOS' 76 members are developing nations from Africa, Asia-Pacific, Latin America, and Oceania.

number of years without any practical outcome. This chapter analyzes these two main issues, critical to space sustainability from the perspective of developing countries but only partially covered by the draft guidelines presented at COPUOS on June 2014, and proposes a way forward for a holistic and comprehensive approach within the UN.

## IMPLEMENTATION OF SPACE TECHNOLOGIES FOR SUSTAINABLE DEVELOPMENT

The international community is now moving towards the Post-2015 Development Agenda, currently involved in a re-examination of the implementation of the Millennium Development Goals, and the launch of new sustainable development targets. In this context, it is imperative to validate to the international community the vital role of space technology in development, particularly within the context of the outcome of the Rio+20 Conference and the emerging Post-2015 Development Agenda, and taking into account the most pressing issues of developing nations. It is commonly accepted that the world is entering an era of shared global utilities from space, and as a result sustainable development is increasingly reliant on space activities that support a myriad of applications and utilities on Earth.

To face mounting skepticism on the United Nations development proposals and in order to support more effective and strengthened implementation of the Post-2015 Development Agenda, the international community will require advanced tools with transformative power. Such tools should not only should have the highest transformative potential, but also be universally applicable to all areas of development in the post-2015 framework – be it health, agriculture, food security, education or disaster risk reduction.

In this context, the United Nations Conference on Sustainable Development decided to establish a universal, intergovernmental, high-level political forum, building on the strengths, experiences, resources and inclusive participation modalities of the Commission on Sustainable Development, and subsequently replacing the Commission. The high-level political forum (HLPF) would follow up on the implementation of sustainable development and avoid overlap with existing structures, bodies and entities in a cost-effective manner. To give a high-level political profile, the UNGA decided that the meetings of the forum under its auspices shall be convened every four years at the level of Heads of State and Government for a period of two days, at the beginning of the session of the Assembly.

To construct a global approach on the sustainability issue, the Post-2015 Development Agenda shall follow up and review progress in the implementation of all the outcomes of the major UN conferences and summits in the economic, social and environmental fields. This is probably one of the first times that the UN applies a holistic approach to one particular issue. Therefore it is necessary to improve cooperation and coordination within the UN system on sustainable development programs and policies by sharing best practices and experiences relating to the implementation of sustainable development goals.

The theme for the 2014 HPLF forum was "Achieving the Millennium Development Goals and charting the way for an ambitious post-2015 development agenda, including the Sustainable Development Goals." One of the main outcomes of Rio+20, was the agreement by Member States to initiate a process to develop a set of sustainable development goals (SDGs). The Rio+20 outcome document stated that the SDGs should be limited in number, inspirational, and easy to communicate. The goals should address all three dimensions of sustainable development in a balanced way and be coherent with and integrated into the UN development agenda beyond 2015.[5]

As recognized in key UN COPUOS resolutions, space technology provides means that can transform traditional approaches into a balanced integration of the three sustainable development elements: social equitability, inclusive and sound economic growth, and protection of environmental features and habitats. We may also deduce that the previous elements rely on a fourth element dealing with security and peace in which space applications are vital.

Among the most visible space applications are those in the agricultural and industrial sectors, which are proven engines of economic and social growth. UN entities make full use of space-derived data and information in their efforts to promote sustainable agriculture and advanced technological development particularly in programs of precision farming. Understanding the complex relationships within agricultural and industrial systems, UN entities make efforts not only to promote the use of geospatial information

---

[5]    The High-level Political Forum on Sustainable Development held its first meeting on Tuesday, 24 September 2013. The meeting was held under the auspices of the General Assembly and gathered Presidents, Prime Ministers, senior officials and representatives of stakeholders. The outcome of the meeting is a summary of the President of the General Assembly (A/68/588).

resources, but also to develop capacities and tools for the active and sustainable participation of Member States in this process.

Space technology has also proven its usefulness in the context of government services, with public health a prime example of a sector in which the use of satellite communications and remote sensing is both a reality and a need. This technology offers appropriate and affordable tools that are needed for achieving universal health coverage, especially in remote and rural areas. Satellite communications are an integral part of an overall health information infrastructure and need to be used intelligently and in partnership between the public and private sector. One of the key space applications with regard to health is health mapping (e.g. of the environment, diseases, movement of people, health facilities), which is used by decision makers to identify populations at risk, assess health-care coverage, guide health sector investments, highlight the geographic spread of diseases, stratify risk factors, assess resource allocation, plan and target interventions, support the monitoring and analysis of trends, and support advocacy and fundraising.

The establishment and strengthening of sustainable and standards-driven spatial data infrastructures should be recognized as means of implementation of development goals and objectives. Concerted efforts are required to ensure continuous monitoring and assessment of the environment in meeting sustainable development objectives at all levels. It is therefore essential to increase the number of countries with enhanced autonomous capabilities to access and use remote sensing and in-situ geospatial data and information in support of decision making processes at national, regional, and international levels.

## EQUITABLE ACCESS TO THE SPECTRUM AND GEOSTATIONARY ORBIT

In order to benefit from the application of space assets in a fair and responsible manner, it is necessary to recognize that space is a limited natural resource in terms of certain classes of orbits and the electromagnetic frequency spectrum available for such applications. The allocation and utilisation of orbital slots in the geostationary orbit continues to be an issue of concern for many countries, especially those without the direct means to access outer space. According to Peter Martinez, chair of the COPUOS Working Group on Long-term Sustainability of Space Activities, the dominant issue for developing countries with respect to the fair and responsible use of space in the domain of satellite communication is "access to spectrum and orbital slots in geostationary orbit." As expressed by Latin America countries in COPUOS, equitable access to the geostationary orbit (GEO)[6] is a determinant factor to achieving space sustainability.

Despite the legal provision on the ITU Constitution and other international instruments, access to existing frequency bands is at present governed by the principle of "first come, first served." That approach, while suited to developed countries, may disadvantage developing countries, especially those that do not yet have access to that orbit. This practice restricts and sometimes even prevents access to and use of certain frequency bands and orbital slots. Additionally, it is exacerbated by the disadvantage of developing countries in coordinating negotiations due to lack of resources, limited expertise, among others. The existing coordination procedures that apply to the non-planned bands are designed to overcome that difficulty, but they are not necessarily suitable for meeting all needs.

In this regard, Article 44, paragraph 196.2, of the ITU Constitution as amended by the Plenipotentiary Conference, held in Minneapolis, United States of America, in 1998, states quite clearly:

> In using frequency bands for radio services, Member States shall bear in mind that radio frequencies and any associated orbits, including the geostationary-satellite orbit, are limited natural resources and that they must be used rationally, efficiently and economically, in conformity with the provisions of the Radio Regulations, so that countries or groups of countries may have equitable access to those orbits and frequencies, taking into account the special needs of the developing countries and the geographical situation of particular countries.

It is generally accepted that the Orbit Spectrum Resource (OSR) has three main attributes.

- OSR is a natural resource conditioned by the laws of nature, being part as a *res communis* of the Province of Humankind concept.
- OSR is not renewable - GEO requires a unique combination of orbit parameters, and its reproduction is beyond human capability.
- OSR is scarce. To avoid radio frequency interference, a minimum angular separation between GEO satellites is required, especially

---

[6]  See document A/AC.105/C.2/2000/CRP.3/Rev.1 for a compendium of documentation relating to the geostationary orbit.

when they share operating frequencies and covered zones. As a result, the number of satellites that can be located in GEO is limited.

We can therefore deduce that the conditions of use that must be met as provided by article 44 of the ITU Constitution are:

- *Rational*: to avoid saturation
- *Economic*: to optimize the cost associated with the use of OSR to reduce final cost of service users
- *Efficient*: to taking advantage as much as possible of a granted OSR
- *Equitable*: to guarantee non-discriminatory access to all the countries.

But how saturated is GEO? Recent efforts reveal the status of GEO. During the 2008 International Astronautical Congress Glasgow the real situation of GEO was demonstrated by contrasting radio space station of space network with satellites present and active at the same orbital position. The outcome was set forth in a Comparative Table.[7] The main conclusions were astonishing:

- A total of 165 notified space networks (10 percent of the total 880) at 83 different orbital positions had no satellite at their respective positions. At least one out of every five notified space networks was not in operation.
- A total of 23 satellites had no notified space networks at their orbital positions.

In response to these findings, Czech astronomer Dr. Luboš Perek stated[8] "Can COPUOS and ITU cooperate to use the Comparative Table to access the real use of the GEO and to turn the practice of bypassing rules into a new set of rules better corresponding to existing situation and to the actual needs of users of the GEO?"[9]

---

[7] The comparative table has been prepared and published in a paper of the 23rd IAA-IISL Scientific-Legal Roundtable in Glasgow, UK, 2008: "'Paper Satellites' - Problems of Policy, Regulation and Economics" dated 30 June 2008.

[8] IAC memoirs Glasgow, 2008.

[9] W. Rathgeber,K.-U. Schrogl.R.A. Williamson (Editors). *The fair and Responsible Use of Space*, Springer Wien: New York, 2010.

A second interesting effort is the GEO Occupancy Analyzer Tool (GOAT),[10] a historical analysis of GEO occupancy. GOAT is a supporting tool for proposed analysis recommended by ITU resolution 80. This spreadsheet-oriented visual basic application performs historic analysis of GEO occupancy by researching a satellite database. GOAT might be used as a data mining tool to provide required information and data for analyzing the basic principles ruling GEO exploitation and their associated variables and models. GOAT and its data base needs to be enriched with updated information.

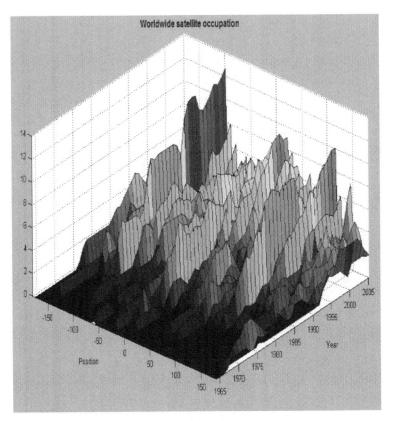

**Figure 1**
**GOAT Worldwide Satellite Occupation**
By position and year presented by Colombian delegation at the 45th
STSC COPUOS, 2008. www.oosa.org

---

[10]  See www.itu. Historical analysis of the occupancy of the geostationary orbit.

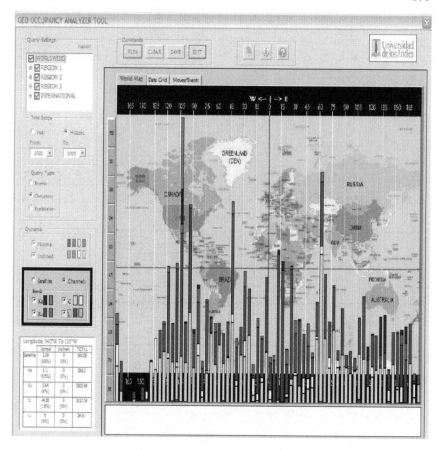

**Figure 2**
**GOAT-Based Map of GEO Occupancy Worldwide**
By country and region. Presented by the Colombian delegation at the
45th STSC COPUOS – www.oosa.org

## PROPOSAL OF A COORDINATION MECHANISM FOR ORBITAL POSITIONS

In 1996, the Colombian delegation submitted to the COPUOS Legal Subcommittee at its 35th session a working paper entitled "Some considerations concerning the utilization of the geostationary orbit" (A/AC.105/C.2/L.200 and Corr.1). In this document, the delegation recommended a coordination mechanism that could be applied to the management of frequencies and orbital positions in GEO. The proposal reached consensus within the framework of the Legal Affairs Subcommittee at its 39ᵗʰ session, (A/AC.105/738, Annex III), was then adopted by the COPUOS plenary in its 43ʳᵈ (A/55/20), and endorsed by the General Assembly in its resolution 55/122.

The main elements of the proposed coordination mechanism are as follows:

A)  Concerning comparable requests:
    Where coordination is required between countries with a view to the utilization of satellite orbits, including the geostationary satellite orbit, the countries concerned take into account the fact that access to that orbit must take place, inter alia, in an equitable manner and according to the ITU Radio Regulations. Consequently, in the case of comparable requests for access to the spectrum/orbit resource by a country already having access to the orbit/spectrum resource and a developing country or another country seeking it, the country already having such access should take all practicable steps to enable the developing country or other country to have equitable access to the requested orbit/spectrum resource.

B)  On the legal framework of the request:
    Countries wishing to use frequencies and satellite orbits, including the geostationary satellite orbit, in the above-mentioned cases file such requests according to the relevant provisions of the ITU Radio Regulations, taking into account resolution 18 of the ITU Plenipotentiary Conference (Kyoto, 1994) and resolution 49 of the ITU World Radiocommunication Conference (Geneva, 1997) in order to guarantee effective use of the orbit/spectrum resource.

By endorsing this proposal, the UNGA recognized that in accordance with article 44 of the ITU Constitution, the satellite orbits and radio frequency spectrum are limited natural resources that must be used rationally, efficiently, economically and equitably. Furthermore, that it is necessary to facilitate equitable access to the GEO orbit and spectrum resource. The current regulations for access to frequencies and satellite orbits may give rise to situations involving difficult processes of coordination among developed as well as developing countries.

As a result of the COPUOS decision, the World Radio Communication Conference (Geneva, 2007), updated Resolution 80 on "Due diligence in applying the principles embodied in the Constitution" (Rev.WRC-07) by "considering that the Legal Subcommittee of the Committee on the Peaceful Uses of Outer Space of the United Nations General Assembly has drawn up recommendations in this respect," then deciding "to instruct the Radio Communication Sector, in accordance with No. 1 of Article 12 of the Constitution, to carry out studies on procedures for measurement and analysis of the application of the basic principles contained in Article 44 of the Constitution" and inviting "administrations to contribute to the studies referred to in resolves 1 and to the work of the RRB as detailed in resolves 2."

From the preceding, we may conclude that in order to secure the sustainability of space activities there is a need to establish a legal regime for regulating access to and the use of the geostationary orbit, which is a limited natural resource and has *sui generis* characteristics. Such a regime could be only possible if the UN and its specialized agencies, particularly the ITU, work in an integrated fashion in order to guarantee equitable access to GEO for all States, taking particular account of the needs of developing countries.

## EFFICIENT, RATIONAL, ECONOMICAL, AND EQUITABLE USE OF SPACE

The issue of the sustainability of space activities prompts the following question: after more than 50 years of exploitation of GEO and its associated OSR, has the resource been used efficiently, rationally, economically, and equitably? There is a need to answer this question from the space sustainability perspective in order to facilitate access to the orbit/spectrum resource for developing countries, as well as for countries that still lack the ability to access. The purpose is to ensure equitable access between those countries already having access to it and those seeking it. Finally, space sustainability should be understood by the unpopular notion that future spacefaring nations may pay for the current close-saturation of space resources.

Further development on this issue is needed. A next step could be convening the working group of Item 6 of the agenda of the COPUOS Legal Subcommittee on the issue of equitable access to the geostationary orbit in accordance with the Subcommittee's normal procedure. The result should enrich the LTS guidelines and therefore elicit more support from developing nations.

## MOVING FORWARD: THE NEED FOR A HOLISTIC APPROACH WITHIN THE UN

Ongoing UN development efforts are benefitting from improved coordination and examination of sustainable development from a holistic perspective. Understanding the value of better linking space technology with development goals would help address a major reservation expressed by developing nations with the LTS effort. At the same time, this also presents an opportunity for reducing the limitations of scattered space-related functions within the UN, which will help drive forward space sustainability discussions and improve progress on issues such as access to the geostationary orbit and spectrum, as discussed before.

At present, there are five principal fora at which overarching space issues are discussed within the UN: (1) UNCOPUOS in Vienna, (2) the Conference on Disarmament in Geneva, (3) the UN General Assembly in New York and several of its committees, such as the Disarmament and International Security Committee and the Special Political and Decolonisation Committee, (4) the United Nations Educational, Scientific, and Cultural Organization in Paris, and (5) the ITU in Geneva. A UN space policy would help improve coordination among these fora and better harmonize UN space activities, thus going beyond the present UN Coordination of Outer Space Activities.

The United Nations is critically reliant on space systems for its day-to-day operations and effectiveness. At least 25 United Nations entities and the World Bank Group routinely use space applications. They make important and sometimes essential contributions to the work of the UN, including in the implementation of recommendations of major world conferences in efforts towards sustainable development and in the implementation of the United Nations Millennium Declaration. However, UN space activities are fragmented geographically and thematically. Such efforts would benefit from greater interdisciplinary cooperation among various public sector institutions and agencies to maximise synergies. There is also the need to optimize inter-institutional cooperation, build and develop human capital, and promote enhance space awareness at all UN levels.

As a result, coordination, cooperation and synergy are essential for those activities to be effectively carried out by the UN system. The annual Inter-Agency Meeting on Outer Space Activities serves as the focal point for inter-agency coordination and cooperation and preventing duplication of efforts related to the use of space applications by the United Nations.

The United Nations has up to now pursued a highly decentralized approach to space among its agencies and organizations. This is not considered a tenable option for the future. The management of the development issue demonstrates the need for a UN Space Governance in order to regain an important place in the global space context and move past current arrangements that are not fully satisfactory. A far more proactive approach is necessary to underpin and sustain the UN's ability to play its role in the rapidly evolving space arena of the 21$^{st}$ century.

•••

## CIRO ARÉVALO-YEPES

After serving for seven years in the Colombian Embassy in Vienna, Ambassador Ciro Arévalo was elected Chairman of the Vienna based United Nations Committee on the Peaceful Uses of Outer Space (COPUOS) from 2008-2010. In 2010, he was designated President of the Latin America and Caribbean Regional Group in the International Astronautical Federation, GRULA-IAF and became a member of the International Academy of Astronautics, IAA.

Recently he became member of the World Economic Forum Network of Global Agenda Council. Ambassador Arévalo is a recognised expert on Legal and Space Policy issues and has often been an authority consulted by different governments and international organizations. Since 2010, he became adviser to the Director General of the International Atomic Energy Agency, IAEA in the Standing Advisory Group on Technical Assistance and Cooperation (SAGTAC). He served as a diplomat during twenty-three years in the Colombian Foreign Service. He worked in the Colombian Embassy to the United Nations both in Geneva and Vienna; Consul General in Australia, and Plenipotentiary Minister among other postings. He was chairman of the Atomic Energy Regional Organization, ORA/ARCAL. He chaired the LOC of the Fourth Space Conference of the Americas.

He is a member of the International Institute of Space Law, IISL and of the Advisory Committee of Secure World Foundation, SWF. Member of the American Institute of Aviation and Astronautics, AIAA, and of the European Space Policy Research and Academic Network (ESPRAN) of the European Space Policy Institute, ESPI. He has authored a number of papers and articles on space policy, space and society, and space and education. He speaks fluent Spanish, English, French and German.

## BIBLIOGRAPHY

Arévalo, C. (2010). Hacia una política espacial de las Naciones Unidas. En General Assembly of the United Nations. *Committee for the Peaceful Uses of Outer Space (Copuos). Document A/AC.105/L.278.* Document prepared in the role of president of copuos 2008-2009.

Arévalo, C., Froelich, A., Martinez, P., Peter, N., Suzuki, K. (2010). The need for a United Nations space policy. *Elsevier/Space Policy*, 26, 3-8. Recuperado de: www.elsevier.com/locate/spacepol

Arnould, J. (2011). *Icarus' Second Chance: The Basis and Perspectives of Space Ethics.* Vienna-New York: Springer Wien.

Bohórquez-Tapia, L. A. & Eakin, H. (s. f.). Conflict and Collaboration in Defining the 'Desired State': The Case of Cozumel, Mexico. En Collaborative Resilience: Moving Through Crisis to Opportunity. Centro de Investigaciones de la Caña de azúcar de Colombia. Recuperado de: http://www.cenicana.org

Christol, C. Q. (1991). *Space Law: Past, Present, and Future.* Deventer: Kluwer. Contribution of the Committee on the Peaceful Uses of Outer Space to the United Nations Conference on Sustainable Development: Harnessing space derived geospatial data for sustainable development. General Assembly Doc. A/Ac,105/993. Recuperado de: www.oosa.unvienna.org

Doyle, K. (2012). Elon Musk on SpaceX, Tesla, and Why Space Solar Power Must Die. *Popular Mechanics.* Available from: http://www.popularmechanics.com/how-to/blog/elon-musk-on-spacex-teslaand-why-space-solar-power-must-die-13386162

eLearningAfrica. ElearningAfrica 2014. Available from: http://www.elearning-africa.com

Documents relating to Space and Socioeconomic Development in the context of Rio+20 and the Post-2015 Development Agenda http://www.unoosa.org/oosa/en/COPUOS/sustdev.html

Goldstein, B. E. (ed.) (2012). *Collaborative Resilience: Moving Through Crisis to Opportunity.* Cambridge: Massachusetts Institute of Technology Press.

International Academy of Astronautics (iaa) (2011). *Summit Declaration.* Available from: http://iaaweb.org/iaa/Communication/iaa_Summit_Declaration.Pdf

iea (2011). *Solar Energy Perspectives: Executive Summary.* Available from: http://www.iea.org/Textbase/npsum/solar2011sum.pdf

Jodoin, S. & Cordonier Segger, Naciones Unidas (2013). *A new global partnership: Eradicate poverty and transform economies through sustainable development.* Available from: http://www.post2015hlp.org/wp-content/uploads/2013/05/UN-Report.pdf

Lele, A. (ed.) (2012). *Decoding the International Code of Conduct for Outer Space Activities.* Washington: Pentagon Press.

M. C. (eds.) (2013). *Sustainable Development, International Criminal Justice, and Treaty Implementation.* New York: Cambridge University Press.

Mankins, John C. (2012). *sps-alpha: "The First Practical Solar Power Satellite via Arbitrarily Large Phased Array"* (A 2011-2012 nasa niac Phase 1Project). Available from: http://www.nasa.gov/pdf/716070main_Mankins_2011_PhI_sps_Alpha.pdf

Moran, M. S. (2000). Technology and techniques for remote sensing in agriculture. *Assoc. Appl. Biol. and Rem. Sens. Soc.// Conf. on Remote Sensing in Agriculture.*

Naciones Unidas (2013). *A new global partnership: Erradicate poverty and transform economies through sustainable development.* Available from: http://www.post2015hlp.org/wp-content/uploads/2013/05/UN-Report.pdf

nasa (s.f.). *Science News.* Recuperado de: www.science.nasa.gov/science

Pelling, M. & High, C. (2005) Understanding adaption: What can social capital offer assessments of adaptive capacity? *Global Environmental Change* (15), pp. 308-319.

Randolph, J. (2012). Creating the Climate Change Resilient Community. En Goldstein, B. E. (ed.). *Collaborative Resilience: Moving Through Crisis to Opportunity.* S. d.

Rathgeber, W., Schrogl, K. & Williamson, R. A. (eds.) (2010). *The Fair and Responsible Use of Space: An International Perspective.* Vienna-NewYork: SpringerWien.

Santos, F. D. (2011). *Humans on Earth. From Origins to Possible Futures.* Springer-Verlag, Berlin Heidelberg, 2012.

Tompkins, E. L. & Adger, W. N. (2003). *Defining response capacity to enhance climate change policy. Working Paper 39.* Norwich: Tyndall Centre for Climate Change Research.

unisdr. (s. f.). *Hyogo Framework for Action (hfa).* Available from: http://www.unisdr.org/we/coordinate/hfa

Chapter 22

# SPACE SYSTEMS FOR SUSTAINABLE DEVELOPMENT ON EARTH

Filipe Duarte Santos, Ph.D.
University of Lisbon

## Sustainable Development

Since the launching of the first artificial satellite, Sputnik, by the Soviet Union in 1957, thousands of satellites have been launched into orbit around the Earth. Currently satellites provide invaluable services to mankind. Communication satellites are essential to bring people together through various forms of telecommunications, such as cell-phones, internet access, radio and television. Navigational satellites provide systems for land, air and maritime traffic management. Weather satellites became indispensable tools for monitoring Earth's weather and climate and for weather forecasting, including extreme weather warning.

Another very important application of satellites is the observation and continuous monitoring of the global environment. Earth Observation (EO) satellites have been monitoring the environment from space for more than half a century now, revealing its beauty, complexity and its exposure and vulnerability to rapidly growing human-induced stresses. EO based geo-information services contribute to socioeconomic development and environmental protection on a global scale. In conclusion, it is well recognized that in our time the use of outer space based systems contributes decisively to the well-being of humanity and to its sustainable development.

Consequently, the space environment is increasingly used by an growing number of States and private sector entities for a very diverse range of outer space activities. This growing use also creates a problem of sustainability. In fact, the long term sustainability of outer space activities is currently in danger due to the proliferation of space debris, the increasing probability of collisions and the congestion of orbital positions and of the radio frequency spectrum, particularly in the Low Earth Orbit (LEO) and Geostationary Orbit (GEO) environments. Our objective here is to review and discuss how space systems contribute to sustainable development on Earth and also to discuss briefly the long-term sustainability of outer space activities.

Sustainable development is a relatively recent concept, introduced in the 1980s and defined by the United Nations World Commission on the Environment and Development in 1987 as "development which meets the needs of the present without compromising the ability of future generations to meet their own needs." This definition, however, did not satisfy everyone, and other definitions arose until gradually it became clear that sustainable development is not a concept of a strictly scientific nature that can be defined without ambiguities, because opinions differ on what precisely should count among the human needs for the application of the principle of intergenerational equity.

These needs can be categorized into the social, economic, and environmental realms, but the relative importance of the different components is a matter of opinion. Sustainable development is currently a meeting point for the debate about the state of the world and how to respond to the social, economic, environmental, and institutional challenges we are facing. Despite the differences of opinion, over the last few decades the principles of sustainable development have been progressively adopted by world leaders following a series of Earth Summits (Rio, 1992; Johannesburg, 2002; Rio 2012), and development targets such as the United Nations Millennium Development Goals have been defined during these summits, and agreed upon by policy makers and business sectors in order to improve quality of life, protect the environment, and fight global poverty and hunger.

## THE SQUARE OF UNSUSTAINABILITY

Although we are still very far from achieving a global sustainable development, it is possible to identify the main drivers of unsustainability. These can be organized into four leading groups, and it will be shown that space systems play a crucial role in addressing the problems raised by these four groups of drivers that constitute the "square of unsustainability" (Fig. 1).

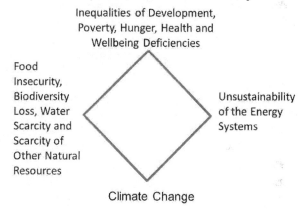

**Figure 1**
**The Square of Unsustainability**
A way to organize the challenges facing sustainable development.

The first group plays a prominent role among the others. It includes the inequalities of development, poverty, especially extreme and severe poverty, hunger, and health and wellbeing deficiencies.

The second group encompasses food insecurity, biodiversity loss, water scarcity and scarcity of other natural resources.

The third is the unsustainability of energy systems, as sustainability of energy systems requires secure access to the energy sources, affordable pricing and environmental compatibility. This last criterion is not satisfied by fossil fuels that represent about 80% of the world primary energy sources because their combustion leads to the emission of carbon dioxide ($CO_2$), which is a greenhouse gas and now understood to be a critical component driving climate change.

Which leads to the fourth driver, anthropogenic climate change generated by the enhancement of the Earth's greenhouse effect which results from the human emissions of greenhouse gases, mainly $CO_2$ from fossil fuel burning and land use changes, especially deforestation, methane ($CH_4$) and nitrous oxide ($N_2O$).

All four unsustainability groups of drivers are strongly interconnected and interdependent. For society to attain some form of sustainable development on a systemic basis, the four groups have to be addressed simultaneously and in an integrated way, and the magnitude and difficulty of this task reveals the perilous state the world is in.

Let us review briefly the global situation regarding the four groups in the square of unsustainability.


## INCOME INEQUALITIES, POVERTY AND FOOD INSECURITY

Since World War II we have witnessed a remarkable acceleration in social and economic development worldwide, which has lifted hundreds of millions of people out of poverty, and improved the quality of life of a great many more. However, poverty is still very significant, and a major impediment in the way to sustainable development.

According to World Bank statistics for 2008, about 50% of the world population live on less than $2.5 a day, and about 80% with less than $10 a day (World Bank, 2008). In the same year of 2008, according to UNICEF, 8.8 million children died before their fifth birthday, mostly due to poverty-induced malnutrition and disease (UNICEF, 2009). There are also, however, reasons for cautious optimism in some indicators, since the global child mortality rate declined by more than one third between 1990 and 2010.

People living in poverty are focused on survival, and therefore give a low priority to environmental issues which are outside of their immediate needs.

## POVERTY AND HUNGER

Poverty is the main cause of hunger. According to the United Nations Food and Agriculture Organization (FAO) the number of hungry people has increased since 1995-97 and reached a peak of 1.02 billion in 2009. In 2010 it decreased to 925 million, which corresponds to 13.1% of the world population, or almost 1 in 7 people. One of the main reasons for the recent trend in malnutrition is the increase in food prices since 2004, with major peaks in 2008 and 2011. The latest report from FAO (FAO, 2014) estimates that the number of people during the period 2012-2014 who were chronically undernourished was 805 million, which represents a significant improvement relative to the end of the last decade. However, food insecurity remains unacceptably high in some regions of the world, as for instance in Sub-Saharan Africa where more than one in four people do not have enough to eat.

There are many reasons why food prices are increasing. One of them is climate change, which leads to more frequent extreme weather and climate events such as droughts, floods and storms, and thus to crop failures. A further reason is the competition between energy and food through the cultivation of biofuel crops, which may take valuable farmland out of food production. The global population is also growing too fast for agricultural production to keep up, and in the developing countries, particularly in those with emerging economies, citizens are consuming greater quantities of higher quality food, which is a very welcome development but requires more water, energy and good quality soils. The rising price of oil makes it more expensive to produce and ship food products, and food prices are also increasing because of decades of neglect of agriculture and agricultural research, especially in hunger-prone regions.

To increase global food production in a sustainable way, improved management of the world's agricultural resources is required. The use of space satellite imagery data at different spatial, spectral and temporal resolutions can contribute significantly to achieving that goal. Space systems applications coupled with sampled *in-situ* observations can be used for sustainable agriculture management and development, including crop system analyses, land cover type analyses, species composition, vegetation structure, net ecosystem productivity, rainfall estimation, soil moisture assessment, soil erosion, soil salinity and alkalinity analyses, integrated agricultural drought assessment and management, desertification monitoring, assessment of land productivity, biomass and carbon estimates of vegetation and soils.

A notable application of satellite data is the Famine Early Warning System Network (http://www.fews.net/), which was initially set up in Sub-Saharan Africa and now operates in other arid environments in developing countries. This system uses satellite images in conjunction with ground-based information to predict and critically to mitigate famines.


## REFUGEES

Another emerging problem is the increasing number of refugees forced to flee their homes across the world due to conflicts, civil wars and natural disasters. According to the United Nations Refugee Agency (UNHCR), the global refugee figure at the end of 2013 reached 51.2 million, the highest number since the World War II. About half the world's refugees are children. There is also a sharp increase in the number of people internally displaced in their own countries, mainly because of civil war and other violent conflicts.

And while there has been significant progress against poverty, income inequalities have been increasing worldwide since the middle of the nineteenth century, and more than 80% of the world's population live in countries where income differentials are widening. The world's richest 20% account for 75% of the total percentage of world income (World Bank, 2008).


## HEALTH, WATER AND URBANIZATION

In the health sector, inequalities between countries are growing at an alarming rate. Life expectancy, for instance, varies from less than 45 years in some sub-Saharan countries to more than 80 years in some OECD countries. The reasons for this profound disparity are well known: poverty, lack of adequate health infrastructures and medical services, lack of or inefficiency in the control of epidemics, and insufficient financing of pharmaceutical and medical research for the specific diseases of the tropical regions, where many of the developing countries are located. Satellites are an essential tool for providing clinical health care at a distance through telemedicine, such as emergency services, general health care, telenursing, telepharmacy, telerehabilitation, telepsychiatry, and teleradiology. These technologies improve access to medical services, especially in distant rural communities and also in emergency situations.

According to the United Nations, water scarcity already affects every continent and 40% of world population (http://www.unwater.org/home/en/). The situation is growing worse due to population growth, urbanization, and to the increase in domestic and industrial water use. Agriculture is the main user of water worldwide, accounting for 70% of all the freshwater withdrawn from rivers, lakes and aquifers. The FAO estimate that by 2025, 1.8 billion people will be living in countries or regions with absolute water shortage, where water resources per person fall below the recommended level of 500 cubic metres per year, the amount of water a person needs for a healthy and hygienic living. Further, about two-thirds of the world's population could be living under water-stressed conditions by 2025.

Satellite imagery data are increasingly used for water resources monitoring and development plans. This methodology is based on coordinated, sustained observations of the water cycle at multiple scales, including river base flow and peak flow, and requirements for irrigated agriculture, power plant cooling and domestic usage. Satellite applications can measure rainfall, water movement and the height of water in rivers, lakes and wetlands, and they can also serve to identify surface and underground water resources in drought-prone regions.

Urbanization has been one of the outstanding features of world development since the beginning of the 20th century. The percentage of the global population living in urban areas was about 2% in 1800, 13% in 1900, 30% in 1950, 47% in 2000, and over 50% since 2007 (UNH, 2011). Current estimates indicate that it will reach 5,000 million people in 2028, corresponding to about 60% of the world population, and 7,000 million in 2050. Growing cities will be the main drivers of the world economy in the next decades, and it is necessary to be prepared for the challenges that will arise from increasing demand for natural resources in the urban areas, especially water and energy, and the need for capital to invest in new housing, office buildings, and port capacity. The EO satellite-based production of standardized mapping is one of the most effective ways to collect digital reference data for urban planning and to monitor urban development, and the same technique can be used for a wide range of specialized urban applications addressing key environmental and urban health issues such as land surface temperature, urban heat islands, air quality and air pollution, urban waste management, hydrologic modelling and flood management.

## DEFORESTATION AND BIODIVERSITY LOSS

Environmental monitoring using satellites has grown very strongly in the past few years, and this trend is expected to continue. A very important example is the application of satellite imagery for forest monitoring and conservation. Deforestation is a significant source of greenhouse gases, and reduces biodiversity. The history of deforestation begins with the emergence of agriculture, and is responsible for a 20–25% reduction of the global forest area. At the present time, there are only three large forest areas with a significant spatial continuity: the river basins of the Amazon and the Orinoco in South America; the northern part of Eurasia, from Scandinavia to the Far East; and the Congo River basin in Africa. The other vast forest regions have disappeared or have been reduced to scattered groups of relatively small areas, as is the case in Central and Southern Europe, the Anatolian Peninsula, India, Indonesia, China, Madagascar, and the Atlantic coast of Brazil.

It is difficult to make a precise assessment of the historical evolution of forested areas on a regional and global scale because different definitions and concepts of forestry have been used, and the quality of the data has a strong spatial and temporal variability. Currently, the situation has greatly improved with access to remote-sensing data, which provides a much more reliable form of monitoring. There are global databases containing satellite images for the past 30 years, making it possible to detect trends and areas where deforestation occurs. According to FAO, around 13 million hectares of forest were converted to other uses or lost through natural causes each year between 2000 and 2010, compared to 16 million hectares per year during the preceding decade. This results not only in biodiversity loss, but also contributes to global warming by releasing 7-14% of the total amount of anthropogenic $CO_2$ emissions into the atmosphere (Harris, 2012) and hampering further $CO_2$ storage.

Currently, the forests in the temperate zones of North America and Eurasia have stabilized or are increasing, while the tropical forests continue to decline rapidly. Using high-resolution satellite images it has been possible to estimate that the average annual rate of tropical forest destruction between 2000 and 2005 was 5.4 million hectares, and that about half of the deforested land is located in Brazil. Approximately 60 to 70% of deforestation in the Amazon is the result of clearing for cattle pasture, and next is large-scale intensive agriculture, especially soybeans, small-scale subsistence agriculture and construction of roads and other infrastructure. By combining hundreds of images from satellite coverage with software analysis, experts can analyse patterns of deforestation down to a single tree, and calculate the emissions resulting from removing trees that would otherwise sequester carbon dioxide. Thus, satellites can assist the United Nations Collaborative Programme on Reducing Emissions from Deforestation and Forest Degradation in Developing Countries (UN-REDD) to achieve its objectives.

The Millennium Ecosystem Assessment published in 2005 reports that ecosystems have declined more rapidly and extensively over the past 50 years than at any other comparable time in human history. Left unchecked, this degradation jeopardizes the world's biodiversity and becomes a significant risk factor for ecosystem services and a threat to long term economic sustainability. Over the last 30 years, biodiversity has decreased in a generalised way, but the freshwater ecosystems have been the most seriously affected as the result of pollution and overexploitation of water resources. The decrease has been less pronounced in the marine and forest ecosystems, although it is still very significant there. Biodiversity loss is particularly high in developing countries in the tropical regions (Jenkins, 2003), but it is important to bear in mind that this is not new, as most of the world's so-called 'natural' systems have been disturbed more or less significantly by direct or indirect human actions for the last 50,000 years.

Data from Earth Observation (EO) satellites highlight land cover spatial changes, both natural and man-made. Combined with botanical data gathered on the ground, information derived from processing of satellite imagery can be used for establishing inventories and calculating indicators of fragmentation, acreages and corridors that are vital for preserving biodiversity. Using satellite imagery to monitor the disruption of natural habitats, which are largely due to human factors such as farming practices, resource exploitation and building, helps to focus conservation strategies. There are several types of remotely-sensed information that can contribute to a better understanding of the direct and indirect drivers of spatiotemporal biodiversity patterns, particularly with respect to the state of the landscape, including topography, land cover and temporal variability. Detailed, broad-scale information about landscape and vegetation structure, combined with maps of human-induced land cover changes such as road networks, urbanized areas, and agricultural areas, can improve the understanding of how activities like logging and poaching affect biodiversity patterns.

## COASTAL ZONES

Coastal zones are simultaneously one of the Earth regions with the greatest biological productivity and with the highest population densities. About 40% of the world's population live in a 100 km wide strip along the coasts, while this constitutes only 20% of the total land surface. This enormous and expanding population inevitably exerts a great pressure on coastal resources, and consequently a large number of coastal habitats in various regions of the world, especially coastal wetlands, mangroves, and coral reefs, are in danger of being destroyed due to overexploitation of resources, pollution, sedimentation, and erosion, in addition to urban and industrial development, and tourism activities. The sustainable management of these coastal zones is thus one of the main challenges that humanity faces at the beginning of the 21st century.

EO satellites can support information collection for the sustainable development and management of coastal areas. This includes coastal water quality monitoring, ship and ship-wake detection, particularly for fisheries protection, oil spill detection, red tide monitoring, mapping of reclamation activities, sustainable coastal land use planning and protection, monitoring the human occupation and activities in the coastal zones, marine habitat mapping, sea grass extent maps, coral reef health maps, assessments of coastal biodiversity trends, risks, and conservation status, and the enforcement of regulations within marine protected areas. The current EO systems can provide the majority of this information in scales ranging from global to local. EO satellites are also used for the sustainable management of marine resources by providing inventories of marine resource stocks and major ecosystems, sea surface temperature, salinity, and pH and $pCO_2$, phytoplankton species composition, and their productivity.

## CLIMATE CHANGE AND EXTREME EVENTS

Climate change is one of the major environmental risks facing humanity in the $21^{st}$ century. At present there is a strong consensus in the scientific community that anthropogenic emissions of greenhouse gases are intensifying the natural greenhouse effect in the atmosphere. These emissions are causing changes in the climate that will very likely intensify during this century.

The signs that climate change is happening are becoming ever more obvious and unequivocal. According to the IPCC (Intergovernmental Panel on Climate Change) $5^{th}$ Assessment Report (IPCC, 2014), the global average surface temperature (land and ocean) has increased by 0.8°C since pre-industrial times, and by 1.0°C over land alone. In the Arctic, the average surface temperature has been increasing at rates about twice the global value. Every single one of the 13 years from 1997 to 2009 was one of the 14 years with the highest global average surface temperature since 1850, the year when temperatures first began to be regularly measured with thermometers. The ice sheets are losing mass, glaciers are shrinking globally, sea ice cover is reducing in the Arctic, snow cover is decreasing, and permafrost is thawing in the Northern Hemisphere. Ice is being lost from many of the components of the cryosphere, although there are significant regional differences in the rates of loss. The temperature of the upper ocean layer down to 75 m is increasing more than 0.1°C per decade.

As a consequence of this increased greenhouse effect, more than 90% of the extra energy stored by the Earth between 1971 and 2010, has gone into warming the ocean. Due to thermal dilatation, melting of mountain glaciers and to a lesser extent melting of the polar ice sheets, global averaged sea level is rising: it increased between 14 and 20 cm in the $20^{th}$ century and since 1993 is increasing at an annual rate between 2.8 to 3.6 mm.

Regarding the future, there are two main methods for making global mean sea level rise projections. The models used are based either on the modelling of geophysical processes in the various systems (upper layers of the ocean, glaciers, ice sheets and storing of land water resources), or semi-empirical models. The latter project the future behaviour of global mean sea level by using statistical relationships between observations on past behaviour, and the mean global temperature of the atmosphere at the surface [GRI 10, RAH 07] or the radiative forcing [JEV 10]. The models based on geophysical processes project for the period of 2081–2100, compared to 1986–2005, an average increase in global mean sea level between 0.26 and 0.98 m.

From these models we can infer that if we do not succeed in reducing current global annual anthropogenic emissions of greenhouse gases by 60–80% before 2050 at the latest, it is very probable that the global mean sea level rise will reach more than 0.5 m at the end of the century.

A very important aspect of anthropogenic climate change, one which has a strong potential impact in many human activities and socio-economic sectors, is the increase in the frequency and intensity of extreme weather and climate events. Research indicates that the percentage of very strong tropical storms, such as the recent hurricane Sandy in the North Atlantic, is increasing due mainly to the increase in the average sea surface temperature (Elsner, 2008).

Earth observation from space over the past 50 years has fundamentally changed our understanding and knowledge of the Earth system (Figure 2).

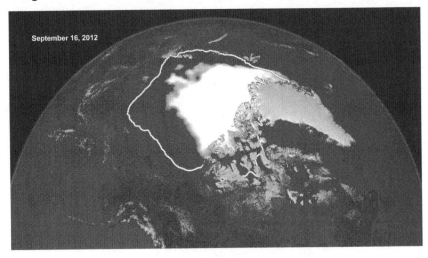

**Figure 2**
**Satellite View of the Artic Polar Region**
**September 16, 2012.**

Satellite data reveal how the new record low Arctic sea ice extent, from Sept. 16, 2012, compares to the average minimum extent over the past 30 years (white line). Sea ice extent maps are derived from data captured by the Scanning Multichannel Microwave Radiometer aboard NASA's Nimbus-7 satellite and the Special Sensor Microwave Imager on multiple satellites from the Defense Meteorological Satellite Program. Credit: NASA/Goddard Scientific Visualization Studio. http://www.nasa.gov/topics/earth/features/2012-seaicemin.html

With increasingly sophisticated space systems it is now possible to obtain quantitative measurements of temperatures in the atmosphere, concentrations of atmospheric gases, precipitation, wind speed, elevations of land and water, water movement, type of soils and vegetation cover. Satellite observations yield continually updated knowledge of the state of the atmosphere, helping meteorologists to devise models that project the weather into the future with much improved accuracy compared to pre-satellite forecasts. Seven-day weather forecasts have more than doubled in accuracy over the past three decades, particularly in the southern hemisphere.

Satellites have been used to monitor the stratospheric ozone layer, which blocks damaging ultraviolet light from reaching the Earth's surface, and to monitor atmospheric aerosol loading. Furthermore, they have contributed decisively to improve our understanding of the climate system and of climate change through the monitoring of the atmosphere, sea and land surface temperatures, ice sheet flows, Arctic sea ice extension, the El Niño-Southern Oscillation and Earth's carbon cycle. Recently, an ensemble of satellite altimetry, interferometry, and gravimetry data sets has been used to conclude that since 1992 melting of the polar ice sheets contributed, on average, 0.59 mm to the annual rate of global sea level rise.

Extreme precipitation and floods are very likely to become more frequent due to climate change. Floods are just one example of various types of disasters where satellites have proved very useful for humanitarian relief by rapidly mapping and assessing local emergency situations and reconstruction activities. Space systems are used extensively today in the management of disasters related to geophysical, meteorological, hydrological and climate events, and they are becoming more integral to reducing reaction time and providing accurate information to rescue and disaster control operations. Satellites are also very useful in disasters for communications, remote sensing and mapping. Meteorological and storm warning satellite technology can also help in predicting water related disasters and setting up precautionary activities.

## CONCLUSION

Many more examples could be given on the importance of space systems for sustainable development, and furthermore it should be emphasized that satellites are currently indispensable to our way of life. One single day without civilian satellites would bring chaos to the global human society, and for longer periods of time it would bring the collapse of our civilization as we know it. Some of the consequences would be ... no GPS, no cellphones, no internet access through satellites, no land, air and maritime traffic control, no financial markets, and severe impairment of most weather forecast services. It is therefore of the utmost importance to guarantee the long term sustainability of outer space activities to support the future sustainability of human society.

•••

## FILIPE DUARTE SANTOS, PH.D.

Mr. Santos holds a M.Sc. in Geophysics by the University of Lisbon and a Ph.D. in Theoretical Physics by the University of London. Presently he is a researcher in Geophysical Sciences and Global Change and Sustainable Development in the Department of Physics of the Faculty of Sciences of the University of Lisbon.

He is since 1979 full professor of Physics, Geophysics and Environment at the University of Lisbon and the Director of the PhD program on Climate Change and Sustainable Development Policies, which involves the University of Lisbon and the New University of Lisbon (http://alteracoesclimaticas.ics.ulisboa.pt/).

His academic career includes being visiting professor at the following Universities: Wisconsin, North Carolina and Indiana, Stanford, Harvard, in the USA, Munich in Germany, Surrey in the UK and Vrije Univ. in the Netherlands.

Mr. Santos was first Vice-President to the UN COPUOS and delegate to the Conference of Parties of the UNFCC. He is a member of the National Council for the Environment and Sustainable Development and member of the Lisbon Academy of Sciences. He was also the Coordinator for the "Area de Desarollo Sostenible, Cambio Global y Ecosistemas del Programa CYTED (Ciencia y Tecnologia para el Desarollo)" from 2007 to 2011 and was Review Editor for the 5th Assessment Report of the IPCC. Member of ESAC – European Academies Science Advisory Council, in representation of the Lisbon Academy of Sciences.

Author of more than one hundred articles in areas of Nuclear Physics, Nuclear Astrophysics and Global Change, Mr. Santos has recently published a considerable number of reports and articles covering climate change and its impacts. He is PI of several research projects in the area of climate change and environment.

# REFERENCES

COPUOS, 2011, United Nations Committee on the Peaceful Uses of Outer
Space, Document A/AC.105/L.281/Add.4
COPUOS, 2014, United Nations Committee on the Peaceful Uses of Outer
Space, Document A/AC.105/C.1/L.339
Elsner, J. B., James P. Kossin and Thomas H. Jagger, 2008, The increasing
intensity of the strongest tropical cyclones, *Nature* 455, 92-
95, doi:10.1038/nature07234
FAO, 2014, The state of food insecurity in the world, Food and Agriculture
Organization of the United Nations.
Grinsted, A., J. C. Moore and S. Jevrejeva, 2010, Reconstructing sea level from
paleo and projected temperatures 200 to 2100 AD, *Climate Dynamics*, 34,
p. 461-472, 2010.Harris, N.L. et al., Baseline map of carbon emissions
from deforestation in tropical regions, Science, 336, 1573-1576, DOI:
10.1126/science.1217962.
IPCC, 2014, Intergovernmental Panel on Climate Change, 5[th]Assessment
Report, WGI, WGII and WGIII, http://www.ipcc.ch/report/ar5/.
Jenkins, M., 2003, Prospects for biodiversity, Science, 302, 1175-1177
Jevrejeva, S., J. C. Moore and A. Grinsted , 2010, How will sea level respond
to changes in natural and anthropogenic forcings by 2100?, *Geophysical
Research Letters*, 37, L07703
MEA, 2005, Millennium Ecosystem Assessment, Ecosystems and Human
Well-being: Synthesis, Island Press, Washington, DC
Rahmstorf, S., 2007, A semi-empirical approach to projecting future sea-level
rise, *Science*, 315, p. 368-370.
Santos, F. D., Humans on Earth. From Origins to Possible Futures, The
Frontiers Collection, Springer, 2011.
Sugden, A., et al., 2015, Forest health in a changing world, Special Section,
Science, 349, 800-836
UNHCR, 2014, War's Human Cost, UNHCR Global Trends 2013
UNICEF, 2009, The state of the world's children, Special Edition.
World Bank, 2008, World Bank Development Indicators.

Chapter 23

# Conclusion: Multilateral Initiatives for Space Sustainability

**Ray Williamson**
Secure World Foundation

## BIG IDEAS

This decade has proved a very exciting and productive time for space activities despite the recent economic struggles of the world community. The rise of so-called New Space, built on major advancements in electronics, materials, propulsion and control, is enhancing the economics of space activities. Private, human spaceflight is on the horizon, and those activities, together with the improvements in practical applications and the dramatic findings of space science research, have created increased interest in the benefits that space systems can bring to humanity.

At the same time, there is growing unease over whether the world community can continue to use the space environment for the long term. Sustainability of the space environment is at stake, and without proactive management of this commons we will likely experience many negative impacts that could even roll back many recent advances in services and scientific knowledge the world community has made because of its effective use of space-based resources. As the previous chapters have made quite clear, there are several multilateral efforts aimed at achieving sustainability of the space environment about Earth, and for many experts in the field, reaching or approximating this goal is one of the key challenges facing spacefaring nations and companies today.

Although experts may not agree on a precise definition of what is meant by "space sustainability" or how to sustain or achieve it, a broad working definition might be: "Space sustainability is ensuring that all humanity can continue to use outer space for peaceful purposes and socioeconomic benefit, now and in the long term."[1]

As noted in the chapters by Doetsch, Brachet, Martinez, and Hodgkins, et al., the impetus to focus on space sustainability derives from several related threads of thought:

1) Increasing awareness that in-space operations need to be managed to reduce the harmful impacts caused by orbital debris generated by past space activities, and by crowding in certain orbits on space systems.

2) Increasing concern over the potential of harmful frequency interference to degrade the performance of existing space communications systems.

3) Growing unease about the potential for the use of space weapons to clutter outer space with debris to the point that it is essentially unusable, thereby depriving Earth of critical space-based services.

4) The knowledge that all spacefaring States have a responsibility to play their part in mitigating the risks of space activities to the space environment.

5) The emergent realization that humanity needs to take an active role in managing Earth's varied environments and sustaining its many resources in order to continue to survive. Because the scale of

---

[1]   Secure World Foundation, "Space Sustainability, A Practical Guide," p. 4. http://swfound.org/media/121399/swf_space_sustainability-a_practical_guide_2014__1_.pdf.

human economic activity is now global, and because space
systems make a significant and increasing contribution to that
effort, it is crucial to maintain their ability to operate with as little
interference as possible.

These lines of thought have converged to the inescapable awareness
that collective action is necessary in the form of research, discussions and
negotiations leading to choices, mitigations, and initiatives. These
concerns have led to three independent multilateral initiatives, the UN
COPUOS Working Group on the Long Term Sustainability of Space
Activities; the European Union proposal for an International Code of
Conduct in Outer Space; and the UN Group of Governmental Experts'
report on space transparency and confidence-building measures.


## ORBITAL DEBRIS

Anxieties about the sustainability of space activities have been a concern to
debris analysts for some time. In the 1980s, that apprehension was framed
as a realization that although near-Earth outer space was vast, the rate of
debris generation in normal launch and operational procedures, as practiced
at that time, would eventually lead to a condition in which debris would
routinely threaten working spacecraft.[2] As Kessler and Cour-Palais showed
in 1978,[3] depositing debris in orbit would likely eventually lead to a chain
reaction of collisions, creating sufficient debris to prevent the use of near-
Earth space, especially in orbits already crowded with active spacecraft.

In the early 1990s, several far sighted individuals at NASA, the
European Space Agency (ESA) and other space agencies decided to start
the Interagency Debris Coordinating Committee (IADC) in order to share
information on debris and States' efforts to mitigate the future generation
of debris. This organization, which now has members from 12 space
agencies, has been key to the development and promulgation of voluntary
debris mitigation guidelines and standards, that have been adopted by most
of the world's space agencies and private space entities.[4] Its efforts have

---

[2]  U.S. Congress, Office of Technology Assessment, "Orbiting Debris, a Space
     Environmental Problem," OTA-BP-ISC-72, October 1990,
     http://ota.fas.org/reports/9033.pdf.
[3]  Donald J. Kessler and Burton G. Cour-Palais (1978). "Collision Frequency of
     Artificial Satellites: The Creation of a Debris Belt". *Journal of Geophysical
     Research* 83: 2637–2646.
[4]  IADC, http://www.iadc-online.org/

led to a decrease in the amount of debris generated each year from routine space launch and other operations, despite a sharp increase in the number of space actors. The 2007 Space Debris Guidelines adopted by COPUOS in 2007[5] are based on those IADC guidelines.

More recently, space experts have also become increasingly troubled about crowding in key orbits, especially in the sun-synchronous orbital regime favored by Earth observation spacecraft and in the geosynchronous orbits of communication satellites and some fixed point weather satellites. Crowding raises the risk of inadvertent collision and reduces the maneuverability of active satellites.

Further, with increasing crowding comes the potential for unintended radio frequency interference. Indeed, frequency interference, both unintentional and intentional, is a growing concern to many satellite system operators.

The recent history of the space sustainability concept as it was developed within the U.N. Committee on the Peaceful Uses of Outer Space (COPUOS) is fully detailed in the chapters authored by Deutsch and Brachet. The need to take action to ensure States' ability to continue to operate in outer space is well understood in the international community, even if precisely what actions to take to ensure that end are not. Further, the need for collective action is clear.

Despite agreement over the need to find multilateral solutions, there is presently little willingness within the international community for attempting to craft a treaty to tackle the issue. Non-binding agreements over international guidelines and principles regarding space activities seem to be the best that can be achieved at present and in the foreseeable future. The following sections briefly review the three major efforts to establish such agreements.


## THE COPUOS WORKING GROUP ON THE LONG TERM SUSTAINABILITY OF SPACE ACTIVITIES (LTS)

COPUOS was moved to tackle part of the issue through its efforts in the Subcommittee on Science and Technology toward reducing the generation of space debris, which resulted in a non-binding set of guidelines for the

---

[5]   United Nations Space Debris Mitigation Guidelines of the Committee on the Peaceful Uses of Outer Space, 2007, http://orbitaldebris.jsc.nasa.gov/library/Space%20Debris%20Mitigation%20Guideli nes_COPUOS.pdf

mitigation of space debris. Later, thanks to the efforts of COPUOS chairmen Karl Deutsch and Gerard Brachet the discussion was broadened beyond space debris to examine the totality of space operations and the threats they face. COPUOS decided to open a Working Group on the Long Term Sustainability of Outer Space and chose Dr. Peter Martinez of South Africa to chair it.

Martinez's chapter in this book reviews the history of the creation and progress of the LTS Working Group. The Group, which had originally been scheduled to complete its work in 2014, but has now been extended by two years to 2016 in order to resolve a series of difficulties resulting from working across different work cultures, languages, and levels of space experience. Added to those difficulties is the reluctance of some states, especially those with very advanced technologies and operational practices, for reasons of national security to reveal very much detail about those practices. Hence, we see that the discussions over LTS can touch on very sensitive matters of national security and operational capabilities.

Over the years, COPUOS working groups have developed varying working styles depending on the subject matter and the political sensitivities they provoke. Most groups generally need additional expertise not present within the COPUOS delegations as, for example, in the case of the Working Group on Near Earth Objects (NEOs). The additional expertise was provided by the so-called Action Team-14, an informal group of experts on different aspects of NEOs, which included individuals from academia, government agencies, and from COPUOS Permanent Observers including Secure World Foundation.

The formation of action teams developed out of the 1999 UNISPACE III conference. In the case of the LTS, several States insisted that all experts be formally nominated by States, perhaps in fear that private entities would dominate a more informal process. Experts were nominated to four Expert Groups:

A. Sustainable space utilization supporting sustainable development on Earth;[6]
B. Space debris, space operations and tools to support space situational awareness sharing;[7]
C. Space weather;[8] [9]

---

[6] Report of Expert Group A:
http://www.unoosa.org/pdf/limited/c1/AC105_C1_2014_CRP13E.pdf.
[7] Report of Expert Group B:
http://www.unoosa.org/pdf/limited/l/AC105_2014_CRP14E.pdf.

D. Regulatory regimes and guidance for new actors in the space
arena.[10]

The Expert Groups worked for more than two years, both within
COPUOS-related meetings and at other times, to develop reports
containing guidelines for their areas of expertise. Some were similar in
content across the groups, but most very specific to the subject of the
working groups. These guidelines were then used by the Working Group to
develop a much smaller set of consolidated guidelines, which could be
presented to COPUOS as a whole for consideration.

With the understandable sharp focus on producing an overall set of
guidelines for operating in outer space that all States can agree to, the
original output of the Expert Groups is often forgotten. Yet these reports
compose a rich compendium of information about operating in outer space
and include proposals for needed research, fact finding, and operational
improvements. By themselves they constitute an important contribution to
future space operations, and are quite worthy of wide study.

The discussion on LTS has helped many delegates, who may not be
familiar with space operations, to increase their knowledge. Indeed, the
physics of outer space are generally unfamiliar to most people, confined as
we are to live on Earth's surface under the protective layer of the
atmosphere, and this causes difficulties of comprehension for non-
specialists. Outer space is truly a different realm than we are used to in
everyday experience. Once objects reach orbit, Newton's Laws of Motion
are experienced in their raw form, unencumbered by any distorting effects
of the atmosphere. Spacecraft and space debris can both travel at velocities
greater than 10 kilometers per second, and at those speeds even very small
pieces of debris can cause serious damage to a functional spacecraft as we
saw in the chapter by Diegelman. The absence of an atmosphere also
means that debris, whether created by impact, explosions, or fragmentation
of aging spacecraft and old rocket bodies, can remain in orbit for many
years, especially in the higher orbits where they constitute a continuing
threat to operating spacecraft.

As noted in Martinez's contribution here, the LTS Working Group still
has a lot to do to complete its mandate, but it has already made

---

[8]   Report of Expert Group C:
      http://www.unoosa.org/pdf/limited/c1/AC105_C1_2014_CRP15E.pdf.
[9]   In the interest of full disclosure, the current author represented Secure World
      Foundation on Expert Group C.
[10]  Report of Expert Group D:
      http://www.unoosa.org/pdf/limited/c1/AC105_C1_2014_CRP16E.pdf.

considerable progress by helping States to focus on the many complex issues surrounding the development of a satisfactory set of guidelines for space operations.

## THE POTENTIAL FOR VIOLENT CONFLICT IN OUTER SPACE

Seldom discussed openly in meetings of COPOUS and its subcommittees is the potential for violent conflict in outer space. States increasingly depend on space systems to support military and intelligence operations, and some foresee skirmishes or even war as a very real possibility in some future time. Such considerations highlight a serious structural issue within the UN. COPUOS is devoted to the promotion and discussion of peaceful activities in outer space, and members take this very seriously, while questions of weapons activity in outer space are supposed to be taken up in the Conference on Disarmament (CD), which has as one of its many agenda items the Prevention of an Arms Race in Outer Space. The CD is "the single multilateral disarmament negotiating forum of the international community,"[11] and is therefore the entity that should be taking up such a question.

Weapons of mass destruction (WMD) are already banned from outer space by the 1967 Treaty on Outer Space. Nevertheless, that leaves many other potential weapons whether based in orbit or at ground-based facilities that could severely disrupt and degrade the outer space environment, especially antisatellite (ASAT) weapons. Destructive conflict would inevitably create thousands of pieces of debris that would stay in orbit for up to hundreds to thousands of years. The Chinese demonstrated this possibility in a test in January 2007 by destroying one of their own defunct weather satellites, which created more than 2,000 pieces of trackable debris in the already crowded low Earth orbit (LEO), and likely many thousands more that cannot be easily tracked from ground based optical or radar telescopes.[12] Much of this debris will continue to threaten spacecraft for hundreds of years.

Several states have made strenuous efforts to introduce discussion of the topic of space weapons into the working agenda of the Conference on

---

[11] Conference on Disarmament,
http://www.unog.ch/80256EE600585943/(httpPages)/BF18ABFEFE5D344DC1256F3100311CE9?OpenDocument.
[12] NASA, Orbital Debris Program Office, 2008.
http://orbitaldebris.jsc.nasa.gov/library/SatelliteFragHistory/TM-2008-214779.pdf.

Disarmament (CD). However, unlike COPUOS, which develops agenda topics that continue from year to year until the subject is considered exhausted, the CD's agenda of work is newly set each year. Because like COPUOS, the Committee works on a consensus basis, it is possible for a state to block any negotiating action within the Committee for that year. In fact, this has happened for nearly two decades, and the serious question of how to deal with or perhaps prevent violent conflict in outer space is not addressed by the United Nations, with the exception of numerous declarations that spacefaring States should avoid conflict in outer space.

Proponents wishing to find a way out of this institutional impasse have come up with two quite different multilateral efforts to move forward. The European Union has offered "an International Code of Conduct (ICOC) for Outer Space Activities as a transparency and confidence building measure…designed to enhance the safety, security and sustainability of activities in outer space."[13] In the EU's view, this proposal can be used as a mechanism for negotiating a new non-treaty instrument that spacefaring States could sign on to, outside of the United Nations.

Within the United Nations, the UN Secretary General can convene a Group of Governmental Experts to examine emerging issues and make recommendations to the Secretary General. In January 2011, the General Assembly adopted a resolution introduced by Russia in 2010 (A/RES/65/68) to establish a Group of Governmental Experts (GGE) to study possible transparency and confidence-building measures in outer space and to make recommendations for future action.

It is also worthy of note that an historic joint meeting of the First and Fourth Committees of the UN General Assembly was held in October 2015. Since the CD reports to First Committee and COPUOS reports to Fourth Committee, this joint meeting provided at least symbolic commitment to build bridges of dialogue between the UN's discussion of military security issues in space and its discussion of peaceful uses. It will take time to see if there will be substance to add to the symbolism.

## THE PROPOSED INTERNATIONAL CODE OF CONDUCT

As summarized in Gerard Brachet's contribution to this book, the European Union's proposal for adopting an International Code of Conduct arose in part out of frustration within Europe at the inaction of the United Nations

---

[13]   European External Action Service, http://eeas.europa.eu/non-proliferation-and-disarmament/outer-space-activities/index_en.htm.

Conference on Disarmament (CD) in addressing the issues that potential violent conflict in outer space present to the international community. It also arose in recognition that promoting good behavior in outer space generally would build confidence among space actors and increase transparency, thereby reducing the potential for conflict. As expressed on the website of the EEAS:

> "By promoting the draft Code of Conduct, the EU supports the notion that voluntary 'rules of the road,' grounded in best practices among space actors, offer a pragmatic approach to achieving, and strengthening, adherence to norms of behaviour in space. The Code of Conduct aims at enhancing safety, security and sustainability in space by emphasising that space activities should involve a high degree of care, due diligence and transparency, with the aim of building confidence among space actors worldwide."[14]

In many ways the contents of the proposed ICOC promoted by the EU simply serves as a reaffirmation of the international treaties on space that were negotiated and ratified in the 1960s and '70s. Indeed, as noted above, the EU itself sees this instrument as reinforcing existing norms. Despite this, the ICOC has generated significant controversy, not only over some of the proposed text, but also the methods by which it was introduced to the international community. Several states expressed concern that the EU's diplomatic efforts lacked professionalism.

After the States of the European Union reached agreement on a text, representatives of the EU External Action Service (EEAS) first introduced that text to the wider community a few days before they convened a side meeting in UN offices in Vienna. This meeting preceded the 2012 meeting of COPUOS. Reaction to the initial draft text was mixed, with very few states welcoming it. Several countries, especially some spacefaring states openly complained that by consulting only with the major space powers during the initial consultation process the EU had slighted them and their potential input to the draft text. Several expressed the opinion that such a meeting should not be held in the same venue as COPUOS, while not involving the United Nations except as a convenient meeting place. Nevertheless, despite the rocky start, 95 States have now attended one or more of the three multilateral consultations in Kiev (May 2013), Bangkok

---

[14]  Ibid.

(November 2013), and Luxembourg (May 2014). They were moved into the negotiating phase beginning in July 2015.

After the introductory meeting, the UN Institute for Disarmament Research (UNIDIR), the UN-affiliated organization devoted to assisting the CD grapple with disarmament issues, was chosen to assist in organizing and providing logistics for a set of regional meetings devoted to informing delegates from States about the draft code and gaining acceptance of the effort. UNIDIR also organized regional seminars in Malaysia, Ethiopia, Mexico, and Kazakhstan. Members of UNIDIR staff have also taken part in other meetings focused on seeking multilateral approaches to transparency and confidence-building measures.

During or after the consultation meetings, States have suggested a wide variety of changes to the text to address concerns they have about the draft, and EEAS officials have made numerous changes, small and large. The EU hopes to conclude the negotiation phase of the effort and soon offer a final draft for signing although it is still unclear just how many States are willing to sign on to the draft.

One potential stumbling block for several States is the draft's provision on States' right to self-defense. The right to self-defense by States is recognized twice in the draft ICOC. It occurs first in the statement of principles as the second principle:

> "The responsibility of states to refrain from the threat or use of force against the territorial integrity or political independence of any state, or in any manner inconsistent with the purposes of the Charter of the United Nations, and the inherent right of states to individual or collective self-defence as recognised in the Charter of the United Nations."[15]

The right of self-defense appears the second time quite explicitly as one of the resolutions. Subscribing States resolve to, "refrain from any action which brings about, directly or indirectly, damage, or destruction, of space objects unless such action is justified...by the Charter of the United Nations, including the inherent right of individual or collective self-defence."

Despite the fact that the Article 51 of the Charter of the United Nations guarantees the right to States of self-defense, some States argue that intentional interference or destruction with space objects should never be allowed and that outer space should only be for peaceful purposes. Thus,

---

[15]  EEAS, Draft International Code of Conduct for Outer Space Activities,

provisions about self-defense are thus very much out of place in this code. For example, Becerra and Acevedo of Venezuela argue that, "The articles of ICOC reflect the right of individual or collective self-defence, which is not acceptable to Latin America and the Caribbean since it contradicts the principle of the peaceful use of outer space, and opens the possibilities to increase the use of force, weapons and the militarisation of the outer space arena."[16]

Furthermore, many of the newer or less advanced spacefaring states especially express concern that they might find their potential for engaging in space activities limited by the code, not necessarily because of legal restrictions, but perhaps because the adopted code would call for higher and more expensive practices in launch and operational standards. They do not wish to be priced out of space activities, nor do they want their right to access space abridged by this code. This feeds into an inherent suspicion that the Code was written by the West and for the benefit of the West.

Hence, the draft ICOC has generated considerable discussion and many pages of argument for inclusion or deletion of various phrases and terms. These deliberations are in many ways similar to those within COPUOS regarding the Working Group on LTS. Although the ICOC is a much broader document derived from the treaties on outer space and basic principles of good behavior, rather than from guidelines based on operational experience.

The EU has made significant changes in the draft ICOC compared to the initial 2008 draft, and now the process is entering a difficult negotiation stage. An initial session held in New York City in July 2015 proved difficult and contentious. The negotiation process will take additional time but may allow States interested in signing on to the Code to offer changes that satisfy their needs and concerns. It may also provide the platform for spreading additional information about the need for an international instrument that defines and promulgates good behavior in outer space.


## THE GROUP OF GOVERNMENTAL EXPERTS (GGE)

As noted above, in large part because of the inaction of the Conference on Disarmament on the issue of weapons in space, the UN Secretary General

---

[16]  Roberto Becerra and Romina Acevedo, ICoC: Perspectives from Latin American and the Caribbean. In Rajeswari Pillai Rajagopalan and Daniel A. Porras (Editors), Awaiting Launch: Perspectives on the Draft ICOC for Outer Space Activities, 2014, http://www.orfonline.org.

convened a group of governmental experts to inform him concerning ways to increase transparency and confidence among States with regard to their space activities, and to make recommendations for member States to consider. The GGE's goal was to craft a set of transparency and confidence building measures (TCBMs) for space activities that could serve as foundational measures for reducing the risks of misunderstanding, mistrust and miscalculations among states in space activities.

The resulting GGE chaired by Russian expert Victor Vasiliev was composed of experts from 15 different countries, and had its first meeting in New York in July 2012. Its second meeting was in Geneva in April 2013, and a final one in New York in July 2013. Members included the representatives from the five permanent members of the Security Council and ten nominated from a geographically diverse selection of countries. According to the terms of reference of the GGE, experts were expected to provide their best politically neutral advice, unaffiliated with citizenship. By focusing primarily on TCBMs with a relatively small group of experts, the group was able to achieve consensus on a limited, but quite significant set of recommendations that provide a firm basis for achieving greater transparency among States in space activities.[17]

The crucial question now for the international community to consider is how, if at all, will the GGE recommendations be implemented? These recommendations are in the hands of the UN Office for Disarmament Affairs (ODA), and three years after the GGE report was delivered to the Secretary General it is still unclear what that office plans to do to promulgate or promote the GGE recommendations. The ODA webpage notes that, "Following extensive and in-depth discussions, the experts agreed upon a set of substantive TCBMs for outer space activities and recommended that States consider and implement them on a voluntary basis." It also notes that "the GGE recommended establishing coordination between the Office for Disarmament Affairs, the Office for Outer Space Affairs and other appropriate UN entities,"[18] but the ODA webpage is silent on any efforts by that office to do so, or by states to begin to implement the GGE's recommendations.

Hence, it is evident that implementation of the GGE recommendations will take a concerted effort by the international community. While there is a lot that a state can do on its own, finding the will within individual states to do so may be difficult without concerted effort from UNODA and other

---

[17]   United Nations, Report of the Group of Governmental Experts, A/68/189, 2013.
        http://www.un.org/ga/search/view_doc.asp?symbol=A/68/189.
[18]   http://www.un.org/disarmament/topics/outerspace/

entities with the power to convene meetings on TCBMs and to promote adoption of them.    The various authors contributing to this volume are unanimous in agreement that this is a matter of both urgency and significance.

## Multilateral Efforts

The three multilateral efforts reviewed here demonstrate that fairly high levels of government in spacefaring States are interested in addressing the long term sustainability of space activities. How successful they will be depends on a variety of factors, including the approach of States to national security, costs of any changes in operational practices, the private sector's willingness to participate, and the inclination of States to adopt and comply with guidelines that are, after all, voluntary.

- **National security**
  The agencies charged with defense of States are generally reluctant to assent to any procedural changes perceived by them to limit options in outer space, especially if following such guidelines might diminish perceived asymmetric military advantage. On the other hand, if potential rivals also agree to adhere to the same restrictions, they are much more willing to agree, and even to champion such restrictions.

- **Private sector operators**
  Only a few States today have a strong private sector presence in space activities, but private investment in space systems is growing each year. Companies always have concerns about the economic bottom line and whether adopted guidelines will negatively affect them and thus may attempt to counter or weaken them. Because states are responsible for the activities in outer space of their citizens, they generally require some sort of license to operate in space, and those licenses carry restrictions on private sector space operations. Companies may lobby for relaxation of license requirements, but ultimately the State will ensure compliance with those requirements.

- **Voluntary nature of the guidelines**
  Critics of these processes raise concerns that voluntary guidelines and non-binding resolutions will not be sufficient to assure the

long term goal of a sustainable space environment. At root is the issue of just how seriously individuals and organizations take concerns over the long term sustainability of space activities. The case of orbital debris is instructive. It took many years and considerable research, operational experience, and technical assessments for space agencies to begin to take action on orbital debris, despite many calls for changing operational practices to avoid creating new orbital debris. Experience with the IADC and UN debris guidelines suggest that gaining 100 percent adherence to the guidelines will be impossible. For example, not every state that participated in the development of operational debris guidelines has lived up to them one-hundred percent. Nevertheless, since spacefaring agencies began to develop operational debris guidelines, the percentage increase in debris created in space operations has decreased despite an increase in the numbers of space actors, suggesting that voluntary measures are far better than none at all.

- **PROTECTING THE RIGHTS OF SMALLER STATES TO ACCESS OUTER SPACE.**

  As noted earlier, the smaller spacefaring States and those that have not yet entered the space arena have worried that they may be inadvertently discouraged from starting a space effort because adopted guidelines or principles might make their entry too expensive. The more capable States can help alleviate this worry by offering technical assistance, both in imparting best practices and transferring relevant technology. The success of so-called New Space activities, which depend in large part on the availability of low-cost components, has demonstrated that cost need not be an insurmountable barrier for reaching space to carry out scientific observations and experiments, and to pursue space-based activities.

- **DEFINING SPACE SUSTAINABILITY.**

  The lack of a precise definition of the term has served as a barrier for some experts, who feel that in order to tackle the problem they need to understand exactly what it means both as a concept and operationally. Indeed, as Peter Martinez notes in his chapter, "attempts to define it precisely have led to much discussion and debate nuanced by the political, cultural, developmental, linguistic and conceptual differences among different countries." Fortunately, as the LTS process within COPUOS has demonstrated, it is possible to have a healthy, productive debate about which operational practices might assist with maintaining or achieving sustainability and which might achieve the opposite.

The LTS process tends to render concerns about the precise definition of space sustainability moot because that process, by focusing on the creation of practical steps to achieve a sustainable condition in outer space, function to define the desired endpoint in terms of procedures and actions, rather than its abstract concepts.

Hence, how we define space sustainability is less important than a shared commitment to taking the necessary actions to doing so. This applies at many levels, including that of individual organizations and companies, as well as nations and multi-nation partnerships, and of course also at the global level of the key international standards and regulatory bodies. It is up to all of us to preserve and protect this critical shared resource, as it is also our opportunity and obligation to enhance the value and the benefits that all humanity receives from our presence in space.

From launch to LEO, from GEO to the moon, to Mars and beyond, the creation and operation of the tools and technologies that bring space benefits to everyone on Earth is a collaborative effort among experts and organizations across the spectrum of the global economy and the global governments.

Reaching and abiding by the necessary agreements on a set of actionable operational guidelines, even where the end goal is not clearly defined ahead of time, is an issue at root of language, culture and operational practices, one that requires dedication and a strong commitment to finding a basic consensus set of guidelines.

It is precisely that debate and interchange of ideas about best practices in the space community that provides a fruitful basis for understanding more clearly just what space sustainability is from an operational standpoint. In the end, this bottom-up approach to negotiating may well prove highly successful in achieving the long term sustainability of space activities.

All three of these efforts have underscored the need for collective action to reduce the danger of collisions in orbit, and helped the diplomatic community to be much more aware of the need to keep the issues of outer space in their thinking and not to take the critical capabilities of space-based systems for granted.  All three efforts have already served a useful purpose in raising issues critical for assuring the long-term sustainability of outer space activities, and causing states to grapple with their meaning and applicability over the long term.  These multilateral processes can be very difficult, frustrating and tedious and often result in achieving less than any single State might wish.  Nevertheless, when successful, they lead to a result to which most States can adhere.  The important chapter by Filipé Duarte Santos reminds us why it is so important to pursue the long term sustainability of outer space.  As he notes, space systems contribute immeasurably to planetary management and therefore to the sustainability of Earth itself.

•••

## AND AGAIN, OUR THANKS

As always, we would like to once again thank the distinguished group of authors who have contributed generously of their thoughts and efforts for sharing their work with us in this volume.  We hope that their insights and experiences have been valuable for all our readers, and that this volume will contribute in a small way to the sustained success of humanity's amazing journeys into space.

## RAY WILLIAMSON

Ray A. Williamson is retired from Secure World Foundation, where he served as Executive Director between 2007 and 2012 and Senior Advisor until 2014. Previously, he was Research Professor of Space Policy and International Affairs in the Space Policy Institute, The George Washington University. From 1979 to 1995, he served first as Senior Analyst and later as Senior Associate for the U.S. Congress, Office of Technology Assessment. Prior to employment with OTA, Dr. Williamson was Assistant Dean at St. John's College, Annapolis, Maryland.

He is editor of Apogeo, a magazine of space technology for human and environmental security. He is also a member of the International Academy of Astronautics, serving on Commission Five: Space Policies, Law & Economics, and a member of the editorial board of the Space Policy Journal. Dr. Williamson is the author or editor of ten books on space policy, historic preservation and the astronomical knowledge and ritual of the American Indian. He has authored more than 150 articles on space policy, space security, and remote sensing. Between 2002 and 2005, he served on the U.S. NOAA Advisory Committee on Commercial Remote Sensing (ACCRES).

Dr. Williamson received his B.A. in physics from the Johns Hopkins University and his Ph.D. in astronomy from the University of Maryland.

# INDEX

## 5

59th International Astronautical
Congress, 38

## A

Abiodun, Adigun Ade, 49
Advanced manufacturing, 222
Aerospace Technology Working
Group (ATWG), III, I, V, 431
Africa, 13, 14, 38, 40, 50, 59, 80, 85,
198, 201, 372, 391, 392, 394, 407
Age of Enlightenment, 302
Agreement on the Rescue of
Astronauts, 66
Air Force Research Laboratory
(AFRL), 280
Air Force's Space and Missile System
Center Development Planning
Directorate, 280
Airbus Defence & Space, 225, 362
Alenia Spazio, 325
Alsbury, Mike, 364
Anzaldua, Alfred, III, 343, 359
Alphasat, 224
Amazon, 394
Analytical Graphics, Inc. (AGI), 99
Ansari X Prize, 289
anthropogenic climate change, 390,
397
Apollo, XI, 21, 299, 314

Arctic, 215, 226, 397, 399
Arévalo-Yepes Ciro, III, 371, 384
Ariane 5, 221, 237, 242, 244, 248,
249, 262
Arms Export Control Act, 150
Armstrong, Neil, 21
Asteroids, 12, 21, 110
Athena Global, 11
atmospheric aerosol loading, 399
Augmented Geostationary
Laboratory Equipment (Eagle),,
280
Australia, VII

## B

Baikonur Cosmodrome, 255
Beason, Doug, 355
Bigelow Aerospace, 110, 111, 127,
339, 345, 364
Bigelow, Robert, 111
biodiversity, 389, 394, 395, 396
bi-propellant, 222
Blair, Brad, III, 343, 359
Bloomberg, 362
Blue Origin, 339, 364
Blum, Michael, III, 337
Boeing, 292, 293, 322, 323, 327, 362
Bostrum, Nick, 185
Brachet, Gérard, 31, 46, 49, 407, 410
Bradley, A.M., & Wein, L.M., 124
Branson, Richard, 364

Brazilian Geosynchronous Satellite, 284
Brazilian Space Agency, 204, 257, 265
Brundtland Commission, 17, 230
Brundtland, Gro Harlem, 230
Bucket Wheel Excavator, 316, 317, 318, 319, 321, 327
Buenneke, Richard H., I, 53, 61, 73
Bureau of European Policy Advisors (BEPA), 214

## C

Cabrera, Enrique Pacheco, 52
California, 134, 135, 211, 275, 286, 292, 293, 303, 304, 336
California Institute of Technology, 286, 292
Canadian Space Agency, 359
Cancer, 305, 306
carbon footprint, X, 243, 304
Carroll, Joe, 347, 349, 350, 356
CelesTrak, 100
cellularization, 348
Center for Public Integrity, 259, 266
Center for Research in Air and Space Law at McGill University, 105
Center of International and Security Studies at the University of Maryland, 75, 117
Chatterjee, Joyeeta, II, 157, 181
Cheyabinsk, 12
China, 33, 34, 37, 40, 44, 69, 71, 76, 80, 85, 86, 88, 89, 92, 94, 101, 115, 128, 136, 177, 196, 197, 205, 206, 207, 210, 237, 238, 252, 256, 265, 266, 267, 268, 303, 347, 394
Chixchulub, 12, 14
Chow, Tiffany, II, 121, 134, 201
Citron, Bob, 325, 326, 327
Clean Air Act, 258
Clean Space, II, 215, 218, 219, 222, 226, 227, 232, 253
CleanSat, 223, 224, 225, 227
CleanSpace One, 365
Climate change, 396
Club of Rome, 16
CNES, 31, 34, 46, 96, 140, 148, 155, 156, 232, 254, 267
Coastal zones, 396

Code of Conduct (CoC) on Outer Space Activities, 199
Code of Conduct for Outer Space Activities, 41, 63
Cold War, VI, 114, 122, 128, 354
Commercial Control List (CCL), 150
Commercial Resupply Services (CRS), 283
Commercial Space Operations Center (ComSpOC), 99
Committee on Earth Observation Satellites, 23, 46
Common Pool Resources (CPRs), 129
Conference on Disarmament, 36, 37, 38, 39, 40, 43, 44, 56, 79, 81, 82, 85, 88, 91, 104, 114, 117, 131, 382, 409, 410, 411, 413
Congo River, 394
Convention on International Liability for Damage Caused by Space Objects (Liability Convention), 66
Convention on International Liability for Damage Caused by Space Objects of 1971, 132
Convention on Registration of Objects Launched into Outer Space, 66, 86, 162, 177
COPUOS, I, 22, 24, 26, 27, 34 - 39, 41, 46, 47, 48, 49, 51, 52, 54, 55, 56, 57, 59, 64, 65, 69 - 73, 77, 80, 81, 82, 84, 85, 88, 90, 91, 93, 94, 95, 96, 97, 101, 103, 104, 106 - 109, 113 - 115, 128, 145, 152, 158, 159, 164, 167, 169, 177, 196, 197, 200, 201, 209, 216, 217, 354, 360, 371 - 385, 400, 405 - 411, 413, 417
Cornell University, 286
Cosmic Study on Space Traffic Management, 34, 107, 166
Cosmos 2251, 23, 126, 127, 347
COTS Space Act Agreement, 283
Cox, Dr. Ken, V, 311
cryosphere, 397
CubeSat Kit, 366
Curiosity, 309
Cygnus, 283

# D

Daily Mail English Channel Crossing Prize, 289

DalBello, Richard, 98

Darwin, Charles, 307, 308, 309

data mining, 378

Database and Information System Characterizing Objects in Space (DISCOS, 101

Debris remediation, 140, 148, 149

Debris Removal, 102, 159, 347, 349, 351, 357

Declaration of Legal Principles Governing the Activities of States and the Exploration and Use of Outer Space of 1963, 65

Deep Space Industries, 110

Defense Advanced Research Projects Agency (DARPA), 100, 278, 348

Deforestation, 394, 395

Delaval, Jessica, II, 213, 228

deorbited, 124, 286, 352

Desalination, 303

Design for Demise (D4D), 223

Deutsch Prize, 289

Diegelman, Thomas, III, 295, 312, 408

DigitalGlobe, 285

Doetsch, Karl, I, 11, 30, 49

Duarte Santos, Filipe, 52, 387, 400

Dunlop, David, III, 343, 360, 361

# E

e.Deorbit, 224, 225, 227

Earth, V

earth imaging, 338

Earth Observation (EO), 388, 395

Earth sensing, 338

Earth Summit 1992, 17

Earth Summit 2002, 18

Earth Summit 2012, 18

Eco-design, 219, 220

École Polytechnique Fédérale de Lausanne (EPFL), 365

ECOSAT, 218, 220

Educational Launch of Nanosatellites (ELaNa) Initiative, 279

Efficient Space Procurement (ESP), 279

El Niño-Southern Oscillation, 399

ElectroDynamic Debris Eliminator (EDDE), 349

Electrodynamic Technologies, LLC, 349

Embraer, 284

Environmental Management System (EMS), 231

Environmental Safety Standard for Space Technologies, 232, 256

Envisat, X, 157, 158, 160, 166

EO satellites, 396

European Centre for Space Law (ECSL), 106

European External Action Service (EEAS), 39

European Organization for the Development and Construction of Space Vehicle Launchers (ELDO), 161

European Space Agency (ESA), II, X, 34, 36, 38, 46, 95, 96, 101, 139, 140, 141, 144, 155, 156, 157, 160, 166, 174, 176, 213, 214, 215, 216, 217, 218, 219, 220, 222, 223, 224, 225, 226, 227, 232, 238, 253, 262, 265, 266, 267, 356, 405

European Union, 36, 38, 39, 41, 44, 77, 79, 86, 94, 98, 128, 200, 214, 218, 253, 405, 410, 411

Eutelsat, 36, 38

Expert Group A: Sustainable Space Utilization, 52

Expert Group B, 53, 55, 84, 92, 93, 95, 96, 97, 102, 103, 407

Expert Group C: Space Weather, 53

Expert Group D: Regulatory Regimes, 53

Export Administration Regulations (EAR), 150

# F

Famine Early Warning System Network, 392

Federal Aviation Administration
    (FAA), V, 110, 252
Federal Law on Safety of Territories
    along Launch Trajectories, 254,
    255
Firefly Space Systems, 337, 338, 341
FlightGlobal, 275
Food and Agricultural Organization
    (FAO), 305, 391, 393, 394
Framework Policy on Sustainable
    Development, 218
France, VII

# G

Gagarin, Yuri, 21
Galaxy Evolution Explorer (Galex),
    286
Galileo, 282, 356
GAMA/Herschel-Atlas/DINGO, 286
General Assembly Resolution 68/50,
    63
General Assembly's First Committee,
    78
GeneSat, 287
genetically modified organisms
    (GMOs, 304
GeoEye, 285
George C. Marshall Institute, 99
George Washington Military Law
    Society, 105
Geostationary Earth Orbit (GEO), IX,
    X, 3, 24, 33, 42, 96, 101, 102, 122,
    123, 125, 126, 132, 139, 216, 217,
    280, 345, 346, 348, 349, 352, 353,
    358, 376, 377, 378, 379, 380, 381,
    388, 417
Globalstar, 126
Glover, Jerry, 304
Goddard Space Flight Center, 103,
    277, 349
Google Lunar X Prize, 288
Goroka tribe, New Guinea, 304
Governmental Group of Experts
    (GGE), 39
GPS, 20, 33, 73, 345, 352, 355, 399
green propellants, 219, 232, 234,
    235, 245, 246, 247, 258, 261, 265
Green propulsion, 222
greenhouse effect, 250, 390, 396,
    397

Grinius, Marius, 37
Ground Service Vehicles (GSVs), 320
Group of Governmental Experts
    (GGE), 46, 62, 63, 78, 84, 197,
    410, 413
Group of Latin American and
    Caribbean states (GRULAC)., 109

# H

Haley, Andrew G., 184
Hardin, Garrett, 129
Harvard Manual of International
    Law Applicable to Air and Missile
    Warfare, 105
Heinlein, Robert, 337
Helium3 (He3), 314, 315, 316, 319,
    323, 327
Hitchens, Theresa, I, 37, 75, 117
Hodgkins, Ken, I, 61, 73, 404
Homo sapiens, 13, 14, 15, 184, 185,
    190
Horikawa, Dr. Yasushi, 70
House Committee on Science, Space
    and Technology, 112
Hubble, IV, XI, 103, 278, 286
Hubble Space Telescope, XI, 103

# I

IAASS White Book  (IAASS, 2009),
    234
IADC Debris Mitigation Guidelines,
    34, 108
Ice Age, 15
Indian Space Research Organisation,
    18
Industrial Revolution, 15
Innovative Partnerships Program
    (IPP), 287
Institute of Air and Space Law of
    McGill University, 355
Intelsat, 36, 98, 166, 348
Inter-Agency Debris Coordinating
    Committee (IADC), 101
Inter-Agency Meeting on Outer
    Space Activities, 27, 382
Inter-Agency Space Debris
    Coordination Committee (IADC,
    33, 128, 143, 158, 216

International Academy of
Astronautics, 7, 12, 34, 46, 59,
106, 107, 155, 167, 384, 385, 419
International Civil Aviation
Organization, 107, 234
International Code of Conduct for
Outer Space, 39, 41, 77, 200, 385,
412
International Code of Conduct for
Space Activities, 128
International Court of Justice, 139,
152, 153, 165
International Institute of Space Law
(IISL), 106, 111, 155
International Maritime Satellite
Organization (INMARSAT), 176
International Space Station, X, 73,
111, 140, 159, 208, 265, 299, 329,
336, 344, 349
International Space University, VIII
International Telecommunication
Union, 32, 82, 91, 109, 125, 372
International Telecommunications
Satellite Organization (ITSO), 357
International Telecommunications
Union, 33, 36
IPCC (Intergovernmental Panel on
Climate Change), 397
Iridium, 23, 121, 126, 127, 135, 136,
347
Isle of Man, VIII
ISO 14001, 231
ISRO, 18, 34, 216
International Institute of Space
Commerce, III, I, VIII, 431
International Space Station (ISS),
140, 265, 283, 299, 306, 307, 312,
329, 330, 332, 338, 349, 350
International Space University (ISU),
I, III, VII, VIII, 431
ITAR, X, 149, 150, 151
*ITU Constitution and the Radio
Regulations*, 40
ITU Radio Regulations, 380

**J**

James Webb Space Telescope, 272,
286
JAXA, 34, 216

Joint Space Operations Center, 98,
355

**K**

Kayser-Threde, 225
Keck Institute of Space Studies, 286
Kelso, T.S., 100
Kennedy, President John F., 274, 304
Kessler, Donald, 345
Kessler syndrome, 153, 344, 345
KickSat', 288
KickStarter, 288
Kistler Aerospace Corporation, 325,
326, 327
Kistler, Walter, 325, 326, 327

**L**

Lachs, Manfred, 142, 147, 156, 173
Latin America and Caribbean
Regional Group (GRULAC),, 372
Launch vehicles, 236
Lavrov, Sergei, 37
Legal Subcommittee, 24, 44, 69, 81,
104, 106, 107, 109, 113, 139, 144,
146, 154, 156, 165, 379, 381
Levin, Eugene, 347
Life Cycle Assessment (LCA), 218,
220
Liou, Jer-Chyi, 344, 347
List of private spaceflight
companies, 362
Long-Term Sustainability of Outer
Space Activities, 38, 39, 59, 69,
77, 83, 84
Low Earth Orbit (LEO), 21, 23, 33,
96, 102, 122, 123, 125, 126, 127,
132, 133, 139, 157, 159, 216, 217,
222, 223, 227, 257, 278, 315, 319,
320, 324, 327, 338, 339, 344, 346,
347, 349, 350, 351, 353, 354, 356,
388, 409, 417
Lunar Transportation Systems
(LTS), 314

**M**

Macauley, Molly, 123, 124, 135

MacDonald, Detwiller, and
    Associates', 348
Mankins, John C., 346
Mann, Ian, 53
Marchisio, Sergio, 53
Mars, 12, 272, 287, 304, 309, 417
Marshall, Will, 366
Martinez, Dr. Peter, I, 38, 47, 50, 59,
    70, 201, 375, 407, 408, 416
McCandless, Bruce, I
McDowell, Dr. Jonathan, 86, 92, 100
Measures for the Administration of
    Registration of Objects Launched
    into Outer Space, 256
Mendel, Gregor Johann, 308, 309
Microgravity Material Processing,
    331
Micrometeorites, 306, 346
Millennium Development Goals, 373,
    374, 389
Millennium Ecosystem Assessment,
    395
Minimum Functional Habitat
    Element (MFHE), 322, 324
Ministry of Ecology, 254, 255
Mitsubishi, 325, 362
mono-propellant, 222
Multilateral Initiatives, III, 95, 403
Murray, Sen. Patty, 113
Mutation-Driven Evolution, 308

**N**

NanoRacks, 366
nano-satellites, 348
nanosats, 354, 357
NASA, III, IV, V, IX, XI, 12, 19, 34,
    103, 111, 112, 153, 156, 163, 191,
    216, 231, 252, 253, 271, 272, 274,
    275, 276, 277, 279, 280, 282, 283,
    285, 286, 287, 288, 289, 290, 292,
    295, 299, 305, 306, 309, 312, 322,
    323, 325, 327, 336, 340, 344, 345,
    347, 349, 359, 360, 405, 409
NASA Ames Research Center, 288,
    366
Nosanov, Jeffrey, III, 271, 292
National Environmental Policy Act
    (NEPA), 232, 253

National Environmental Satellite
    Data and Information Service
    (NESDIS), 73
National Geospatial-Intelligence
    Agency (NGA), 284
National Policy for the Development
    of Space Activities (PNDAE), 257
National Program for Space
    Activities (PNAE), 204
National Science Foundation (NSF),
    279
National security, 415
National Space Policy of the United
    States of America, 273
Neanderthals, 14, 15
Near Earth Objects (NEOs), 11, 12,
    19, 21, 23, 357 407
Nei, Masatoshi, 307, 308
Nelson, Michael, 53
Neosat, 222, 224
NewSpace Global, 366
Nixon, Richard M., 298
Northrop Grumman, 362
nuclear power, 67, 68, 73

**O**

Obara, Takahiro, 53
Occupancy Analyzer Tool (GOAT,
    378
Olson, Mancur, 128
Oneida Indian Tribe, 6
Opensource, 290
Orbcomm, 126
Orbital Debris, II, III, IX, X, 32, 34, 37,
    73, 103, 137, 138, 156, 159, 166,
    167, 344, 345, 346, 347, 348, 349,
    350, 351, 352, 354, 355, 356, 357,
    358, 404, 405, 409, 416
Orbital Debris Program Office, 365
Orbital Mechanics, 352
Orbital Sciences, 277, 280, 364
Organisation for Economic Co-
    operation and Development
    (OECD), 214
Orinoco, 394
Orteig Prize, 289
Ostrom, Elinor, 129
Our Common Future, 17
Outer Space Treaty, 40, 56, 65, 66,
    103, 109, 111, 113, 123, 125, 141,

143, 154, 155, 161, 162, 168, 169, 170, 171, 173, 174, 175, 176, 177, 178, 179
overshoot, 16
Oviedo Convention on Human Rights and Biomedicine (Oviedo Convention)., 188

## P

Paradigm, 275, 276, 292
Patten, Norah, III, 271, 292
Pearson, Jerome, 347, 349, 351
People's Republic of China (PRC) Environmental Protection Law, 256
PharmaSat, 287
Planetary Resources, 110, 112
Portelli, Claudio, 53
Posey, Rep. Bill, 113
post humans, 184, 185, 186, 188, 190
Potter, Michael, III, 271, 292
Prevention of an Arms Race in Outer Space (PAROS), 36, 37, 39
Principles Relating to the Remote Sensing of the Earth from Outer Space, 67
Private Finance Initiative (PFI), 275, 281
Privatization, 176, 286
propellants, X, 219, 230, 231, 232, 233, 234, 235, 236, 237, 238, 239, 241, 242, 243, 245, 246, 247, 248, 256, 260, 261, 262, 263, 264, 350, 359
Protein Crystal Experiments, 330
Prudhoe Bay, Alaska, 321
Public Private Partnerships (PPPs), 281

## Q

Qatar, 303, 304

## R

radio frequencies, 108, 376
REACH (Registration, Evaluation and Authorization of Chemicals,

also REACh), 218, 220, 221, 222, 254
ResearchGate, 290
Resolution on Transparency and Confidence Building Measures (TCBMs), 39
Ricco, Antonio, 288
Rio Declaration on Environment and Development, 17
Rio+20, 18, 27, 373, 374, 385
Robinson, George, II, 183, 191
RoHS, 218, 221
Russian Academy of Sciences Keldysh Institute of Applied Mathematics, 101

## S

salt, 296, 297, 300, 310
Samson, Victoria, II, 195, 211
SatelliteToday, 285
Schmitt, Harrison "Jack", 318
Scientific and Technical Subcommittee, 38, 44, 49, 50, 51, 54, 55, 59, 69, 70, 77, 78, 81, 82, 83, 84, 92, 93, 97, 102, 103, 159, 164, 200, 216, 372
Sea-viewing Wide Field-of-view Sensor (SeaWiFs), 277
Secretary-General Ban Ki-Moon, 198
Secure World Foundation, III, I, V, VI, IX, 3, 7, 38, 105, 117, 121, 130, 134, 135, 195, 198, 201, 211, 371, 384, 403, 404, 407, 408, 419, 431
Siebold, Pete, 364
Simpson, Michael, 3, 7, 11
SkyCube, 289
Skylab, 286
Skynet, 5, 275, 276, 277, 282
Smith, Lesley Jane, II, 137, 155
Solar Power Satellites (SPS), 345
solar storms, 32
Solar System Research Institute, 287
Soviet Cosmos 954, 132
Space Activities of the Russian Federation in 2013-2020, 207
space commerce, VIII, 295, 296, 297, 298, 299, 300, 301, 303, 304, 306, 308, 310, 359
Space Debris, II, IX, 22, 23, 25, 33 - 38, 48, 52, 53, 55, 64, 68, 69, 72,

73, 80, 81, 82, 92, 94, 95, 96, 97,
101, 103, 109, 110, 121 - 129, 132
- 135, 138 - 141, 143, 144, 146,
148, 155, 156, 158 - 168, 170,
171, 176, 179, 180, 200, 206, 215
- 220, 223, 224, 225, 227, 235,
350, 352, 354, 356, 358, 360, 371,
372, 388, 406, 408
Space Debris Mitigation Guidelines,
22, 25, 34, 35, 37, 69, 94, 95, 96,
144, 158, 159, 164, 216, 217, 406
Space Manufactured Silicon Carbide,
331
Space Resource Exploitation and
Utilization Act, 112
Space Security Index, 28, 138
space situational awareness (SSA),
129
Space Shuttle, XI, 163, 236, 283, 305,
325, 327, 336, 364
Space Shuttle – STS 7, 364
space sustainability, 5, 6, 47, 48, 51,
52, 55, 56, 57, 58, 121, 122, 129,
130, 132, 133, 136, 193, 201, 233,
371, 373, 376, 381, 382, 404, 406,
416, 417
Space Venturers Holdings, LLC (SVA,
333
SPACEHAB, 323, 325, 326, 327
SpaceShip 2, 364
SpaceX, IV, 283, 339, 364, 385
Sputnik, 20, 35, 159, 230, 273, 387
Star Technology and Research, Inc.,
349
State of Registry, 158, 174, 177, 179
Steer, Dr. Cassandra, 105
Stott, Christopher, III, 271, 293, 295
Strategic Command's Joint Space
Operations Center, 71, 98
sustainable development, 11, 16, 17,
18, 19, 23, 25, 26, 52, 67, 69, 70,
83, 201, 206, 215, 218, 226, 230,
372, 373, 374, 375, 382, 385, 386,
388, 389, 390, 396, 399, 407
Sustainable Space Development
Board (SSDB), 233, 264

## T

Tallinn Manual on International Law
Applicable to Cyber Warfare, 105

Taylor, Thomas C., 313, 326
Techamerica Space Enterprise
Council, 99
Technology, V
Telebras, 284
Tether Applications, Inc.,, 347, 349
Thales Alenia Space, 225, 362
*The Limits to Growth*, 16, 17
*The Right Stuff*, 271
*The Space Millennium: Vienna
Declaration on Space and Human
Development*, 25
*The Space Report, 2012*, 274
The Space Resource Exploitation
and Utilization Act of 2015, 112
Tragedy of the Commons, IX, 129,
135, 356
transhumans, 184, 185, 186, 188
Transparency and confidence-
building measures (TCBMs), 62,
195
Tunguska event, 12

## U

U.S. Department of State, 94, 354
U.S. Space Surveillance Network
(SSN), 98
UN Committee on the Peaceful Uses
of Outer Space, 7, 22, 31, 41, 48,
62, 73, 354
UN Debris Mitigation Guidelines,
101, 115
UN General Assembly, 25, 26, 35, 39,
40, 56, 77, 176, 199, 382, 410
UN Institute for Disarmament
Research (UNIDIR), 412
UN Treaty on Disarmament, 150
UNGA Resolution 62/101, 145
Union of Concerned Scientists, 100,
122, 135
Union of Concerned Scientists (UCS),
100
Unispace 1, 24
Unispace II, 24
Unispace III, 25, 26
United Nations, 5, 17, 18, 25, 26, 27,
32, 33, 34, 35, 40, 43, 44, 46, 48,
50, 55, 59, 62, 63, 67, 71, 73, 77,
78, 79, 80, 85, 86, 87, 94, 97, 109,
117, 128, 131, 142, 158, 159, 160,

161, 162, 164, 169, 170, 171, 174,
176, 177, 178, 196, 197, 198, 201,
207, 216, 305, 360, 371, 373, 381,
382, 383, 384, 385, 388, 389, 391,
392, 393, 395, 406, 410, 411, 412,
414
United Nations Convention on
Biological Diversity, 18
United Nations Framework
Convention on Climate Change,
17
United States Munitions List, USML,
150
Universal Declaration on the Human
Genome and Human Rights
(UDHGHR), 188
University of Wisconsin at Madison,
317
Urbanization, 392, 393
US Air Force, 280, 325
US Department of Transport (DOT),
242
Uwingu, 290

**V**

Valencia-Bell, Ferran, 262
Vasiliev, Victor, 40, 414

Virgin Galactic, 237, 261, 364
Visiona, 284
volume, V

**W**

Wedgwood, Emma, 307
Weeden, Brian, II, 100, 121, 134,
140, 149
Weizmann Institute of Science, 286
Western Switzerland University of
Applied Sciences, 366
White House Office of Science and
Technology Policy, 112
Wicht, Anthony, 53
Williamson, Ray, III, 403, 419
Wiskerchen, Dr. Michael, III, 329
Wolfe, Tom, 271
World Bank, 382, 390, 392
World Conference on Sustainable
Development Rio+20, 372

**Z**

Zhdanovich, Olga, III, 229, 265
Ziglar, Zig, 300

# SPACE FOR THE 21ST CENTURY
·
## DISCOVERY
## INNOVATION
## SUSTAINABILITY

AN AEROSPACE TECHNOLOGY WORKING GROUP BOOK
VOLUME 5

IN PARTNERSHIP WITH

SECURE WORLD FOUNDATION

AND

THE INTERNATIONAL SPACE UNIVERSITY

AND

THE INTERNATIONAL INSTITUTE OF SPACE COMMERCE

31877561R00282